Springer Proceedings in Business and Economics

Springer Proceedings in Business and Economics brings the most current research presented at conferences and workshops to a global readership. The series features volumes (in electronic and print formats) of selected contributions from conferences in all areas of economics, business, management, and finance. In addition to an overall evaluation by the publisher of the topical interest, scientific quality, and timeliness of each volume, each contribution is refereed to standards comparable to those of leading journals, resulting in authoritative contributions to the respective fields. Springer's production and distribution infrastructure ensures rapid publication and wide circulation of the latest developments in the most compelling and promising areas of research today.

The editorial development of volumes may be managed using Springer Nature's innovative EquinOCS, a proven online conference proceedings submission, management and review system. This system is designed to ensure an efficient timeline for your publication, making Springer Proceedings in Business and Economics the premier series to publish your workshop or conference volume.

This book series is indexed in SCOPUS.

João Reis • Tiago Pinho • Vítor Barbosa •
Luís Barreto • Sandrina B. Moreira • Pedro Pardal •
João Nabais • Carlos Mata • Ana Mendes
Editors

Driving Quality Management and Sustainability in VUCA Environments

4th International Conference on Quality Innovation and Sustainability (ICQIS), Setúbal, Portugal, May 22-23, 2023

Springer

Editors
João Reis
Industrial Engineering
Universidade Lusófona
Lisbon, Portugal

Vítor Barbosa
Escola Superior de Ciências Empresariais
Instituto Politécnico de Setúbal
Setúbal, Portugal

Sandrina B. Moreira
Escola Superior de Ciências Empresariais
Instituto Politécnico de Setúbal
Setúbal, Portugal

João Nabais
Escola Superior de Ciências Empresariais
Instituto Politécnico de Setúbal
Setúbal, Portugal

Ana Mendes
Escola Superior de Ciências Empresariais
Instituto Politécnico de Setúbal
Setúbal, Portugal

Tiago Pinho
Escola Superior de Ciências Empresariais
Instituto Politécnico de Setúbal
Setúbal, Portugal

Luís Barreto
Escola Superior de Ciências Empresariais
Instituto Politécnico de Viana do Castelo
Viana do Castelo, Portugal

Pedro Pardal
Escola Superior de Ciências Empresariais
Instituto Politécnico de Setúbal
Setúbal, Portugal

Carlos Mata
Escola Superior de Ciências Empresariais
Instituto Politécnico de Setúbal
Setúbal, Portugal

ISSN 2198-7246 ISSN 2198-7254 (electronic)
Springer Proceedings in Business and Economics
ISBN 978-3-031-52722-7 ISBN 978-3-031-52723-4 (eBook)
https://doi.org/10.1007/978-3-031-52723-4

This Springer imprint is published by the registered company Springer Nature Switzerland AG
The registered company address is: Gewerbestrasse 11, 6330 Cham, Switzerland

Preface

Introduction

The 4th International Conference on Quality Innovation and Sustainability (ICQIS) took place between 22 and 23 May 2023 and was organized at the Polytechnic Institute of Setúbal, Portugal. The ICQIS Conferences have been a meeting forum for researchers and interested professionals on Quality Innovation, with a special focus on Sustainability. Thus, the ICQIS23 aimed to foster an inclusive environment that offered everyone a unique opportunity to connect, collaborate, exchange innovative ideas, and showcase their groundbreaking research to a global audience.

Over the course of the last three years, the host universities of ICQIS have played a pivotal role in uniting scientists with the fundamental principles of Quality Innovation and Sustainability. By establishing a robust connection and encouraging advancements in these domains, they have inspired us to follow a similar path. Notably, since 2022, Springer Proceedings in Business and Economics (SPBE) has been engaged with the conference [1] and has emerged as a key player in disseminating valuable scientific knowledge to scholars, professionals, and inquisitive readers interested in the subjects covered. ICQIS is also establishing a new tradition, laying the groundwork for the convergence of master and doctoral students in Quality Innovation and Sustainability with researchers, professionals, and academics from various corners of the world.

In line with other well-established conferences [2, 3], we share the belief that providing networking opportunities from an early stage is essential for the academic development of young researchers as they explore uncharted grounds of knowledge. Our commitment lies in continuously supporting the entire academic-scientific community, facilitating ongoing discussions, and encouraging the publication of valuable contributions that enrich the collective body of knowledge.

General Overview

The International Conference on Quality Innovation and Sustainability (ICQIS) has quickly established itself as a prominent event, now reaching its fourth edition. Despite its relatively young age, the conference has already gained substantial recognition from the global community, and we expect its influence to grow even further, attracting increased participation in the future. In this edition, we have meticulously selected 21 high-quality book chapters that address a wide range of compelling topics, including Product Service System Development, Electronic Commerce in Small- and Medium-Sized Companies, Sustainability in Supply Chain, Lean Methodology, Sustainable Development Goals (SDGs), Reverse Logistics, Local Development Strategies (LDSs), and Innovation in Higher Education Institutions (HEIs).

In the following section, we provide a succinct yet comprehensive summary of each book chapter, offering a glimpse into the valuable insights and research presented in this proceedings book.

Outline of the Volume

This book's first chapter investigates Circular Economy promotion, the R principles application, and sustainability policy by non-profit and public utility organizations in Setúbal, Portugal. It focuses on Social Economy and Sustainable Development Goals (SDGs) in alignment with the 2030 Agenda for Sustainable Development.

The second chapter optimizes industrial drug processing in the pharmaceutical sector by enhancing integrated equipment for active pharmaceutical ingredients (API) production. The proposed equipment offers significant ergonomic, hygiene, and practical advantages compared to the originals, enhancing the company's competitiveness.

Chapter 3 explores the growing market advantage of customized 3D-printed prosthetics and orthotics (P&O), improving patients' well-being. The flexible 3D printing process tailors devices to individual patient needs, enhancing functionality.

The fourth chapter addresses energy poverty, spotlighting energy-efficient building solutions in Portugal and Europe. It presents a pioneering solution through linear programming, streamlining decisions for near-zero energy buildings (nZEB) certification.

The fifth chapter comprehensively analyzes consumer behavior with Augmented Reality (AR) during shopping, revealing a predominantly positive sentiment toward its benefits.

The sixth chapter discusses e-commerce adoption in agricultural SMEs, with a focus on factors shaping their adoption and the role of digital platforms.

Chapter 7 explores managerial innovation, focusing on introducing novel management practices, structures, or techniques to improve organizational goals. The chapter highlights the significant link between applying these innovations and their effects on protected areas. An internal action plan is provided as a tangible illustration of this connection.

Chapter 8 explores the adaptation of small- and medium-sized enterprises (SMEs) to digital platforms.

Moving on to the ninth chapter, a groundbreaking approach for determining cargo capacity in agricultural product transportation is introduced. This approach contributes to sustainability in supply chains, particularly for the transportation of perishable goods over extended distances. By categorizing delivery options using key criteria, this methodology establishes an average cargo capacity standard for trucks. This empowers carriers to efficiently optimize tonnages for specific supply chains, minimizing unnecessary expenses associated with perishable transport by streamlining load capacity management.

The subsequent Chapter 10 critically reviews the profound impact of blockchain on maintenance practices. The authors present a well-thought-out proposal for a customized data structure tailored specifically for ship maintenance. The chapter concludes by outlining potential directions for future research in this dynamic field.

Chapter 11 introduces an innovative methodology that combines TRIZ and Lean tools to revolutionize maintenance management within the ship system of the Portuguese Navy. By harnessing these methodologies, the objective is to enhance operational efficiency, reduce waste, and eliminate non-value activities.

Chapter 12 focuses on a company specializing in manufacturing metallic structures for the aeronautics sector. The goal here is to elevate the intrinsic availability of assembly line equipment to a world-class 90%. This ambitious target is pursued through the strategic application of TPM (Total Productive Maintenance) and Lean tools.

The thirteenth chapter is dedicated to examining the impact of time window duration on the Vehicle Routing Problem with Time Windows (VRPTW). Through a case study, the authors delve into the intricate relationship between various time window intervals and their implications for costs, customer satisfaction, and travel time.

Chapter 14 embarks on a comprehensive investigation of Industry 4.0 through a meticulous literature review, encompassing 132 selected and analyzed papers. The review identifies seven distinct thematic research clusters, including Quality Management Systems, Total Quality Management, Measurement and Analysis, Human Factors, Technology, Human Systems Integration, and Training.

The subsequent Chap. 15 delves into Quality 4.0, presenting it as a multifaceted framework anchored on processes, human engagement, and technology. This framework shapes a forward-looking quality system that underscores the vital importance of digitalization skills and resilient quality frameworks essential for a flexible, data-driven Quality 4.0 setup.

Chapter 16 takes a global perspective on education service quality within Higher Education Institutions (HEIs), spanning institutions in Portugal and Angola. Employing the advanced HEISQUAL framework for assessment, this study reveals variations in student satisfaction levels between the two regions. Employing quantitative methods, the study provides actionable insights aimed at elevating quality performance indicators in Angola.

Chapter 17 introduces a pioneering blockchain-driven corporate communication tool that enhances Corporate Social Responsibility (CSR) reporting. Through a study of 96 articles, the authors demonstrate how blockchain technology bolsters transparency and credibility in CSR reporting, mitigating potential distortions.

In Chap. 18, the focus shifts to the analysis of the 2021-2027 Erasmus+ Program Regulation in alignment with the 17 SDGs. The study identifies key SDGs, highlighting the program's potential to catalyze Sustainable Development through participation and project implementation. The interconnections between SDGs are explored, shedding light on their intricate relationships.

Chapter 19 delves into the realm of Reverse Logistics, exploring its implications on consumer recycling behavior. Utilizing a combination of secondary sources and a questionnaire, the study identifies key factors influencing recycling behavior and categorizes consumers into distinct categories based on their attitudes and behaviors.

In the twentieth chapter, Community-Led Local Development (CLLD) takes the center stage as a catalyst for integrated development in local communities. The study investigates how CLLD can enhance tourism in specific municipalities, emphasizing the need for collaborative stakeholder efforts in effective implementation.

Lastly, the final chapter underscores the paramount importance of innovation in shaping a smart and sustainable future. The study delves into how micro-regions, often overlooked, contribute to regional innovation. Through a focus on Vidzeme University of Applied Sciences in Latvia, the research showcases the impact of Higher Education Institutions-driven innovation on Smart and Sustainable Futures.

Acknowledgments

The editors of this book of proceedings extend heartfelt gratitude to all participants and the entire Quality, Innovation, and Sustainability Community. We would also like to express our sincere appreciation to the following: (1) the diligent reviewers who invested their time and expertise to ensure each book chapter received at least two high-quality reviews promptly; (2) our esteemed Publishing Editor, Nitza Jones, for her unwavering support throughout the entire process; (3) the dedicated Springer team whose hard work made it possible for all selected papers to reach the wider community. Without your invaluable contributions and dedication, this book would not have been possible.

Lisbon, Portugal — João Reis
Setúbal, Portugal — Tiago Pinho
Setúbal, Portugal — Vítor Barbosa
Viana do Castelo, Portugal — Luís Barreto
Setúbal, Portugal — Sandrina B. Moreira
Setúbal, Portugal — Pedro Pardal

Setúbal, Portugal João Nabais
Setúbal, Portugal Carlos Mata
Setúbal, Portugal Ana Mendes

References

[1] de Oliveira Matias, J. C., Pimentel, C. M. O., dos Reis, J. C. G., das Dores, J. M. C. M., & Santos, G. (Eds.). (2023). *Quality innovation and sustainability: 3rd ICQIS, Aveiro University, Portugal, May 3–4, 2022*. Springer Nature. https://doi.org/10.1007/978-3-031-12914-8
[2] Gonçalves dos Reis, J. C., Mendonça Freires, F. G., & Junior Buda, M. V. (Eds). (2023). Industrial engineering and operations management. In *XXIX IJCIEOM, Lisbon, Portugal, June 28–30, 2023*. Springer Mathematics & Statistics. Springer, Cham. Forthcoming.
[3] López, S., Freires, F., Reis, J., & Dores, J. (Eds.) (2022). Industrial Engineering and Operations Management. In *XXVIII IJCIEOM, Mexico City, Mexico, July 17–20, 2022*. Springer Mathematics & Statistics. Springer, Cham. https://doi.org/10.1007/978-3-031-14763-0.

ICQIS 2023 Committees

Conference Chair

João Reis (Chair)	Lusófona University, Portugal
Tiago Pinho (Co-Chair)	Polytechnic Institute of Setúbal, Portugal

Organizing Committee

Ana Mendes	Polytechnic Institute of Setúbal, Portugal
Carlos Mata	Polytechnic Institute of Setúbal, Portugal
João Nabais	Polytechnic Institute of Setúbal, Portugal
João Reis	University of Aveiro, Portugal
Luís Barreto	Polytechnic Institute of Viana do Castelo, Portugal
Pedro Pardal	Polytechnic Institute of Setúbal, Portugal
Sandrina B. Moreira	Polytechnic Institute of Setúbal, Portugal
Susana Galvão	Polytechnic Institute of Setúbal, Portugal
Tiago Pinho	Polytechnic Institute of Setúbal, Portugal
Vítor Barbosa	Polytechnic Institute of Setúbal, Portugal

Editors

João Reis, Lusófona University, Portugal
Tiago Pinho, Polytechnic Institute of Setúbal, Portugal
Vítor Barbosa, Polytechnic Institute of Setúbal, Portugal
Luís Barreto, Polytechnic Institute of Viana do Castelo, Portugal
Sandrina B. Moreira, Polytechnic Institute of Setúbal, Portugal
Pedro Pardal, Polytechnic Institute of Setúbal, Portugal

João Nabais, Polytechnic Institute of Setúbal, Portugal
Carlos Mata, Polytechnic Institute of Setúbal, Portugal
Ana Mendes, Polytechnic Institute of Setúbal, Portugal

Scientific Committee

Ana Mendes	Polytechnic Institute of Setúbal, Portugal
Anabela Alves	Minho University, Portugal
Ângela Neves	Polytechnic Institute of Viseu, Portugal
António Pereira	Polytechnic Institute of Leiria, Portugal
António Abreu	Polytechnic Institute of Lisboa, Portugal
António Amaral	Polytechnic Institute of Porto, Portugal
Claúdia Silva	Aveiro University, Portugal
Dulce Matos Coelho	Polytechnic Institute of Setúbal, Portugal
Dzintra Atstaja	BA School of Business and Finance, Latvia
Federica Murmura	University of Urbino Carlo Bo, Italy
Helena Carvalho	Nova University of Lisboa, Portugal
João Nabais	Polytechnic Institute of Setúbal, Portugal
José Carlos Sá	Polytechnic Institute of Porto, Portugal
Kristina Zgodavova	Technical University of Košice, Slovakia
Luís Barreto	Polytechnic Institute of Viana do Castelo, Portugal
Luis Ferreira	Polytechnic Institute of Porto, Portugal
Luís Barbosa	Federal Center for Technological Education Celso Suckow Fonseca, Brasil
Graça Costa	Polytechnic Institute of Setúbal, Portugal
Maria Teresa Pereira	Polytechnic Institute of Porto, Portugal
Miladin Stefanović	University of Kragujevac, Serbia
Neringa Vilkaitė-Vaitonė	Vilnius Gediminas Technical University, Lithuania
Raul Campilho	Polytechnic Institute of Porto, Portugal
Sandrina B. Moreira	Polytechnic Institute of Setúbal, Portugal
Stevan Stankovski	University of Novi Sad, Serbia
Tamer Haddad	An-Najah National University, Nablus Palestine
Teresa Morgado	Polytechnic Institute of Lisboa, Portugal
Vítor Barbosa	Polytechnic Institute of Setúbal, Portugal

Keynote Speakers

Matevž Obrecht	University of Maribor, Slovenia
Gilberto Santos	Polytechnic Institute of Cávado e do Ave, Portugal

Contents

The Social Economy and the R Principles in Third Sector Organizations in the District of Setúbal

Ana Vizinho, Sandrina B. Moreira, and Raquel Pereira

Abstract This study aims to analyze how nonprofit and public utility organizations in the district of Setúbal promote a circular economy and apply the R principles and sustainability policy.

Motivated by the lack of previous studies that relate to social economy (SE), circular economy (CE), R principles, and sustainable development goals (SDGs) applicable to the social economy, the following question arose: in what way are nonprofit and public utility organizations concerned with the fulfillment of the new Sustainable Development Agenda entitled "Transforming Our World: The 2030 Agenda for Sustainable Development," published in 2015 by the United Nations (UN) and to be implemented by 2030.

A survey was sent to a random sample of 100 organizations, obtained by cross-referencing a list from Social Security in Portugal with the Portuguese Tributary and Customs Authority's list of entities receiving the 0.5% remittance of personal income tax (PIT). The survey questions focused on the ten Rs of sustainability policy and how the SDGs are relevant to their mission.

The results show that there is some knowledge on the subject and attempts to implement it, but the resistance encountered discourages its application. SE can be a pillar for the development of the CE, but this will require measures to encourage stakeholder involvement, through information targeted at this socioeconomic sector.

Keywords Circular economy · Social economy · Sustainability · SDG

A. Vizinho (✉) · R. Pereira
Instituto Politécnico de Setúbal (ESCE/IPS), Setúbal, Portugal
e-mail: ana.vizinho@estsetubal.ips.pt

S. B. Moreira
Instituto Politécnico de Setúbal (CICE, ESCE/IPS) e BRU-IUL (Business Research Unit), Setúbal, Portugal

J. Reis et al. (eds.), *Driving Quality Management and Sustainability in VUCA Environments*, Springer Proceedings in Business and Economics,
https://doi.org/10.1007/978-3-031-52723-4_1

1 Introduction

This study aims to relate the SE to the CE, namely, to understand how SE organizations promote a CE by applying the R principles and the SDGs.

The SE emerges with the Treaty on Social Economy published by Charles Dinoyer in the 1930s. In Portugal, the Republic Constitution reflects this idea by describing the cooperative and social sector in its articles 80 and 82 [1].

In 2013, the Social Economy Framework Law [2] was approved, defining the principles and organizations framed in this sector. The growing concern with supporting citizens, protecting fundamental values and rights such as health, education, and housing, gave rise to the creation of nonprofit organizations whose objective is social welfare.

As of 1992, the focus on people also meant an increased concern about sustainability; on this date, the UN Assembly for the Environment (RIO92) launched the pillars of sustainability that aimed at protecting the environment, making the economy more efficient and sustainable, and changing consumption and production habits. The key words chosen at the time were three Rs: reduce, reuse, and recycle. In 2000, the UN Millennium Declaration was adopted, an ambitious program to save the planet and consequently humanity, referring to the Millennium Goals to be achieved in the next 15 years. In 2012, during RIO + 20, the United Nations Conference on Sustainable Development, new goals called SDGs were approved, entering into force in 2015 with the goal of achieving them by the year 2030.

The R principles of sustainability created the CE concept, ending the concept of a linear economy and giving a new life to all products in a closed loop concept.

These global concerns with economic, social, and environmental sustainability led to the following research question: how SE organizations perceive these concerns, implement them, and incorporate them into their day-to-day life, in the development of their social mission. This question motivated a survey among the SE organizations in the district of Setúbal to understand the extent to which these organizations apply the Rs policy of sustainability and how these organizations contribute to the achievement of the targets of the 2030 Agenda through the application of the SDGs.

The remainder of the paper is organized as follows. Section 2 provides the relevant literature review on the topic. Section 3 describes both the purposes of the empirical study and the methodological framework. Section 4 proceeds with the analysis and discussion of the results, followed by the conclusions in Sect. 5.

2 Literature Review

The SE reflects activities whose main goal is to meet social needs to which the State may be directly or indirectly linked, encompassing a whole sector that incorporates nonprofit but solidarity-based organizations [3]. states that SE is not a by-product:

SE institutions reflect the need for an economy that reconciles the social, economic, and financial dimensions, manages to generate wealth, and is not only evaluated in terms of its financial capital but, above all, by its social capital.

The European Parliament Resolution on "A Model for the Future" (2009) [4] refers to SE as one of the pillars of the European social model, as it "plays an essential role in the European economy by combining profitability and solidarity, creating quality jobs, strengthening social, economic and territorial cohesion, generating social capital, and promoting active citizenship, solidarity and a form of economy with democratic values, putting the human being at the forefront and supporting sustainable development and social, environmental and technological innovation."

It is worth mentioning that SE has a diversity of funding from the private and public sector, donations, associate members, employees, volunteers, etc. creating different dynamics and relationships with the organization.

Regarding sustainability, the word derives from the Latin *sustentare* (Vergílio) or *sustinere* (Cicero); everything that can be supported or maintained is the definition given in the Portuguese Latin dictionary [5]. Sustainability is everyone's responsibility, especially the organizations that make SDG possible, where all of them should be rethought to be made on a more sustainable basis [6]. The emergence of the term sustainability has led many companies to value more sustainable business.

The pillars of sustainability were created at the RIO92, thus emerging the 3R principles, reduce, reuse, and recycle. The Rs are proposed to make the connection between environmental protection and economic efficiency that is linked to consumption habits. Beginning by R, followed by the letter "e," and thus forming a prefix "re," meaning repetition, they lead us to think about environmental sustainability and should be applied and reapplied through loops:

1. Refuse – products that harm nature and do not respect environmental legislation.
2. Rethink – daily attitudes and conscious consumption.
3. Reduce – ecological footprint.
4. Reuse – what would be discarded by giving it another destination.
5. Repair – what can be repaired.
6. Recycle – when it has no use or repair.
7. Reintegrate – returning to nature what came from it (e.g., organic waste).
8. Respect – life and living beings, people, their work and school, and environment.
9. Responsible – for your actions and the impacts they cause.
10. Repass – the knowledge that can make the world a better and sustainable place.

The CE emerges to replace the linear economy model identified with extract, produce, consume, and discard, which promotes the overuse of resources. This new paradigm has a multidisciplinary character and is interconnected with the process of sustainable production and consumption. In the linear process of production, the generation of waste is inevitable since it does not foresee the reuse of waste. In the circular model, on the other hand, there is production planning so that waste is reused in the production process with the aim of avoiding the generation of industrial waste [7]. CE and its relation to a more sustainable world consider not only the reduction of waste but also the limit of natural resources on the planet.

In September 2015, high representatives of the United Nations met with heads of State and government of the 193 member countries in New York and officially adopted the new Sustainable Development Agenda, entitled "Transforming Our World: The 2030 Agenda for Sustainable Development" [8]. The SDGs are an integral part of this Agenda and aim to change mindsets through coordinated actions by governments, business, and civil society to achieve the 17 SDGs and their 169 targets. They came into force in January 2016 and are expected to be met by December 31, 2030. As such, CE as a sustainability paradigm has been acquiring its space in the application of the SDGs, more specifically in some of the goals defined by the UN in 2015.

According to the Sustainable Europe Research Institute, social sustainability for a company is as important as its economic and/or environmental dimension [9]; the social dimension is everyone's responsibility, therefore including companies in all its dimensions. Social sustainability should be concerned with human rights and the fulfillment of the targets mentioned, namely, in the SDGs on hunger, education, gender equality, health, and decent work.

The SE (third sector) in Portugal encompasses nonprofit social organizations of public utility, autonomous in their decisions and relying in voluntary work, with the aim of promoting social change. Their revenues are intended solely for use in the organization itself. In this way, they pursue purposes of general interest in cooperation with the central or local administration. On a financial level, they are supported by public subsidies, membership fees, payment of users for services rendered, and sporadic sale of goods produced.

In Portugal, the government subsidizes institutions that provide solidary support to society. However, several of these entities are at their limit, and in some cases, the waiting time is long, unlike other countries (e.g., the UK, the Netherlands) that use a different model.

Among the main social organizations in Portugal are Cooperatives, Mutual Associations, Benefactor Organizations, Foundations, Charities and Private Social Solidarity Institutions (IPSS in Portuguese). They are of relevant importance in the society in which they operate, contributing to the SE and filling the gaps of the State in society.

3 Methodology

This study aims to investigate how nonprofit and public benefit organizations in the district of Setúbal promote a CE and apply the Rs policy of sustainability.

Analyzing the various options in this field of research, the questionnaire survey methodology proves to be the most correct and the most applied [10]. The goal is to collect valid and reliable information, around which conclusions are drawn that can be generalized, or not, to the universe under study [11]. A questionnaire survey was thus prepared as a data collection technique widely used in research, as a result of its standardization [12]. In particular, it is intended to be a quantitative study analyzed by statistical method on the application of sustainability in various types of SE institutions.

The anonymity and confidentiality of the answers make the respondent more comfortable to express opinions about the study and not be influenced by the researcher [13]. The present self-designed questionnaire is anonymous, includes a brief introduction describing the purpose of the questionnaire [14], and is composed mostly of closed-ended questions. Over time, closed-ended questions have become popular and open-ended questions less and less used [15]. The willingness to respond to questionnaires is often dependent on their size and content [14]. Thus, this questionnaire was straight to the questions that were intended to be analyzed according to the research objectives. The online survey prepared was sent by email.

The survey questions related to the application of the 10R principles are in accordance with the Dutch Environmental Assessment Agency (PBL) and included in the Council of Ministers Resolution No. 190-A/2017 on the Action Plan for Circular Economy in Portugal and the 17 UN SDGs [13]. These questions used the Likert method, with a five-level graded scale, either at the level of agreement, relevance, classification, or application.

As for the sample population object of study, a sample of 100 organizations that develop community support activities for the population was taken from the population of social organizations in the district of Setúbal. The same resulted from the crossing of two lists, one taken from the Social Security website and the other from the Tax Authority, which assigns entities a percentage of PIT, under n° 4 and 6 of article 32 of Law 16/2001 of June 22 [16]. The main difficulty encountered in choosing the sample was the lack of information on the organizations in a national database. Consulting the 2018 Social Economy Sector Survey (SESS) [17], prepared by the National Statistics Institute (INE in Portuguese) in collaboration with the António Sérgio Cooperative for the Social Economy (CASES in Portuguese), it appears that 47.7% of the entities connected to this sector at a national level do not have an Internet presence, in terms of websites or electronic page, which makes it difficult to access their contacts. On the other hand, there are several entities that no longer have a valid email address, and telephone contact is difficult, either because they do not have a listed phone number or because they simply do not answer. This sample includes Cooperatives for the Education and Rehabilitation of Maladjusted Children (CERCI in Portuguese), IPSS, associations, charities, and foundations, which, after contact research, were found to have a valid telephone number and/or email address.

4 Main Results and Reflections

The district of Setúbal is made up of 13 municipalities, and the sample chosen was randomly distributed across them, with a greater response in the municipalities with the largest population, Almada and Setúbal. Emails were sent to the 100 organizations chosen to check if the email addresses were valid. Approximately 7% of the emails sent were returned because that address no longer existed. Thus, the sample considered is composed of 93 valid organizations, with a response rate of 36% (see Fig. 1).

The largest number of answers was given by IPSS (61%), with no answers being obtained from social solidarity associations, mutual associations, and social and parish centers (Fig. 1). Overall, results suggest that although the theme of sustainability is referred to daily, there is still a lot of resistance to the application of sustainable behavior, even in the context of SE. When asked if the organization opts for renewable and natural processes in various aspects presented in Fig. 2, the answers were as follows:

- The organization invests in service innovation and/or internal processes, aiming to introduce renewable energy and resources: with 47% of the answers being mostly positive, it seems that most organizations are committed to resource innovation.
- The organization encourages recycling: there are no respondents with a negative response, and 94.1% of the organizations seem to be concerned with the recycling policy, and people and users are aware of it.
- The organization encourages the reuse of materials: the answers were positive in 88.2%, but there are still 5.9% of the organizations that do not promote this attitude.
- The organization substitutes materials and processes for sustainable ones: although there are situations opposing sustainability, 64.47% promote this substitution.

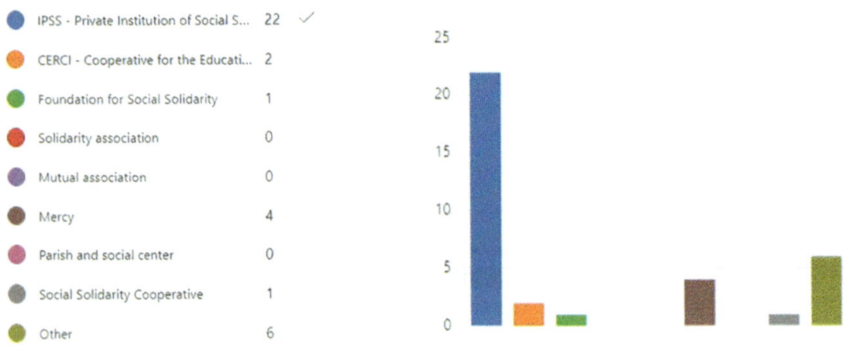

Fig. 1 Organizations that responded to the survey

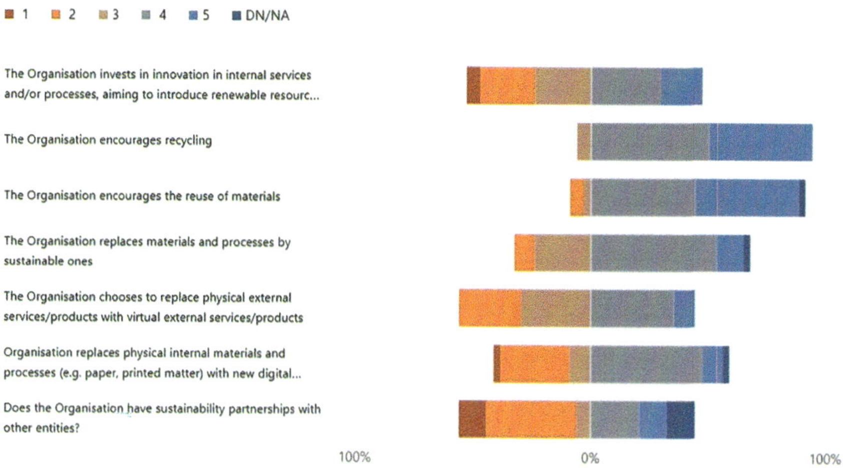

Fig. 2 Renewable and natural processes used by the organization

- The organization chooses to replace physical external services/products by virtual ones: in terms of new technologies, although a positive response, there is still not enough awareness for this attitude, with less than 50% of the organizations (44.1%) applying them.
- The organization replaces physical materials and internal processes (such as paper, printed materials) with new digital technologies (dematerialization): as many suppliers already offer the possibility of sending the invoice by email and consequently the reduction of paper, it is a growing trend as 55.9% of the organizations apply this.
- The organization has sustainability partnerships with other entities. Partnerships are extremely useful in reducing waste and costs; however, 50% of respondents gave a negative response, compared to 32.4% who have partnerships.

In terms of the use of renewable and natural processes, organizations seem to mostly apply recycling and reuse of materials.

The next question focuses on the organizations' perception regarding the society in which it operates and the environment.

Figure 3 shows a neutral trend in questions 2 and 3 and very positive one in questions 4 and 5, with values of 91.1% and 70.6%, respectively, suggesting that respondent organizations are aware of sustainable practices and develop and encourage them.

Considering the sustainability aspects for organizations, Fig. 4 presents varied trends.

A negative trend in some points deserve attention, namely, taps without flow reduction, flushing cisterns without water reducers, lack of drinking water dispensers, non-reuse of vegetable water in food production, and composting. Those mentioned first cause a waste of water, which could be avoided; at the level of

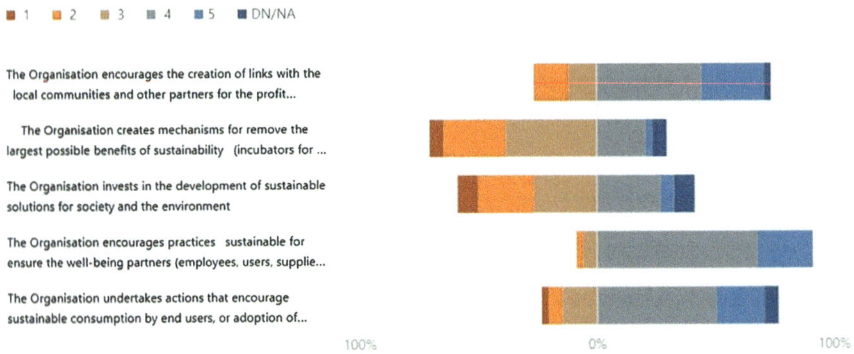

Fig. 3 Opinion of organizations regarding the society in which they operate

composting, this can generate a benefit as fertilizer if the organization has farms or gardens. Moreover, partnerships with the surrounding society could generate a positive exchange for the consumption from local producers. Results also show a huge concern with energy consumption, using energy-saving light bulbs and turning off electrical appliances that are not in use.

Analyzing the respondents' perception to these situations (Fig. 5), the highest percentage of response is at the following levels:

- Refuse: polluting materials (61.8%), carbon emissions (58.8%), and obsolete or damaged materials (26.5%).
- Rethink: plastics (32%), natural resources (26.5%), and recreational objects (23.5%).
- Reduce: carbon emissions (61.8%), plastics (41.2%), and polluting materials (26.5%).
- Reuse: give clothing a second life (38.2%), paper/cardboard (35.3%), and give clothing a second life again (29.3%).
- Repair: give clothing a second life (70%), obsolete or damaged materials (32.4%), and again give clothing a second life (29.4%).
- Recycle: obsolete or damaged materials (41.2%), paper/cardboard (32.4%), school material (23.5%).
- Reintegrate: food waste (91.2%) and natural waste (85.3%), with the other figures not being relevant.
- Respect: natural waste (55.9%), polluting materials (35.3%), and food waste (26.5%).
- Responsible: polluting materials (47.1%), carbon emissions (38.2%), and natural waste (26.5%).
- Repass: polluting materials (50%), carbon emissions (32.4%), and natural waste (20.6%).

Finally, the analysis of the SDGs through their importance and contribution to achieving the targets is reported in Fig. 6.

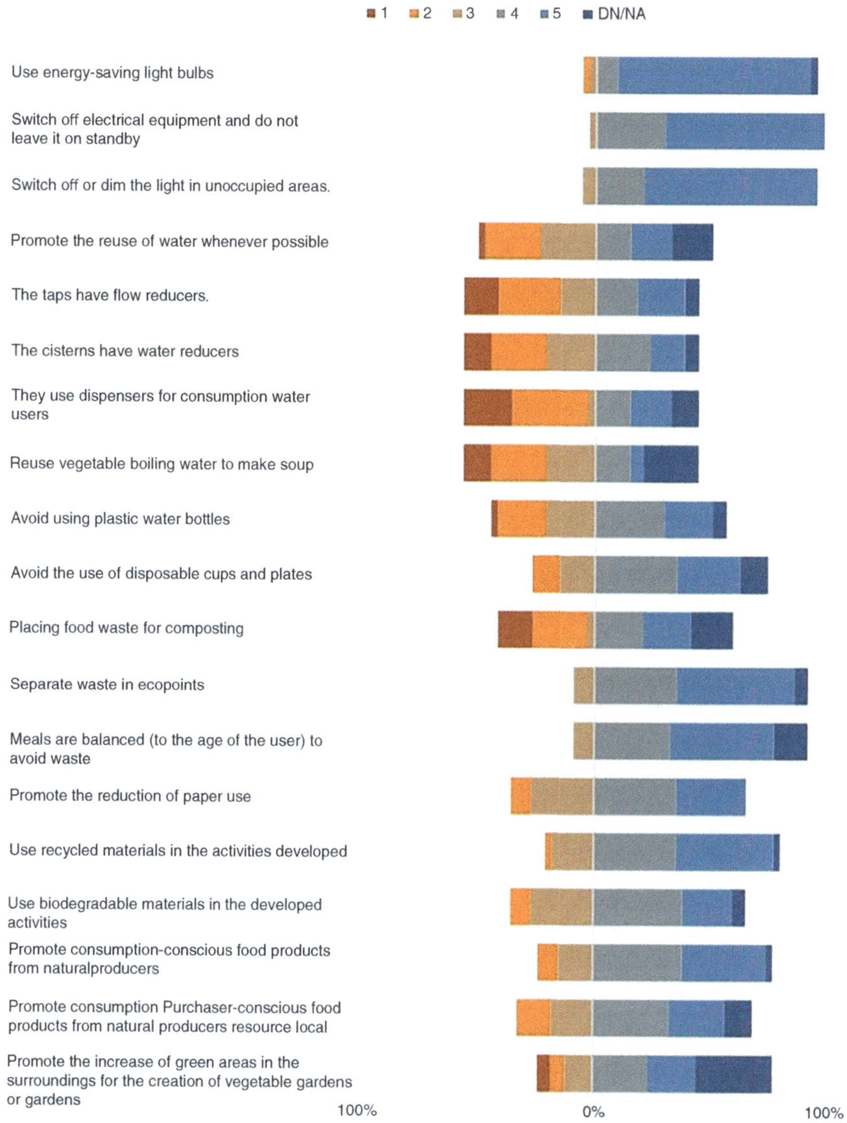

Fig. 4 Map of sustainability questions

When asked about other good practices applied in the organization, references were made to an ecological vegetable garden, concern in replacing equipment by others with better energy levels, avoiding food waste, and other initiatives. As for challenges in promoting sustainability, organizations highlighted lack of information, lack of incentive, limited financial resources, sole promotion of what is known to be achievable, lack of time and support, mindset change, misinformation

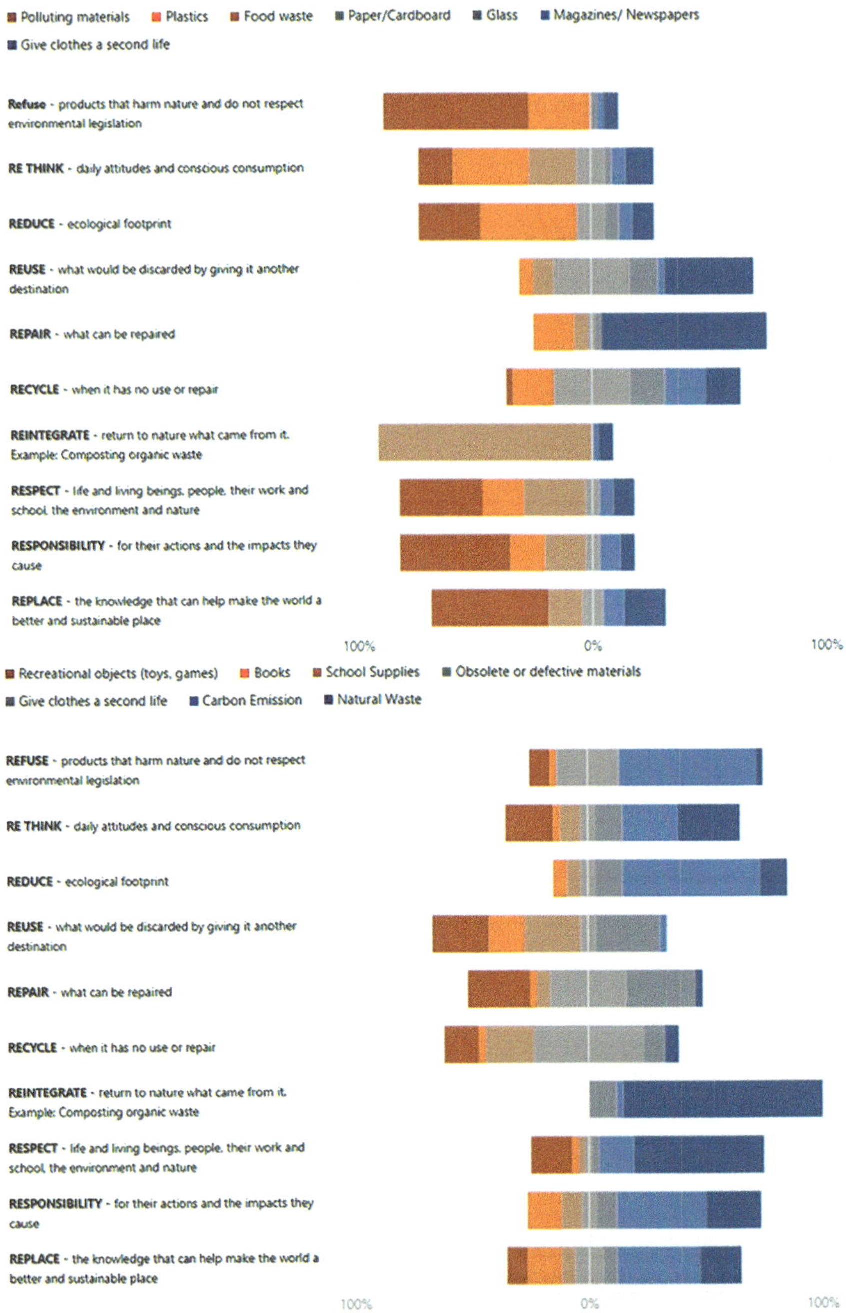

Fig. 5 Answers about the policy of the Rs in different situations

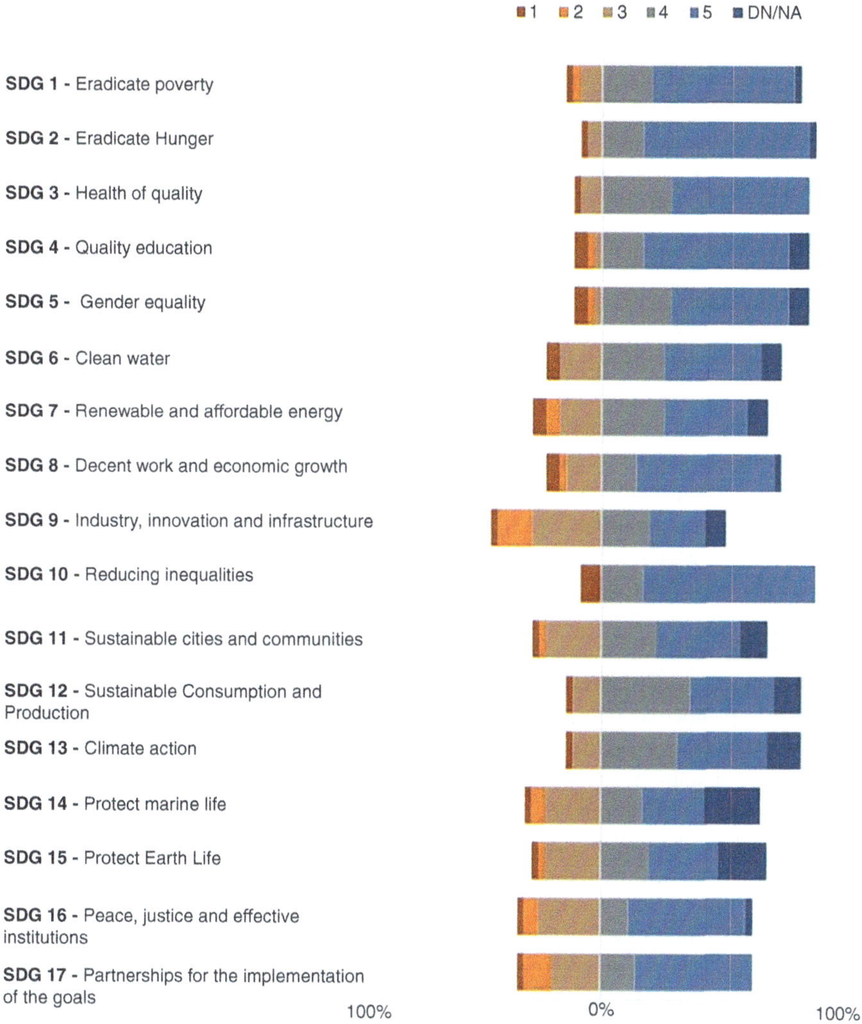

Fig. 6 Responses to the 17 SDGs

among employees, resistance from users, and lack of interest from the family, community, and users.

5 Conclusion

The present study aimed to understand how SE fits into the CE through the application of the R principles and the SDGs.

Surveys are recommended for studies in social sciences, but low response rates are a reality. On the other hand, one does not know who receives online surveys or if they eventually end up in spam. The immense amount of information received today by email causes people to lose interest in answering surveys, even if they consider it important. Though not comparable with this survey questionnaire, the national SESS only got 3550 answers out of a sample of 6019 entities.

The survey results of the present study suggest that there is a lack of interest or knowledge on the topic of sustainability. Despite a response rate of 36%, respondents seem to have some knowledge of the issues, but its implementation is difficult mainly due to resistance of users, families, financial availability, and lack of information/training.

Focusing on the 3R principles, responding organizations seem to place greater emphasis on situations that are heard daily in the media and mainly consider reduce, reuse, and recycle for carbon emissions, plastics, paper, and clothing. When faced with the other seven Rs, organizations continue to give greater importance to the R principles mentioned above and additionally to the Refusal to use polluting materials and the Repair of clothing. From these answers, and due to the low percentage attributed to other choices, one can infer that there is a lack of information about the various ways of treating waste and materials, to which a second life can be given.

Regarding food waste, one can highlight the rethinking of conscious consumption, since very few organizations reuse it, and some still put it in the regular trash. In this regard, protocols could exist between these organizations and local producers, who in partnership could use this waste as fertilizer, and thus the organizations could obtain more sustainable products.

As for energy consumption, there is some concern to reduce it by switching off unnecessary lights and unplugging appliances from the standby state, but as for water consumption, flow reducers for savings are not much applied, which causes waste.

Concerning the SDGs (Agenda 2030), referenced almost daily, directly, or indirectly, we all contribute to their implementation. It is noted in the responses to the questionnaire that hunger, education, gender equality, sustainable production and consumption, and the reduction of inequalities, i.e., the most publicized SDGs, are the ones with the highest percentage of response, although there is some interconnection between them. However, analyzing the SDGs only by the goals may lead to a misperception of their importance because there are only 17, although they include 169 targets that are applicable to various sectors of the economy. Portugal is a motivator of these policies, even though they are not yet fully disseminated in a way that is perceptible to all.

To sum up, the research question that motivated this study, trying to understand the extent to which organizations apply the R principles, how they fit into the SE, and how it is given compliance with the SDG goals, was thus only partially answered. There is some knowledge on the subject, and attempts at implementation, but the resistance encountered discourages its application. The SE can be a pillar for the development of the CE, but for this, it will be necessary to take incentive measures in these organizations that involve stakeholders, employees, users, and volunteers through information directed to this socioeconomic sector.

In view of the institutions of this third sector and with their partnership, future research could understand the behavior of SE organizations by municipality, involving the City Councils (which are one of the main sources of funding), and try to understand the perception that these organizations have of the goals of each of the SDGs. It would also be interesting to study possible partnerships with local producers and to understand how they rate the possibility of a sustainable certification.

References

1. Constituição da República Portuguesa: https://www.parlamento.pt/Legislacao/Paginas/ConstituicaoRepublicaPortuguesa.aspx. Last accessed 29/03/2023.
2. Lei n° 30/2013 da Assembleia da República Diário da República n.° 88/2013, Série I de 2013-05-08, páginas 2727–2728 (Lei de Bases da Economia Social) https://dre.pt/dre/detalhe/lei/30-2013-260892. Last accessed 29/03/2023.
3. Coheur, A. Livro branco 'A economia social... retomar a iniciativa'. Propostas para fazer da economia social um pilar da União Europeia (Social Economy). https://www.cases.pt/livro-branco-economia-social-retomar-a-iniciativa/. Last accessed 29/03/2023.
4. Social Economy European Parliament resolution of 19 February 2009 on Social Economy (2008/2250(INI)) (OJ C, C/76, 25.03.2010, p. 16, CELEX: https://eur-lex.europa.eu/legal-content/EN/TXT/?uri=CELEX:52009IP0062). Last accessed 29/03/2023.
5. Torrinha, F. (1939). *Dicionário Português-Latino*. Ed. Domingos Barreira.
6. Campos, L. M. S., et al. Relatório de Sustentabilidade: Perfil das organizações brasileiras e estrangeiras segundo o padrão Global Reporting Initiative. *Revista Gestão & Produção, 20*(4), 913–926. https://www.scielo.br/j/gp/a/96R88t5VprQhkPy7XzqrBFN/?lang=pt. Last accessed 29/03/2023
7. Souza, N. d. J. (2007). *Desenvolvimento Económico* (5th ed.). Atlas.
8. Transformando nosso mundo: Agenda 2030 -Resolução adotada a 25 de setembro de 2015: https://brasil.un.org/sites/default/files/2020-09/agenda2030-pt-br.pdf. Last accessed 29/03/2023.
9. Sustainable Europe Research Institute.: https://www.seri.at/en/. Last accessed 29/03/2023.
10. Ferreira, M. J., & Campos, P. (2009). O Inquérito Estatístico: uma introdução à elaboração de questionários, amostragem, organização. *Lisboa, Instituto Nacional de Estatística.*: 214 p.
11. Thayer-Hart, N., Dykema, J., Elver, K., Schaeffer, N. C., & Stevenson, J. (2010). Survey fundamentals – A guide to designing and implementing surveys. *Office of Quality Improvement*. 20 p.
12. Borg, W., & Gall, M. (2002). *Educational research: An introduction*. Person Education, Coleção Landranger Map.
13. Bravo, R. S. (1991). *Técnicas de investigação social: Teoria e ejercicios* (7th ed.). Paraninfo.
14. Hill, M.M. & Hill, A.B. Investigação empírica em ciências sociais: Um guia introdutório. (1998).
15. Smith, T. W. (1987). That which we call welfare by any other name would smell sweeter: An analysis of the impact of question wording on response patterns. *Public Opinion Quarterly, 51*, 75–83.
16. Resolução do Conselho de Ministros n.°190-A/2017 - Diário da República n.° 236/2017, 2° Suplemento, Série I de 2017-12-11, páginas 54–73 (2017).
17. Instituto Nacional de Estatística – Inquérito ao Setor da Economia Social:2018. Lisboa: INE, 2020. https://www.ine.pt/xurl/pub/450307417. Last accessed 29/03/2023.

Improvement of the Manufacturing Process of Active Pharmaceutical Ingredients

A. C. P. Soares and R. D. S. Campilho

Abstract Currently, the pharmaceutical industry needs to optimize industrial drug processing procedures. The constant improvement of industrial equipment is an unavoidable strategy to improve the quality of the process, as well as its efficiency and productivity. This work aims to analyze and implement improvements in equipment integrated into production processes in pharmaceutical laboratories to produce active pharmaceutical ingredients (APIs). In the project, studies were carried out using the finite element method (FEM) and traditional structural design, to test the feasibility of the selected solutions. Then, the cost analysis was carried out, and the benefits and results of the new equipment were presented. It was possible to conclude that the proposed equipment demonstrates advantages at an ergonomic, hygienic, and practical level compared to the original ones, which translates into an increase of the company's competitiveness in the pharmaceutical market.

Keywords Pharmaceutical industry · Active pharmaceutical ingredients · Spray drying · Standardization · Process improvement · Finite element method

1 Introduction

The pharmaceutical sector is a specific branch of production that is responsible for creating APIs present in medicines that generate the intended effects [1]. Approving

A. C. P. Soares
ISEP—School of Engineering, Polytechnic of Porto, R. Dr. António Bernardino de Almeida, Porto, Portugal

R. D. S. Campilho (✉)
ISEP—School of Engineering, Polytechnic of Porto, R. Dr. António Bernardino de Almeida, Porto, Portugal

INEGI—Institute of Science and Innovation in Mechanical and Industrial Engineering, Pólo FEUP, Porto, Portugal

J. Reis et al. (eds.), *Driving Quality Management and Sustainability in VUCA Environments*, Springer Proceedings in Business and Economics,
https://doi.org/10.1007/978-3-031-52723-4_2

a drug involves research and preclinical testing of several compounds, which are then submitted to clinical tests that can last for about 6 years. During this time interval, the manufacturing process is also developed [2]. A series of key aspects must be incorporated for the creation of the product, namely [3], the product must fit the current offer for the target market, the financial sustainability of the product is essential, latest advances in manufacturing capability and technologies must be used, the development of new products should be focused on population groups with certain similar characteristics, tests are necessary to guarantee functionality and security, and environmental care, health, and safety are valuable for product development and its fabrication.

The most used operation in the pharmaceutical manufacturing process is spray drying, as it produces drugs with uniformly distributed particles in a polymeric matrix. This technique consists of transforming liquids with dissolved or suspended particles into dry powder. In this method, the APIs are dissolved in solvent and synthesized in a gas flow, usually nitrogen, with high temperature and speed [4]. The spraying process consists of concentration, atomization, contact of liquid with gas at high temperature, drying of the liquid, and finally separation. An important factor for the efficiency of the process is the collection of the product, mainly of small particles. The greater the efficiency of these equipment, the greater the yield of pharmaceutical products that are composed of expensive components, in this case API [4]. Cyclone size and geometry also influences the maximum yield of spraying processes. Computational fluid dynamics (CFD) allow evaluating the performance of cyclones with geometries that differ from the traditional ones [4]. The gas used in the cyclones for the transformation is sent to condensers, to be treated and the solvents to be removed. Thus, it can be reused in the process, reducing its consumption on a large scale [5].

Machine design is vital for the development of API fabrication equipment [6]. Each project needs several resources, namely, information, design tools, equipment, and qualified human resources [7]. The FEM acquires special relevancy within the scope of machine design [8]. The engineering problems that can be analyzed by the FEM include the analysis of stresses, temperatures, fluid flow, and aerodynamics [9]. On the other hand, industrial automation also plays an important role in improving manufacturing production performance, as it is responsible for operating and activating equipment and processes in industries. Therefore, task times and movements must be evaluated and optimized to increase work performance [10]. Freitas [10] employed industrial automation technologies to improve production processes in a pharmaceutical company, aiming to optimize the time of industrial procedures, especially in the packaging stage of the manufactured product. Therefore, human-machine interfaces (HMI) and programmable logic controllers (PLC) communication software were applied to the scale, packer, and baler equipment to eliminate the preparation of machines by workers. This application resulted in a gain of 2 h for product production and allowed a 35% decrease in wasted product, which caused a 47% recovery of costs related to product waste. The review of Chen et al. [11] discussed digital twins' application to pharmaceutical and biopharmaceutical manufacturing. Emphasis was given to latest developments in process analytical technology (PAT), process modelling approaches, and data integration studies. It was concluded that, in pharmaceutical manufacturing, digital twin building blocks, including PAT methods, and integration approaches have been proposed and validated in recent years, although integration of all systems is still difficult due to PAT accuracy, computation and data communication in real time, and data security limitations. In general, scientific works on process improvement in the pharmaceutical industry from a fabrication line approach are virtually inexistent, and thus a successfully implemented work on this topic could provide productivity, cost reduction, and competitiveness benefits in the industry.

This work aims to analyze and implement improvements in equipment integrated into production processes in pharmaceutical laboratories to produce API. In the project, studies were carried out using the FEM and traditional structural design, to test the feasibility of the selected solutions. Then, the cost analysis was carried out, and the benefits and results of the new equipment were presented. The work is divided as follows. Initially, the project to be undertaken is described, along with the characterization of the current design problems, which should be addressed in the proposed equipment. The main section (Equipment Design) describes the final solution, followed by details on the design process, prototype construction, and finally the cost analysis. The conclusions summarize the work done and respective limitations.

2 Methods

2.1 *Project Description*

Due to the growing need of the pharmaceutical market for the improvement and installation of equipment and laboratory modules called spray dryers, a project was initiated with the scope of manufacturing pharmaceutical equipment and integrating them into skids. Skids enable the integration of equipment and piping, which aim to carry out a given process, in this case the spray drying process, in a module limited by a steel structure (Fig. 1). The skid concept is advantageous for the quick and efficient installation of spray dryers in buildings intended for the manufacture of API.

The objective of this project is to model an industrial spray dryer skid that integrates the demonstrated equipment and the necessary interconnections for the operation of the manufacturing process. During this project, an attempt is also made to optimize the equipment manufactured in the company that is inserted in the industrial spray dryer, more specifically the bag filter and the Hepa filters.

2.2 *Problem Characterization*

The current version of the spray dryer module being manufactured (Fig. 2) is about 3 m wide and long and can reach 12 m in height. These dimensions may vary

Fig. 1 Example of a skid module

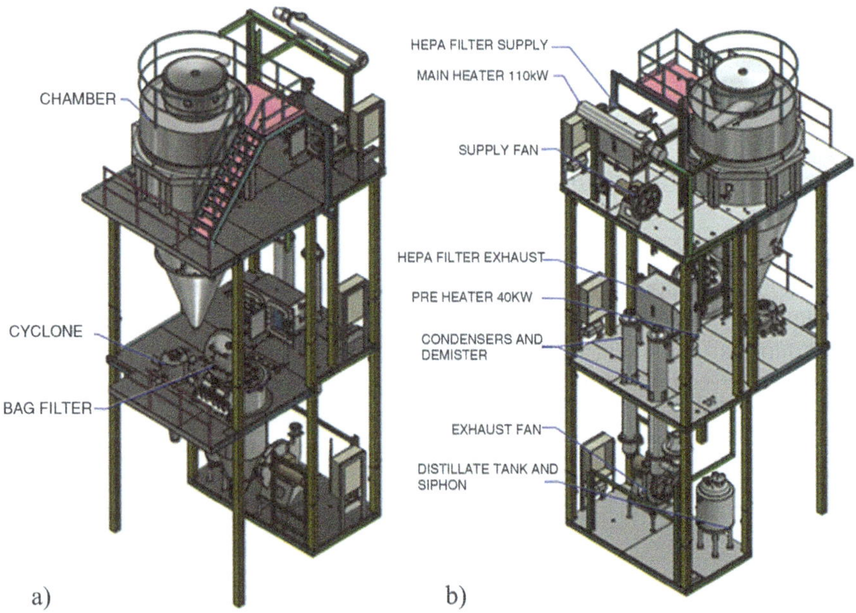

Fig. 2 Current spray dryer: (**a**) clean zone and (**b**) technical zone

depending on the selected camera size. The larger the dimension of this equipment, the larger the module created will be, due to the necessary adjustments to the structural design. The spray dryer comprises a variety of essential equipment for the manufacturing process. The main components are the chamber, the bag filter,

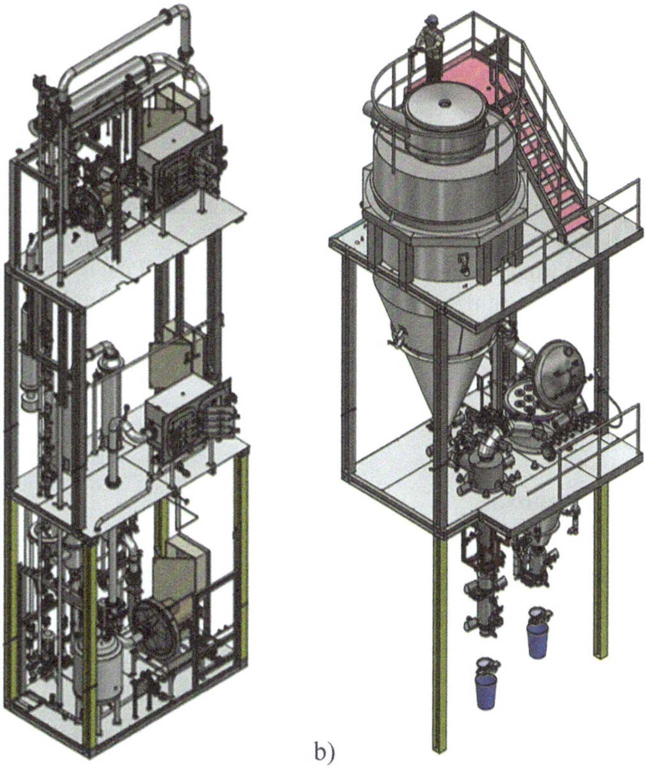

a) b)

Fig. 3 Spray dryer module: (**a**) skid of the technical part and (**b**) skid of the clean part

the cyclone, the Hepa filters, the resistances, the fans, and the condensers, as shown in Fig. 2.

The module is made up of two skids, namely, the clean part and the technical part, as shown in Fig. 3. This division is due to the considerable size and weight of the spray dryer to carry out its transport and installation inside the building as a single piece. The skids are around 12 meters high, the technical module weighs 10 tons, and the clean module weighs 12 tons. The clean zone consists of rooms where operators have access to API production, so these must be completely sealed and sanitized to avoid contamination of people, air, or the final product. In technical areas, access is generally only necessary to carry out qualification tests or to maintain apparatus or equipment. The space is used to place the piping responsible for cleaning the process piping at the end of product processing and draining waste created during the procedure. The manufacture of bag filters and Hepa filters is currently the responsibility of the partner company, which aims to optimize this equipment according to the needs and growing demands of the pharmaceutical market. This equipment initially had design and finishing flaws, with emphasis to the warping caused by the current lid opening system, which compromises the equipment's sealing. The situation with Hepa filters is similar, as the opening system needed to be more practical for the user and the equipment doors suffer flexion due to their weight. Thus, it was necessary to create new designs for the equipment to improve the resistance to high pressure, for better surface finishing and easier handling by the operator. Another difficulty encountered was the construction of modules in the intended buildings. Since these laboratory buildings usually have other pharmaceutical units in operation, access to the site and working conditions is limited. This problem led to the idea of creating the spray dryer modules in skid format, with the advance installation of all the necessary equipment and connections, to optimize their assembly time at the intended location.

3 Equipment Design

3.1 Final Proposal

In this section, the final model of the spray dryer module is described with the changes implemented to the current system, which have the purpose of standardizing these laboratory modules. The new spray dryer manufactured at the company is called SD-653 (Fig. 4). This spray dryer is approximately 3 meters wide and long, and the module reaches 12 meters in height.

The module consists of two skids, namely, the clean part and the technical part, as shown in Fig. 5. This division is due to the considerable size and weight of the spray dryer to transport and install inside the building as a single piece. The clean zone consists of rooms where operators have access to API production with proper personal protection. Thus, these must be completely sealed and sanitized to avoid contamination of people, air, or product. In technical areas, the access is generally only necessary to carry out qualification tests or to do maintenance of equipment. The space is also used for placing the piping responsible for cleaning the process piping after product processing and for draining the waste created during the procedure.

3.2 Design Process

Spray Dryer Compared to the previous module (SD-1255), the most recent SD-653 module is composed of intermediate floors, having a total of five floors. With this distribution of floors, the clean module has intermediate floors that correspond to the technical part of the spray drying process. This solution serves to reduce the number of connections between the two skids, which simplifies the process of preparing the transport of this Spray Dryer module, because it reduces the number

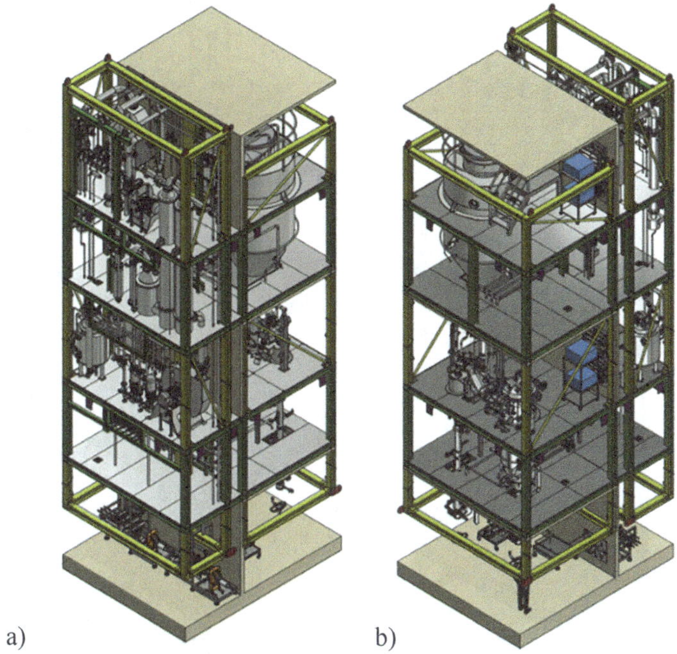

Fig. 4 Module SD-653: (a) technical module and (b) clean module

of pipelines that can be removed and transported in a separate container. At the same time, this solution frees up space in the clean area for operators to move around. The main equipment manufactured at the company to install in the spray dryer are the Hepa filters, located at the fifth and third floors, and the bag filter, located at the second and third floors, and will be subjected to improvement at an ergonomic, hygienic, and technical level, through the alteration of the system opening and tightening of the equipment, to improve the quality of the produced products and the manufacturing process.

Bag Filter For the bag filter, the aspects to be improved are the lid opening system, the interior cleaning system, and the lid tightening system. To improve the opening system, a cap balancer consisting of two springs is installed in the bag filter of the SD-653 module to facilitate handling the lid of this equipment. The bag filter's interior cleaning system was also modified. The updated model is composed not only by spray balls, but also of pop-ups and pneumatic hammers, to carry out a more effective cleaning (Fig. 6). The two pneumatic hammers are used to loosen residual particles from the walls so that the interior can be cleaned afterward. The eight pop-ups have the same function as the spray balls, but these can move forward to wash the equipment, and at the end of the cleaning process, they move back to be aligned with the inner wall of the bag filter. For maintenance access, a cone was placed in the lower area, like what happens with the chambers inserted in the spray

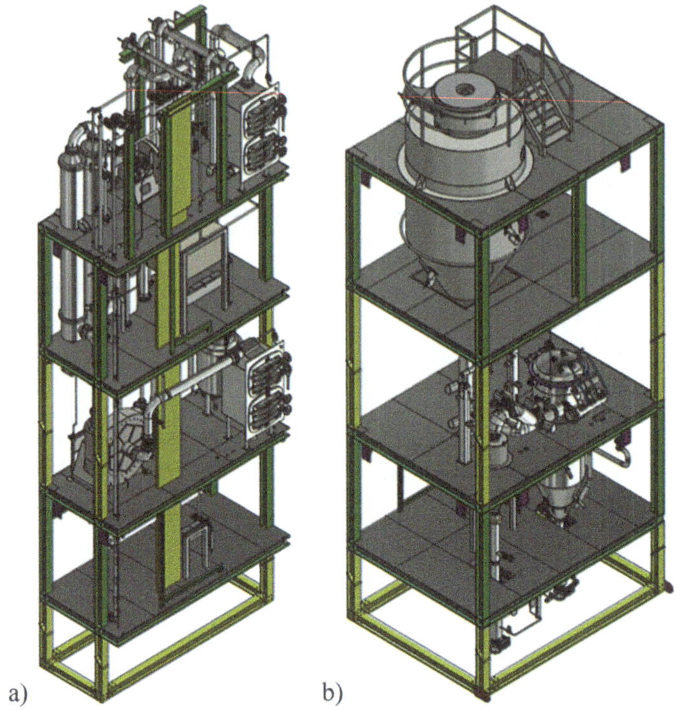

Fig. 5 SD-653 module: (a) technical module and (b) clean module

drying process. This inspection cone was designed to remain suspended and rotated to one side of the cylindrical body in case the equipment is opened. This component is screwed to the bag filter body and has an O-ring for sealing the area between the inspection cone and the body of the equipment. This modification allows access from inside the bag filter through the technical area to carry out any necessary maintenance. As the tightening system had flaws in terms of sealing, hygiene, and ergonomics, it was replaced. In turn, the updated bag filter tightens the cover to the body of the equipment through a triclamp manufactured by the company (in purple color), as shown in Fig. 7. The rotation axes of the clamp tightening ends are positioned next to the cap balancer in the rear area, and the adjustment is made using a handwheel with a trapezoidal threaded screw located in the front part of the bag filter.

Hepa Filters The improvements considered in the Hepa filters consisted of modifying this equipment to be more resistant to increased pressure, by replacing the tightening system. The changes carried out to improve the pressure resistance involved increasing the thickness of the structural sheets and the installation of reinforcements in the doors to restrict their deformation. As the working conditions in this equipment ranged from constant pressure 400 mbar to variation between

Fig. 6 Bag filter's interior cleaning system

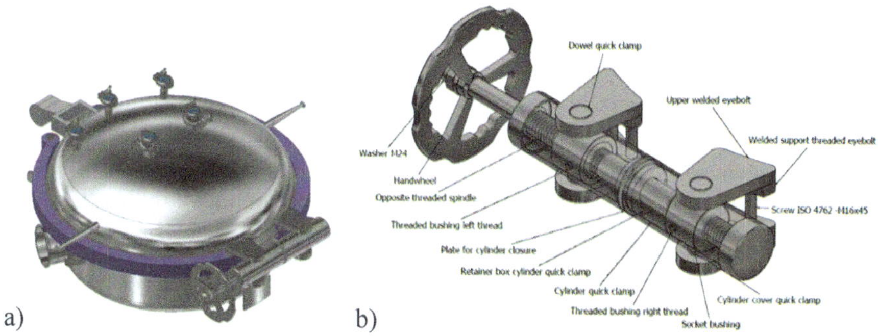

Fig. 7 (**a**) Bag filter's tightening system installation and (**b**) bag filter's tightening system

−500 mbar and 500 mbar, it must be ensured that there are no significant structural deformations that compromise the operation, sealing, and durability of the Hepa filters. In the study, the most extreme cases were considered, which correspond to the application of uniform pressure on the structure and on the equipment door of

Fig. 8 Mesh applied to Hepa filter structure

500 mbar and −500 mbar. The mesh applied to the equipment structure is standard (Fig. 8), and geometric nonlinearities were considered to simplify the analysis. The structure of the equipment is made of stainless steel (316 L), and in this study the materials are modeled as elastic. For both tests, the boundary conditions consisted of clamping the two front panels of the equipment. The pressures were applied to all interior walls of the equipment, either with the inward or outward direction.

As the applied pressures are symmetrical, the von Mises equivalent stress and safety coefficient are equivalent between the −500 and 500 mbar. However, the resulting displacement and deformation proved to be symmetrical between tests. The data resulting from the studies carried out demonstrate that the maximum von Mises equivalent stress corresponds to 210.5 MPa, but it is restricted to a very small area in the contact zone between the cone and the structural body of the Hepa filter. The maximum deformation and the minimum factor of safety are also located at the same area. It is considered that the structure has the capacity to resist the working pressures, since the minimum safety factor is 1.2. The maximum displacement in both tests is 2.14 mm. As this occurs in the central part of the interior walls, it

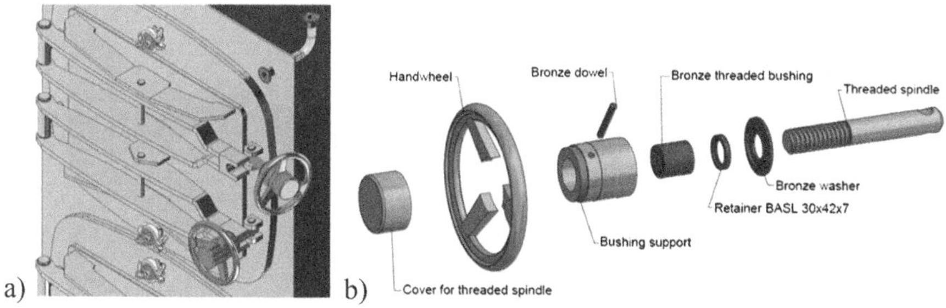

Fig. 9 (**a**) Hepa filter's tightening system installation and (**b**) Hepa filter's tightening system

is considered that the sealing of the equipment is not affected. To optimize the tightening system of the Hepa filters, the tightening system of the old model is replaced by a similar system installed in the bag filter (Fig. 9). The demonstrated set makes opening the equipment more practical and hygienic compared to the screwed tightening system.

3.3 *Prototype Construction*

The spray dryer system was constructed and tested by the company. By handling the lid of the manufactured bag filter, it is visible that the installation of the cap balancer makes opening or closing the equipment a simple and safe task. The weight of the cap is practically nullified by the elastic force exerted by the cap balancer. The bag filter's optimized tightening system (Fig. 10a) also proved to be an improvement in ergonomic, hygienic, and practical terms. This implementation makes the process of opening the equipment faster and does not compromise its sealing, which benefits the working conditions of the operators. As the opening system and the tightening system proved to be efficient, it is expected to be implemented in future bag filter models manufactured by the company. For the approval of the Hepa filters built, shown in Fig. 10b, hydraulic experimental tests were performed to evaluate their behavior under working conditions. The equipment was tested at constant pressures of −500 mbar and 500 mbar for a period of 1 h. The test results showed that this equipment is resistant to the applied pressures, since it did not show leaks, visible deformations, or warping in the structure or doors of the Hepa filters that could compromise their functionality or sealing. The tightening system implemented in this equipment also proved to be more efficient compared to the previous system. This improvement contributed to the hygiene of the equipment and makes the process of opening and closing the equipment faster. Since the results of the changes made to the Hepa filters were satisfactory, these improvements will be implemented in the following models.

a) b)

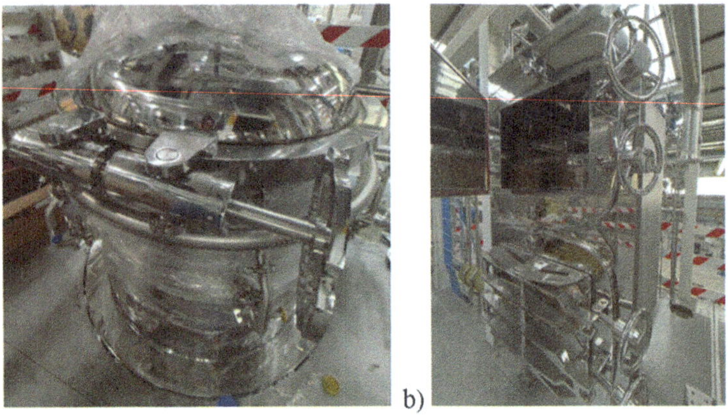

Fig. 10 Manufactured (**a**) bag filter and (**b**) Hepa filter

Table 1 Labor costs and manufacturing processes to manufacture the bag filter

	€/h	Resources	Runtime (h)	Cost (€)
Drawing office	27	1	72	1944
Manual sheet cutting	22	1	3	66
Laser cutting	22	1	6	132
Forming/bending	22	1	33.5	737
Machining	32	2	32.8	2100
Assembly	25	2	355.2	17,760
Polishing	22	1	123	2706

3.4 Cost Analysis

Bag Filter Table 1 represents the estimated costs of labor and manufacturing processes for the execution of a bag filter. The total cost of labor and manufacturing processes reached 25,445 € with a running time of 625.5 h. The costs of materials, purchase of market components, and raw materials are 14,710 €. It is possible to conclude that the bag filter costs 40,155 €, considering the labor and materials used.

Hepa Filter The costs associated with labor and manufacturing processes for each Hepa filter unit are shown in Table 2. The total cost of labor and manufacturing processes for the construction of a single Hepa filter is equivalent to 15,191 € with the runtime of 503.5 h. Thus, the execution of two units of this equipment represents a cost of labor and manufacturing processes of 30,382 €. The purchase of the market components and raw material corresponds to a total cost of 7853 €, which implies the total expenditure on materials for the two equipment corresponding to 15,706 €. As a result, the execution of the two Hepa filter units costs 46,088 €, considering the labor, manufacturing process, and purchased materials.

Table 2 Labor costs and manufacturing processes to manufacture the Hepa filters

	€/h	Resources	Runtime (h)	Cost (€)
Drawing office	27€/h	1	38	1026
Manual sheet cutting	22€/h	1	7,4	163
Laser cutting	22€/h	1	1,6	35
Forming/bending	22€/h	1	2	44
Machining	32€/h	2	13.3	851
Assembly	25€/h	2	120.2	6010
Polishing	22€/h	1	321	7062

Table 3 Manufacturing costs for SD-1255 and SD-653 modules

	SD-1255 (€)	SD-653 (€)
Spray dryer	316,301	208,352
Bag filter	46,000	40,155
Hepa filters	36,980	46,088
Total	399,281	294,595

Comparison of SD-1255 and SD-653 Module Costs It is possible to compare the manufacturing costs between the SD-1255 module and the estimated costs of the SD-653 (Table 3). Based on the presented values, the manufacturing cost to build a module reduced by 116,575 €. This difference is significant, considering the improvements applied to increase the quality of the spray drying process.

4 Conclusions

The main scope of this work was the improvement of an equipment supplied to the pharmaceutical area, to facilitate its handling and cleaning, and to reduce delivery times for spray dryers as well as pharmaceutical equipment used in these units, specifically bag filters and Hepa filters, to offer a faster response to the needs of the pharmaceutical market. The predesign stage was essential to define different solutions for each problem encountered. Through the comparison and analysis of the ideas presented, it was possible to select the most viable solution for the problems reported in the spray dryer module. FEM analyses and traditional structural design were applied to assess the feasibility of improvements made to equipment. After the design process was completed, the proposed equipment can now be considered as viable substitutes for the originals that formed the starting point of this work, with ergonomic, hygienic, and practical advantages. These advantages translated into an increase in the company's competitiveness in the global market due to the improvement of the products made available to the industry. The solutions found for the standardization of spray dryer modules enabled reducing the manufacturing time of these modules, as well as the reduction of costs resulting from raw materials, labor, and manufacturing processes related to the manufacture of the module and its components. Nonetheless, few limitations can be pointed out. In the event of changing the working conditions currently imposed on the spray drying process, it would therefore be essential for the equipment that the bag filter and the Hepa filters could be improved to be resistant to greater variations in pressure and temperature. To reduce the delivery response time of a spray dryer module, strategies should be analyzed to help reduce the time taken to build them. The standardization of pharmaceutical modules allows for the manufacture of piping and equipment for stock. One strategy for reducing response times is to build stock to offer a faster response to requests from the pharmaceutical market.

References

1. Salwin, M., Andrzejewski, M., & Kraslawski, A. (2021). Information asymmetry in product-service system design-example of the pharmaceutical industry. *Procedia Manufacturing, 55*, 282–289.
2. Cervera-Padrell, A. E., Skovby, T., Kiil, S., Gani, R., & Gernaey, K. V. (2012). Active pharmaceutical ingredient (API) production involving continuous processes–A process system engineering (PSE)-assisted design framework. *European Journal of Pharmaceutics and Biopharmaceutics, 82*(2), 437–456.
3. ten Kate, A. J., Piccione, P. M., Westbye, P., & Becker, A. F. A. (2022). An industrial and chemical engineering perspective on the formulation of active ingredients in pharmaceuticals and agrochemicals. *Current Opinion in Chemical Engineering, 36*, 100747.
4. Graham, L. J., Taillon, R., Mullin, J., & Wigle, T. (2010). Pharmaceutical process/equipment design methodology case study: Cyclone design to optimize spray-dried-particle collection efficiency. *Computers & Chemical Engineering, 34*(7), 1041–1048.
5. Baumann, J. M., Adam, M. S., & Wood, J. D. (2021). Engineering advances in spray drying for pharmaceuticals. *Annual Review of Chemical and Biomolecular Engineering, 12*, 217–240.
6. Golenko, A. (2010). *Fundamentals of machine design. A coursebook for Polish and foreign students*. Dolnośląska Biblioteka Cyfrowa.
7. Wang, M., Zhang, Z., & Zhang, Z. (2019). Current research and prospect of cloud application in manufacturing equipment design. *IOP Conference Series: Materials Science and Engineering, 631*(3), 032007.
8. Santos, P. M. M., Campilho, R. D. S. G., & Silva, F. J. G. (2021). A new concept of full-automated equipment for the manufacture of shirt collars and cuffs. *Robotics and Computer-Integrated Manufacturing, 67*(February), 102023.
9. Campilho, R. (2012). *Método de elementos finitos* (Vol. 205 (1), pp. 203–205). Publindústria.
10. Freitas, E. S. (2020). Aplicação de tecnologias de Automação industrial para melhoria de processos de produção: Um estudo de caso. *Revista Eletrônica de Engenharia Elétrica e Engenharia Mecânica, 2*, 91–102.
11. Chen, Y., Yang, O., Sampat, C., Bhalode, P., Ramachandran, R., & Ierapetritou, M. (2020). Digital twins in pharmaceutical and biopharmaceutical manufacturing: A literature review. *Processes, 8*(9), 1088.

Product-Service System Development for 3D Customized Production in Hospitals

Milene Santos, Bruno Soares, and Helena Carvalho

Abstract Customized prosthesis and orthosis (P&O) manufacturing is growing as a market advantage and contributes to the positive impact on patient's wellbeing. Three-dimensional printing allows a flexible manufacturing, adapting different characteristics according to the patient's requirements while increasing the device functionality. Considering the relationship between the customized P&O and the extended attendance that patients need, this manufacturing approach is requiring a new system which can consider not only the product but also an integrated service. Despite the huge potential that product-service system (PSS) could have in healthcare industry, there is a lack of research around the topic. PSS is a supporting network which is related to the product that generates value while satisfying the customer's needs. The objective of this paper is to identify representative PSS models applied to the healthcare context and find a proposal that can be applied to 3D printing customized P&O. Pursuing this objective, a review on PSS business models applied to healthcare was driven. This paper aims the integration of the business models proposed in the literature, to mitigate the limitations presented in the current state-of-the-art.

Keywords Customized · 3D printing · Product-service system · Prosthesis and orthosis

M. Santos (✉)
Department of Mechanical and Industrial Engineering, NOVA School of Science and Technology, NOVA University Lisbon, Caparica, Portugal
e-mail: msa.santos@campus.fct.unl.pt

B. Soares · H. Carvalho
UNIDEMI, Department of Mechanical and Industrial Engineering, NOVA School of Science and Technology, NOVA University Lisbon, Caparica, Portugal

Laboratório Associado de Sistemas Inteligentes, LASI, Guimarães, Portugal

J. Reis et al. (eds.), *Driving Quality Management and Sustainability in VUCA Environments*, Springer Proceedings in Business and Economics,
https://doi.org/10.1007/978-3-031-52723-4_3

1 Introduction

According to the Assistive Technology program, [1] there are an estimated 65 million people globally who live with a part of the body removed following a trauma, disease, or congenital conditions. A prosthetic device is an externally applied device used to replace wholly or in part an absent or deficient limb segment. An orthotic device is an externally applied device used to modify the structural and functional characteristics of the neuromuscular and skeletal system to restore or improve mobility and functioning and to correct deformities [2]. Both devices can restore or improve mobility, functioning, and correct deformities. The appropriate selection of devices improves quality of life and reduces mortality. Personalized assistive devices are becoming an emergent healthcare field where products, such as customized prosthesis and orthosis (P&O), can provide not only a substantial impact on people life's but can be also a huge market advantage.

Three-dimensional printing allows the manufacturing of customized P&O with adapted geometries leading the design flexibility according to the patient's requirements, increasing the device functionality, improving their well-being [3]. At the same time, it increases the physician knowledge regarding the patient anatomical variants, since it continuously monitors their adaptation to the device while reducing the assembly requirements [4], number of surgeries, anesthesia times, and healthcare costs [5]. It is becoming more evident that customized medical devices manufacturing is facing a growing demand, and considering the relationship that should exist between the devices and the extended attendance that the patients need, customized devices manufacturing is requiring a new systems approach which considers not only the product but the integrated service. The PSS approach considers the product, but also a service with supporting networks to satisfy and generate value while satisfying customer need. This type of system could have huge potential in healthcare industry; however, despite all the application promising potential, there is a lack of research around the topic. The objective of this paper is the integration of the business models proposed in the literature to mitigate the business model limitations presented in the current state-of-the-art.

2 Literature Review

2.1 Additive Manufacturing in P&O Production

About half of the world's annual healthcare expenditure is due to patient's injury or damage to tissue and organs [5]· which makes this healthcare area not only very important in the patient point of view but also to involve giving the patient the better possible recovery using the better economical choice. Traditionally, prosthesis and orthosis (P&O) were fabricated using a cast mold (Fig. 1).

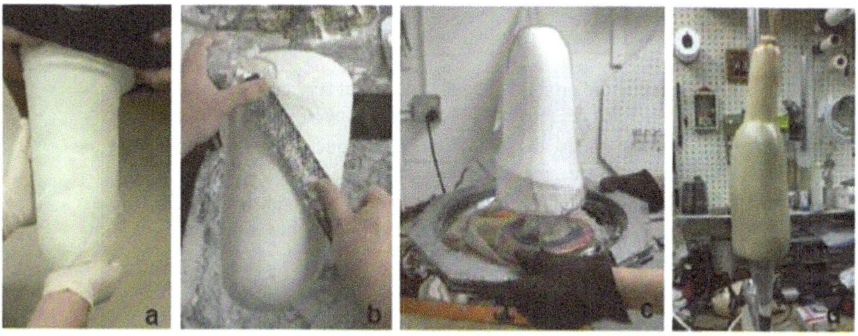

Fig. 1 Traditional transfemoral prosthesis fabrication process [3]

To produce the mold, first the relevant patient anthropometric measurements are obtained by taking the relevant patient anthropometric measurements plaster bandages around the affected body part; then the mold is created with the use of thermoplastic heated and vacuum-forming. Body adjustments to ensure comfort and functionality could be needed. Although the casts creation is the most traditional fabrication method, the procedure has limitations. Waste of materials, high time, labor costs, and product quality depending on the prosthetist or orthotist skills [6] are examples of the process limitations. Moreover, the process normally includes several patient visits which leads to final product delivery time variation from several days to weeks delaying the process and the patient access to the piece [7].

2.2 *Additive Manufacturing in P&O Production*

Three-dimensional printing is a process allied to "computer aided design" (CAD) and is performed layer by layer where 3D objects are formed when the layers are fused in a row in a two-dimensional cross section [8], presented in recent years, utilities of medical 3D printing that have grown exponentially [9]. Several authors highlight the increasing use of 3D printing in the customized P&O sector because it (i) could provide the benefits of the customization, personalization, cost-effectiveness, improved productivity, accessibility, short production time, and assembly simplicity [8]; (ii) allows the production, in one single run, of parts that are extremally difficult or even impossible to fabricate with another technique, completely changing the traditional product life cycle [10]; (iii) facilitates just-in-time, specific, or tailored objects, manufactured on a small scale, but cheaply and without the need for supply chains [11]; and (iv) provides the opportunity to create unique pieces that would make a difference in patient's life. Despite the benefits associated with 3D printing and the P&O manufacture, there is a lack of framework of the procedure that covers the process from the conception to the final adaptable products [6]. The customization factor indicates a close relationship between the

Table 1 PSS types

PSS category	
Product-oriented	Selling the product includes additional services such as after sales, maintenance, repair, reuse and recycling, training, consulting, etc.
Use-oriented	Selling the use or availability of a product which is not owned by the customer (leasing, sharing)
Result-oriented	Selling a result or capability instead of a product (web information replacing directories, selling laundered clothes instead of a washing machine)

Adapted from Baines et al. [15]

product and a differentiated service provided to each patient regarding their unique characteristics.

2.3 Servitization and Product-Service System

The activity of selling the services provided by the product rather than the product itself for a manufacturer is called servitization [12]. Servitization is pointed as the transformation of manufacturers from selling products to providing product-service systems or customer solutions [13].which involves a shift from product-oriented business logic to service-oriented business logic, which is more customer-centric and market-driven [14]. Low level of servitization indicates that product-oriented logic still dominates, and it is more likely to develop services around current products to consolidate these products. High levels of servitization are guided by service-oriented logic, and it is more likely to provide advanced services to support customers' actions [14]. Servitization is the transformation process followed by a product-oriented organization to become a product-service system provider [15]. In the literature, PSS is described in two perspectives:

- A system of products, services, supporting networks, and infrastructures design to satisfy customers' needs and generate value [16].
- A business concept based on a marketable set of services and products/technologies, capable of jointly fulfilling a user's need [17], where there is a shift from commercialization of manufactured products to the combination of a product and a service into an integrated offer [18]·

Though different authors use different subdivisions to describe PSS, there is some convergence on the existence of three different PSS types (Table 1).

Although the interest and potential of PSS, many companies are still facing challenges transitioning from product-centric to service-centric business models, and for this reason, it has been pointed as a promising research field [19]. PSS approach requires the configuration of an appropriate business model that could enhance the development of an innovative and sustainable offer to the customers,

[20] yet the existing tools and methods available are more product-focused, and the design of a PSS, instead, falls in a different domain [20]. Although there are very few tools supporting the development of PSS, there needs to be a unified PSS business model perspective to advance theory on PSS and servitized business models [19]. The focus of a PSS design activity shifts from the definition of new products to the reorganization of existing elements. In the design of a PSS, cultural and social values, people, and technology have the same relevance [21]. Companies have a well-defined product development process, but they need a sufficiently defined service development process and to be better equipped with appropriate approaches, methodologies, and tools to support the development of PSS [22]. Table 2 shows an overview of different authors that deal with PSS design.

2.4 *Product-Service System in Healthcare*

The most prevailing motivation for applying a PSS approach in healthcare industry is because it is considered as a complex social system which value is created through the combination of tangible and intangible goods, high professional skills, regulatory regimes and rules, organizational practices, funding mechanisms, and advanced technologies [23]. Table 3 shows an overview of different authors that study PSS in the healthcare context.

Santos and Wauben [36] conducted a study focus on the role of systems design on the challenges related to the safety and performance of medical equipment that is transferred in the aftermath of a humanitarian crisis, indicating that PSS could be combined to support humanitarian innovation, to explore ways to redesign their system, optimizing the use of resources and enhancing the competitivity and transparency in the way of the medical equipment use and how they service the beneficiary of aid. However, it is important to further elaborate and test the potential contribution of the PSS contribution in this context.

Pourabdollahian and Copani (2015) [28] created a general business model structure for customized production in healthcare considering the hospital as the manufacturer. The main features of this model are (i) location, which describes the possible location of the AM lab; (ii) operation personnel, which describes the provenience of the operation personnel necessary to develop the AM production; (iii) equipment ownership which is provenience of the necessary equipment and technology; (iv) maintenance which is maintenance of the machines; (v) payment modes which is payment regarding the ownership of the machinery or machinery outsourcing; and (vi) target segment which is final product user. According to the author, the development of those business model alternatives can be considered as a tool for a business model design for customized production of prostheses by hospitals. The model allows to develop different positions in terms of the levels of servitization and customization. These different positions enable the hospital to decide the best alternative business model, considering the strategy and objectives that need to be fulfilled. Although the proposed business models are beneficial to

Table 2 PSS design methods

Method	Definition	Authors
Integration definition for function modeling (IDEF0)	Graphical process modeling methodology used to implement systems and engineer software and used in data functional modeling, simulation, object-oriented analysis, and knowledge acquisition	[24]
Service blueprint	Diagram that displays the entire process of service delivery, listing each stage activities performed by the different roles involved	[24][25][26]
Life cycle model	Framework of processes and activities regarding the life cycle that may be organized into stages	[27]
Function model	Graphical representation of an enterprise's function within a defined scope	
Actor network model	Theoretical and methodological approach to social theory where everything in the social and worlds exists in constantly shifting networks of relationships	
Morphological box	Template to generate solutions to complex problems. The approach is to logically decompose the problem into several variables/factors for which solutions or ideas can be identified	[28]
PSS characterization	Characterization approach to characterize a PSS systematically in terms of different parameters	[29]
Business model canvas (BMC)	Strategic management template used for developing new business models and documenting existing ones	[30][31]
Categorization template	Template that summarizes different categories regarding the product life cycle, orientation, spatial proximity level, and relational proximity	[32]
Life cycle thinking	Holistic view of a product, process, or service from production through to consumption or use to end-of-life. Assesses the impact of the product, process, or service in environmental, economic, and social terms through the whole life cycle	[33]
Model-based system engineering (MBSE)	Methodology used to support the requirements, design, analysis, verification, and validation associated with the development of complex systems	[34]
Goal modeling (GM)	Capture stakeholders and system goals, together with social, intentional, and structural relationships, in a way that supports trade-off analysis and decision-making	
Engineering value assessment (EVA)	Review of new or existing products during the design phase to reduce costs and increase functionality to increase the value of the product	[35]

support the hospital in terms of customized production of medical products, the author considers that the study has its own limitations. The developed business model is theoretical and has importance to convert the proposed business model to a practical that can be used by the hospitals.

Table 3 PSS in healthcare

Authors	Research area	Research methodology	Research gap
[36]	Systems design		PSS developments are not specific to the context of companies in the healthcare or humanitarian aid sectors
[28]	Development of a PSS-oriented business model	Development of a PSS-oriented business model	The developed business model is theoretical, and it is essential to convert the proposed business model into a practice that the hospitals can use
[29]	Characterizing product-service systems	Develop a repeatable process for PSS characterization	PSS characteristics proposed may be limited to new PSS development. In the healthcare industry only, the feasibility of the approach is influenced by the facilitation skills of the researcher who performed the facilitation
[23]	PSS in healthcare	Systematic review	The author considers that even though healthcare industry has always been conceived as an area where PSS can have a huge potential, studies on this topic is still limited
[28]	Development of a PSS-oriented business model	Development of a PSS-oriented business model	The developed business model is theoretical, and it is essential to convert the proposed business model into a practice that the hospitals can use

Xing et al. [23] conducted a systematic review regarding the state-of-the-art literature concerned with the application of PSS in the healthcare industry, characterizing the landscape of state-of-the-art research, methods, and practices for PSS integration with various forms of healthcare services. The author considers that even though healthcare industry has always been conceived as an area where PSS can have a huge potential, studies on this topic are still limited.

Yip et al. [29] proposed a new PSS characterization approach that supports the early-stage new PSS development process using 11 case studies in the healthcare industry. Their approach includes five steps as described below and the tools illustrated in Fig. 2:

- *Step 0: Stakeholder identification* – where the stakeholders involved in the product-service system should be listed.
- *Step 1: PSS depiction* – which is the construction of a diagram showing the products, services, and key stakeholders within the new PSS (Fig. 2).
- *Step 2: PSS abstraction* – it is a diagram representing the focus of the new development. The rectangle represents the operating environment, the dotted line circle represents the PSS, the pink circle with the letter P represents the product, and the blue circle with the letter S represents the service.

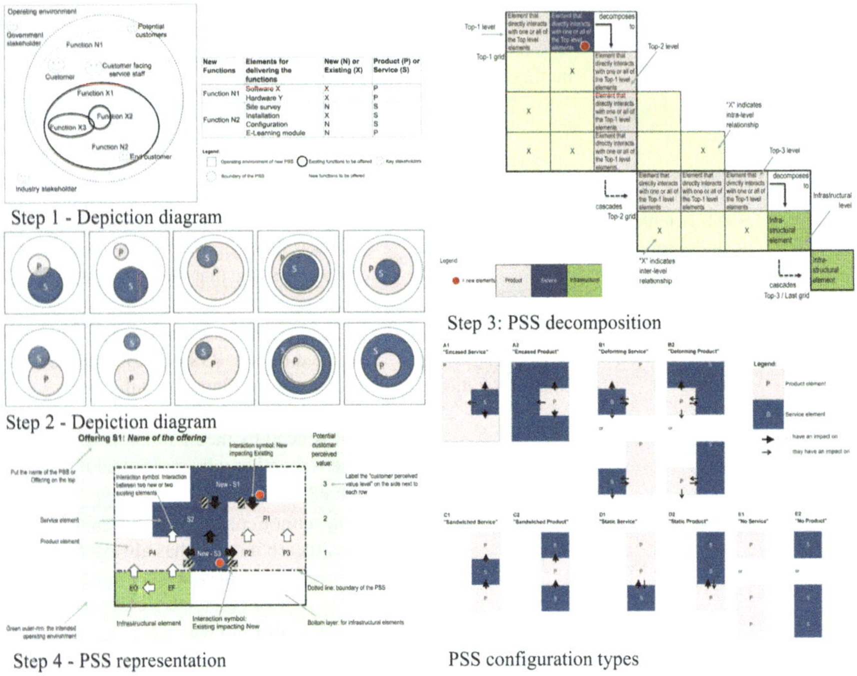

Fig. 2 PSS development process according to Yip et al. Images retrieved from [29]

- *Step 3: PSS decomposition* – development of a progressively series of grids during the analysis of the constituent parts of the PSS.
- *Step 4: PSS representation* – development of a structured diagram that represents the elements and the relationships identified in the PSS decomposition.
- *Step 5: PSS characterization* – determines the "value" of the four PSS characteristics: (i) *potential customer perceived value level* that is represented by the height, or the number of levels, of the PSS representation diagram; (ii) *type and degree of connectivity* that depend on the numbers of black and striped arrows within a PSS representation are counted; (iii) *connectivity number of the PSS* which is represented by the following formula: connectivity number = 2× (total number of black arrows) + total number of striped arrows; and iv) *PSS configuration type* which is represented by five PSS configurations.

Yip et al. [29] work contributes to extend the application of technical systems to tools supporting the engineering design process for PSS. The approach helps visualizing which part of the PSS is of higher potential customer value and how the new PSS will interact with the operating environment, including the key stakeholders. Considering the lack of PSS applied to healthcare context, there is a need to analyze the existing business models in the literature that are applied in the context of PSS.

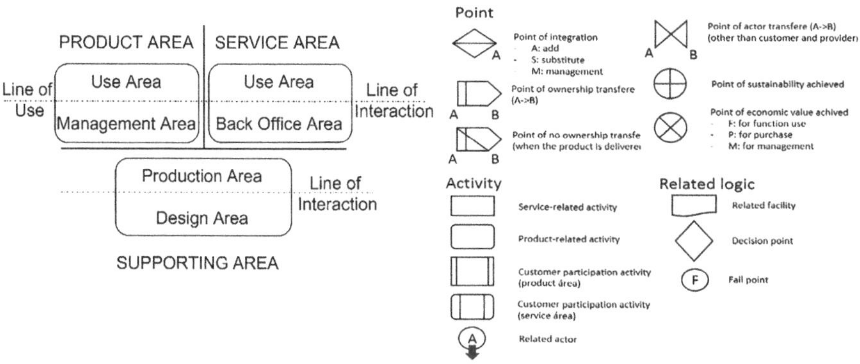

Fig. 3 Product-service blueprint template and symbology [25]

Among the tools and methodologies for the PSS design presented in Table 2, service blueprint (Fig. 3) firstly introduced by Shostack G [37] takes a main role in the PSS detail design [37]. One of the service blueprint's main benefits is that it allows a detailed map of the customer's journey within the service [26] providing the advantage of clear modeling and visualization of the service processes, which turn the service blueprint has a wide use for the representation of services [25].

Geum and Park [25] propose a business model design based on the service blueprint classical model, although it suggested a product-service integration.

The model includes additional areas, lines, and appropriated symbols reflecting the PSS characteristics (Fig. 3). The PSS design structure consists of three main areas: product area, service area, and supporting area. According to [25], there are three fundamental steps:

- *Step 1*: Preparation – allows the PSS development, products, services, infrastructure, and stakeholder identification. This is the first point of the PSS design with the representation of the elements, and activities, of each area, are represented using the boxes, arrows, and symbols.
- *Step 2*: Element's identification – the activities are linked and related; the step makes the relationship between the product area, the service area, and the support area.
- *Step 3*: Building relationships – the activities in each area are linked and related across the value network. This step includes the building of relationships between the product area, the service area, and the supporting area to demonstrate a sequential view of the system.

Regarding the presented models presented for general PSS characterization and PSS in healthcare, the paper proposes an integration of two presented models. Concerning the PSS applied to healthcare, the model presented by Yip et al. [29] shows more detail regarding the PSS elements and considering the identification of the stakeholders involved. The model proposed by Geum and Park [25] presents the benefits of the service blueprint and additionally the product integration.

This paper proposes the characterization of a PSS of a customized 3D printed P&O production considering the hospital as the manufacturer considering both approaches.

3 PSS Business Model Design: An Illustrative Example

In this paper, we present an illustrative example related to the 3D printing production of a knee-positioning orthosis for a patient with cerebral palsy [4], with the proposal of designing a PSS business model. The knee-positioning orthosis case description is available in [3]. Also, it is considered the scenario proposed by [20], in which the hospital is the customized P&O manufacturer.

Take in consideration the contribution of the approaches presented in Sect. 2; this paper intents to use and integrate both tools to design a customized P&O 3D printed PSS: (i) PSS characterization and (ii) product-service blueprint (Fig. 4).

3.1 PSS Characterization [3]

To categorize the present PSS, it was selected and adapted the steps of the template proposed by Yip et al. [29] and Geum and Park [25] to create a PSS template adapted to the illustrative example. According to Yip et al., [29] PSS characterization Step 0 is regarding the stakeholder's identification and Step 1 the construction of a diagram showing the products, services, and key stakeholders within the new PSS, while Geum and Park [25] describe Step 1 as a preparation step that allows the PSS development, products, services, infrastructure, and stakeholder identification.

Gathering the information required in the steps defined by the authors, Table 4 was created with the objective of summarizing the information necessary to develop the PSS.

Considering the information gathered in Table 4, it was possible to build the depiction diagram (Fig. 5) as indicated by Yip et al. [29] Step 1. The depiction

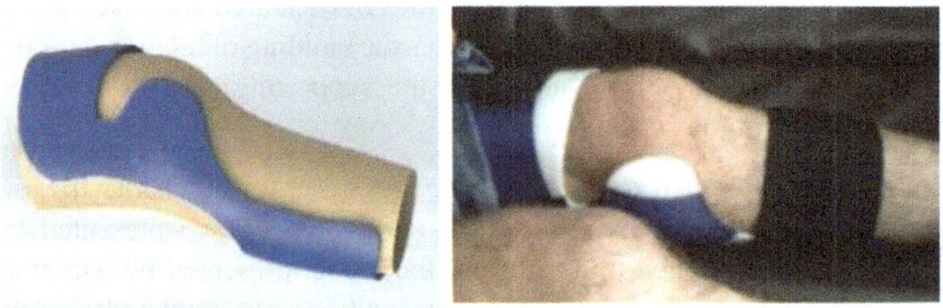

Fig. 4 Three-dimensional printing customized orthosis. (Retrieved from [4])

Table 4 PSS elements defined

New function	Elements for delivering the functions	New (N) Existing (X)	New (N) Existing (E)	Product (P) Service (S)
Customized 3D printed P&O	Medical care	Appointment	E	S
		Consultation	E	S
		Establishing patient needs	E	S
		Medical prescription	E	S
		Device delivery	E	P
		3D anatomical dataAcquisition	N	S
	Customization	Anatomical dataTreatment	N	S
		3D modeling	N	S
		Device fabrication	N	S
		Maintenance and follow-up	N	S
	Management	Device acquisitionRequest	E	S
		Device declarationAnd registration	N	S
		Appointment	E	S

diagram enables to organize the information (defined in Table 2) allowing the PSS visualization in a more schematic way. The rectangle represents the operating environment which in this case is the hospital. Dotted line represents the defined PSS which is the 3D printed customized P&O. The black circle represents the existing products or services which are represented by the medical care and the management. The red circle represents the new product and services that are represented by the customization. The "smiley faces" represent the key stakeholders of the PSS which are the hospital management, healthcare staff, national healthcare system, drugs and health products national authority, and the patients.

To represent a business model design of the illustrative example, according to Geum and Park [25] product-service blueprint template, it is important to create the relationships between the PSS elements identified in Table 4 and the product-service blueprint areas (Fig. 6). The medical care element is related to product and service area, the management element is related to the management and back-office area, and the customization element is related to service, back-office, production, and design area.

The created product-service blueprint is illustrated in Fig. 7 and represents a business model design of the illustrative example. The blueprint is divided into three related areas. The product area represents the product activities flow and is divided by product use line. The service area, related to service provision flow, is divided by the front- and back-office activities. The supporting area portrays the related infrastructures and underlying activities and is divided by the product design and production stage. Considering the PSS elements definition (Table 4) and the depiction diagram, the black shape represents the existing product and services, the

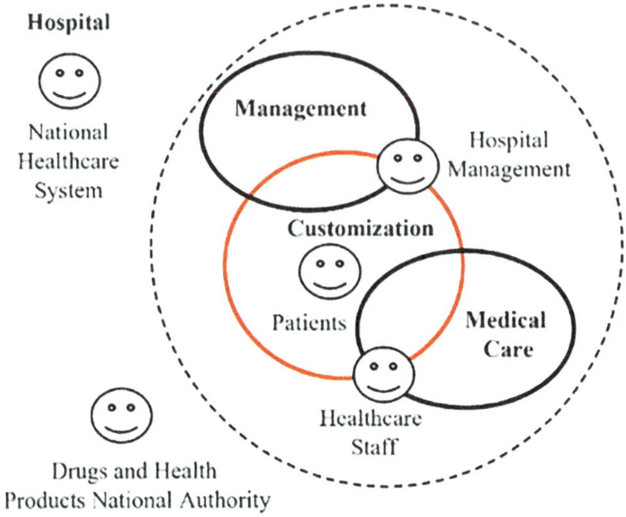

Fig. 5 PSS depiction diagram

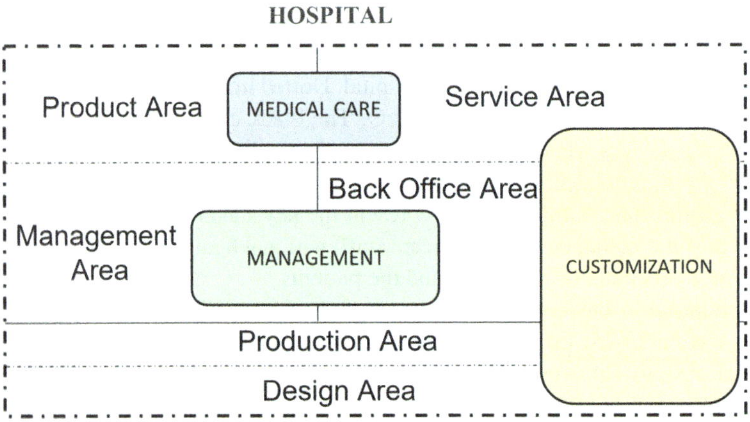

Fig. 6 Relationship between main elements and service blueprint template areas

red shape represents the new service or products, and the "smiley faces" represent the integration of the key stake holders into the PSS product-service blueprint.

4 Conclusion

This paper proposes an integration of two business models to help and support the product-service integration in customized 3D printing industry, considering the hospital as the manufacturer. An illustrative example is used to show how the

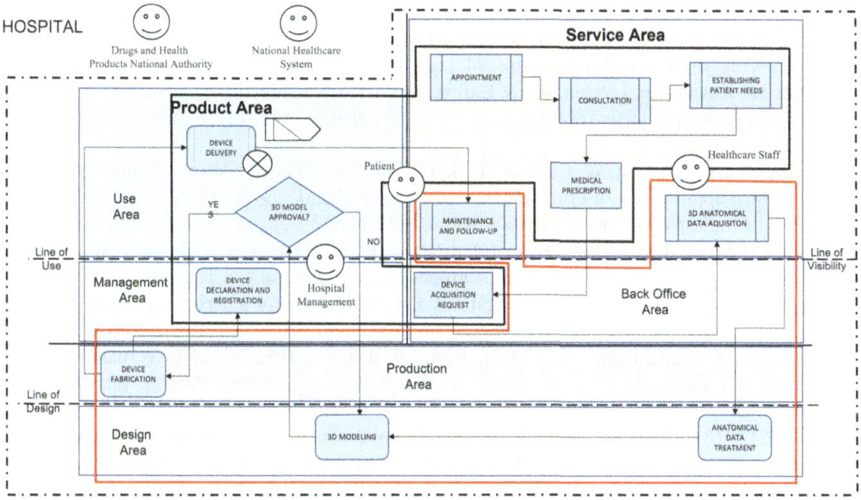

Fig. 7 Product-service blueprint of customized 3D printing P&O for the illustrative example

integration between two different tools enables to organize and categorize the main elements for a PSS for customized 3D printing orthosis. The integration enables a schematic relationship between the product, service, and infrastructure, design of the relationships between elements, and a clear definition of the point of the ownership of the product or service and economical value and the identification of the new or existent product or service with the indication of the key stakeholders. Despite the contribution of this paper, it is required further study to test the model in a practical case of study.

Acknowledgments The authors acknowledge Fundação para a Ciência e a Tecnologia (FCT – MCTES) for its financial support via project UIDB/00667/2020 (UNIDEMI).

References

1. Clinton Health Access Initiative. (2020). *Prostheses*.
2. WHO Standards for Prosthetics and Orthotics. (2017).
3. Kourra, N., Warnett, J. M., Attridge, A., Dibling, G., McLoughlin, J., Muirhead-Allwood, S., King, R., & Williams, M. A. (2018). Computed tomography metrological examination of additive manufactured acetabular hip prosthesis cups. *Additive Manufacturing, 22*, 146–152. https://doi.org/10.1016/j.addma.2018.04.033
4. Leite, M., Soares, B., Lopes, V., Santos, S., & Silva, M. T. (2019). Design for personalized medicine in orthotics and prosthetics. *Procedia CIRP, 84*, 457–461. https://doi.org/10.1016/j.procir.2019.04.254
5. Spencer, S. R., & Kay Watts, L. (2020). Three-dimensional printing in medical and allied health practice: A literature review. *Journal of Medical Imaging and Radiation Sciences, 51, Issue 3*, 489. https://doi.org/10.1016/j.jmir.2020.06.003

6. Wang, Y., Tan, Q., Pu, F., Boone, D., & Zhang, M. (2020). A review of the application of additive manufacturing in prosthetic and orthotic clinics from a biomechanical perspective. *Engineering, 6*(11), 1258–1266. https://doi.org/10.1016/J.ENG.2020.07.019
7. Totah, D., Kovalenko, I., Saez, M., & Barton, K. (2017). Manufacturing choices for ankle-foot orthoses: A multi-objective optimization. *Procedia CIRP, 65*, 145–150. https://doi.org/10.1016/J.PROCIR.2017.04.014
8. Kumar, R., Kumar, M., & Chohan, J. S. (2021). The role of additive manufacturing for biomedical applications: A critical review. *Journal of Manufacturing Processes, 64*, 828–850). Elsevier Ltd. https://doi.org/10.1016/j.jmapro.2021.02.022
9. Andrés-Cano, P., Calvo-Haro, J. A., Fillat-Gomà, F., Andrés-Cano, I., & Perez-Mañanes, R. (2021). Revista Española de Cirugía Ortopédica y Traumatología role of the orthopaedic surgeon in 3D printing: Current applications and legal issues for a personalized medicine. In. *Revista Española de Cirugía Ortopédica y Traumatología, 65*). www.elsevier.es/rot, 138.
10. Singh, S., & Ramakrishna, S. (2017). Biomedical applications of additive manufacturing: Present and future. In *Current opinion in biomedical engineering* (Vol. 2, pp. 105–115). Elsevier B.V. https://doi.org/10.1016/j.cobme.2017.05.006
11. Lee, N. (2016). The lancet technology: 3D printing for instruments, models, and organs? *The Lancet, 388*(10052), 1368. https://doi.org/10.1016/S0140-6736(16)31735-4
12. Kanatlı, M. A., & Karaer, Ö. (2022). Servitization as an alternative business model and its implications on product durability, profitability & environmental impact. *European Journal of Operational Research, 301*(2), 546–560. https://doi.org/10.1016/J.EJOR.2021.10.052
13. Vandermerwe, S., & Rada, J. (1988). Servitization of business: Adding value by adding services. *European Management Journal, 6*(4), 314–324. https://doi.org/10.1016/0263-2373(88)90033-3
14. Wei, Z., Huang, W., Wang, Y., & Sun, L. (2022). When does servitization promote product innovation? The moderating roles of product modularization and organization formalization. *Technovation, 117*, 102594. https://doi.org/10.1016/J.TECHNOVATION.2022.102594
15. Baines, T., Ziaee Bigdeli, A., Bustinza, O. F., Shi, V. G., Baldwin, J., & Ridgway, K. (2017). Servitization: Revisiting the state-of-the-art and research priorities. *International Journal of Operations and Production Management, 37*(2), 256. https://doi.org/10.1108/IJOPM-06-2015-0312
16. Kim, Y. S., & Lee, H. (2020). Process characteristics of product-service systems development: Comparison of seven manufacturing company cases. *Journal of Cleaner Production*, 124971. https://doi.org/10.1016/j.jclepro.2020.124971
17. Goedkoop, M. J., van Halen, C. J. G., te Riele, H. R. M., & Rommens, P. J. M. (1999). Product service systems, ecological and economic basics. In *Report for Dutch Ministries of Environment and Economic Affairs*.
18. Tao, J., & Yu, S. (2019). Developing conceptual PSS models of upper limb exoskeleton based post-stroke rehabilitation in China. *Procedia CIRP, 80*, 750–755. https://doi.org/10.1016/J.PROCIR.2019.01.031
19. Moro, S. R., Cauchick-Miguel, P. A., de Sousa, H., & Mendes, G. (2022). A proposed framework for product-service system business model design. *Journal of Cleaner Production, 376*, 134365. https://doi.org/10.1016/J.JCLEPRO.2022.134365
20. Durugbo, C., Tiwari, A., & Alcock, J. R. (2009). *An infodynamic engine approach to improving the efficiency of information flow in a product-service syste*m. 1st CIRP IPS2 Conference, April.
21. Morelli, N. (2003). Product-service systems, a perspective shift for designers: A case study – The design of a telecentre. *Design Studies, 24*(1), 73. https://doi.org/10.1016/S0142-694X(02)00029-7
22. Marilungo, E., Peruzzini, M., & Germani, M. (2016). Review of product-service system design methods. *IFIP Advances in Information and Communication Technology, 467*. https://doi.org/10.1007/978-3-319-33111-9_25
23. Xing, K., Rapaccini, M., & Visintin, F. (2017). PSS in healthcare: An under-explored field. *Procedia CIRP, 64*, 241–246. https://doi.org/10.1016/j.procir.2017.03.068

24. Morelli, N. (2006). Developing new product service systems (PSS): Methodologies and operational tools. *Journal of Cleaner Production, 14*(17), 1495. https://doi.org/10.1016/j.jclepro.2006.01.023
25. Geum, Y., & Park, Y. (2011). Designing the sustainable product-service integration: A product-service blueprint approach. *Journal of Cleaner Production, 19*(14), 1601–1614. https://doi.org/10.1016/J.JCLEPRO.2011.05.017
26. Silva, H. M., Gonzaga do Nascimento, M. M., de Morais Neves, C., Oliveira, I. V., Cipolla, C. M., Batista de Oliveira, G. C., de Almeida Nascimento, Y., & Ramalho de Oliveira, D. (2021). Service blueprint of comprehensive medication management: A mapping for outpatient clinics. *Research in Social and Administrative Pharmacy, 17*(10), 1727–1736. https://doi.org/10.1016/J.SAPHARM.2021.01.006
27. Nemoto, Y., Kawase, K., Akasaka, F., & Shimomura, Y. (2013). Model-based framework for management of PSS design knowledge. *Proceedings of the international conference on engineering design, ICED, 4 DS75-04.*
28. Pourabdollahian, G., & Copani, G. (2015). Development of a PSS-oriented business model for customized production in healthcare. *Procedia CIRP, 30*, 492–497. https://doi.org/10.1016/j.procir.2015.02.158
29. Yip, M. H., Phaal, R., & Probert, D. R. (2015). Characterising product-service systems in the healthcare industry. *Technology in Society, 43*, 129–143. https://doi.org/10.1016/j.techsoc.2015.05.014
30. Adrodegari, F., Saccani, N., & Kowalkowski, C. (2016). A framework for PSS business models: Formalization and application. *Procedia CIRP, 47*, 519. https://doi.org/10.1016/j.procir.2016.03.073
31. Orellano, M., Neubert, G., Gzara, L., & Le-Dain, M. A. (2017). Business model configuration for PSS: An explorative study. *Procedia CIRP, 64*, 97–102. https://doi.org/10.1016/J.PROCIR.2017.03.008
32. Zanetti, V., Cavalieri, S., & Pezzotta, G. (2016). Additive manufacturing and PSS: A solution life-cycle perspective. *IFAC-PapersOnLine, 49*(12), 1573–1578. https://doi.org/10.1016/j.ifacol.2016.07.804
33. Orellano, M., Lambey-Checchin, C., Medini, K., & Neubert, G. (2018). Towards an integration of lifecycle thinking into PSS business models. *Procedia CIRP, 73*, 291. https://doi.org/10.1016/j.procir.2018.04.017
34. Orellano, M., Medini, K., Lambey-Checchin, C., & Neubert, G. (2019). A system modelling approach to collaborative PSS design. *Procedia CIRP, 83*, 218. https://doi.org/10.1016/j.procir.2019.03.089
35. Rondini, A., Bertoni, M., & Pezzotta, G. (2020). At the origins of product service systems: Supporting the concept assessment with the engineering value assessment method. *CIRP Journal of Manufacturing Science and Technology, 29*, 157. https://doi.org/10.1016/j.cirpj.2018.08.002
36. Santos, A. L., & Wauben, L. S. G. L. (2014). Systems design perspective of healthcare provision in humanitarian aid. *FormAkademisk – Forskningstidsskrift for Design Og Designdidaktikk, 7*(3). https://doi.org/10.7577/formakademisk.790
37. Shostack, G. L. (1982). How to design a service. *European Journal of Marketing, 16*(1), 49–63. https://doi.org/10.1108/EUM0000000004799

An Optimization Model for the Selection of Renovation Actions in a Building

Miguel Moura, Alcinda Barreiras, Maria Teresa Pereira, and Marisa Oliveira

Abstract In Portugal, as in all of Europe, there has been a growing concern about energy poverty and its impact on people's health and well-being. Rehabilitating and making buildings more energy efficient allows for the reduction of energy consumption, that is, transforming existing buildings into near-zero energy buildings (nZEB). Energy consumption in buildings can be reduced by implementing measures to improve energy efficiency. However, the choice of improvement measures to be implemented needs to be optimized since, in general, they require investments, despite causing a decrease in energy costs. The research provides a linear programming optimization approach based on a bi-objective knapsack model to pick the portfolio of actions to undertake while analysing energy consumption reductions and investment payback periods. The model enables the selection of the investments required for the building to achieve nZEB certification. A case study is used to illustrate the model.

Keywords Energy buildings · Energy consumption · Energy efficiency · Linear optimization model

---

M. Moura
Engineering University of Porto (FEUP), Porto, Portugal

A. Barreiras
School of Engineering of Porto (ISEP) - Polytechnic of Porto, Porto, Portugal

M. T. Pereira (✉) M. Oliveira
School of Engineering of Porto (ISEP) - Polytechnic of Porto, Porto, Portugal

Associate Laboratory for Energy, Transports and Aerospace (LAETA-INEGI), Porto, Portugal
e-mail: mtp@isep.ipp.pt

J. Reis et al. (eds.), *Driving Quality Management and Sustainability in VUCA Environments*, Springer Proceedings in Business and Economics,
https://doi.org/10.1007/978-3-031-52723-4_4

1 Introduction

Energy is a key component of technological, economic, and social advancement. Its demand has risen over time, particularly in recent decades, until declining in 2020, owing mainly to the COVID pandemic [1]. The energy consumption of the building sector represents more than 40% of global consumption and one-third of greenhouse gas (GHC) emissions [1]. Buildings with insufficient energy efficiency use more energy than necessary, increasing carbon dioxide (CO_2) emissions [2]. Today's transition to the expansion of renewable energy depends on buildings, and it is necessary programmes to lessen the building sector's reliance on primary energy sources. All buildings, old and new, must adhere to energy efficiency principles. In the European Union (EU), the Energy Performance of Buildings Directive(EPBD) approved in 2002 [3], imposed to all Member States that all new or refurbished buildings placed on the market for sale or lease should have an assessment of energy performance and the respective certificate, which is a mandatory reference in their advertising, carried out by real estate brokerage companies. The revised EPBD [4] and the revised Energy Efficiency Directive (EED) [5] show how important is to improve the energy efficiency of the existing buildings, particularly the ones built before the existence of thermal regulations (1970) [6]. According to this study, approximately 50% of the buildings in the EU were constructed before 1970, so the majority of the EU buildings are energy inefficient [6]. Portugal approved the National Energy and Climate Plan 2030 [7] and the Roadmap to Carbon Neutrality 2050 [8] to meet the objectives of energy efficiency, CO_2 emissions reduction, and primary energy consumption set out in Directive 2018/844/EU [4]. The renovation of buildings was identified as a strategic action to achieve the decarbonization objectives.

The strategy to be used to fulfil the commitment made under RNC 2050 [8] is to convert existing buildings into buildings that are more energy efficient and have better energy performance because a sizeable portion of Portugal's national building stock is ageing and in need of renovation interventions. A bi-objective knapsack model is suggested in this study to aid in choosing renovation measures that take into account the budget. We present a case study of a retrofit project for a service building in Portugal. We then discuss some of the areas in which more research is required and assess the pros and disadvantages of the model.

2 Literature Review

There has been a long-standing worry about increasing buildings' energy efficiency, and various nations have been creating legislation and regulations in this area. As was previously said, in Europe, the EPBD [5] and the EED [6] were published to encourage member states to define policies to stimulate renovations of the building stock. A building is a complex system with many subsystems, such as lighting, water, heating, cooling, ventilation, and envelope. Each subsystem affects the total energy efficiency performance, and the interaction between the subsystems also has a great impact on energy savings [9, 10]. To achieve the objectives defined in the EPBD, it is also necessary to take into account the life cycle of a building in determining the most cost-effective set of measures applied by to a reference building [11, 12]. With the aim of combining energy efficiency and cost-effectiveness, the EPBD established that energy efficiency improvement strategies should be selected based on "cost optimization" [13]. The methodology set out in this directive is based on the principle of cost–benefit analysis and can be evaluated from a financial perspective, which includes costs and rates, or from a macroeconomic perspective, which considers the costs of emitting greenhouse gases. The calculation parameters to obtain the optimal cost are: the discount/interest rate and the annual increase in energy prices [14]. As was previously said, selecting energy efficiency solutions is typically a complex and heterogeneous process that calls for the fusion of several specialities [15]. In order to support the decision-making process for the retrofit of buildings for increased energy efficiency and performance assessment, a number of models and approaches have been created. Which of the current facilities needs to be changed? How many interventions are required to upgrade current buildings on a tight budget? Which interventions to use when there are multiple options [10]? Generally, in order to satisfy sustainability objectives, efforts must be placed on finding a valid solution. Some studies have found a solution based on the cost–benefit analysis of energy efficiency measures in buildings. The net present value (NPV) is one of the most used approaches [16–18]. This analysis is based on assumptions that could affect the results, including changes in the price of equipment and materials, both of which are very variable, and the unpredictability of energy prices. A sensitivity analysis can assist in removing this uncertainty and improving the reliability of a study's findings by enabling continual assessment, which is necessary to obtain the anticipated return [19]. Commission Delegated Regulation (EU) No. 244/2012 [20] that complements the EPBD is based on cost minimization, which is obtained as a function of avoided costs due to energy savings and reduction of CO_2 emissions. In general, the solutions evaluated are based on their energy performance; however, the evaluations may not focus exclusively on energy efficiency and take into account other factors such as environmental costs [21, 22] among others. It should also be taken into account that the energy efficiency classification of a building may enhance it in the market, that is, the market value of an energy-efficient building is higher than a low-performance one [23, 24]. Investigate the influence of energy efficiency measures on the economic value of buildings. For example, [25] calculated the payback period of investments to analyse the impact of investments in efficiency measures. As the cost of energy increases, so does the need for strategies that optimize investment. Finding an optimal solution is difficult because there are various remodelling choices side by side with varied objectives to fulfil. Building energy retrofit is an optimization problem [26]. The problem of choosing energy efficiency measures has been the subject of research over the years, there is a wide variety of measures that can be implemented in a building, and it is not easy to choose [9]. The mix of measures that should be used in a building is specific, even though there is an increasing amount of information and resources available. This is because each building is a distinct entity with its own architectural features, operating temperature zone and use, among other things. The selection of which retrofit measures to implement in a certain project is a multi-objective optimization problem with multiple constraints and limitations [22, 27]. Several review studies with various points of view on improving the energy performance of buildings have surfaced in recent years. Studies that examine the optimization of the building's design and geometry, as well as its energy efficiency for some building types and climate zones, have recently been published [28, 29]. The adaptation of a bi-objective optimization problem based on the knapsack problem, which considers trade-offs between renovation costs and payback periods estimated in accordance with Portuguese law [30] and the estimated annual energy savings of the building, is developed in order to assist decision-makers in determining which retrofit actions to carry out. The TRNSYS software, an energy simulation programme, is used to estimate the building's energy needs and the energy savings corresponding to each of the improvement measures. To demonstrate how the suggested model works in a real-world situation, a case study example is provided.

3 Problem Description

Choosing a building renovation action is a difficult procedure that involves numerous options. There are lots of selections and possibilities, such as replacing equipment with more energy-efficient models, updating exterior features such as window doors, and replacing incandescent lighting with LEDs, among others. On the market, there are numerous materials for structural components and models for technical equipment. As a result, the alternatives are diverse and wide. According to Portuguese regulations [31], energy efficiency measures fall into four categories: building improvements, which include facades and glazed areas, technical equipment replacement, heating/cooling equipment, ventilation, hot water heating, among others, lights and regulation equipment, and finally the installation of renewable energy source systems. The group of building improvements includes, among other things, the examination of constructive pathologies, fungus, and molds. One example is the thermal improvement of horizontal or vertical opaque elements, both internal and external. The installation or replacement of doors with better thermal performance is also another activity, as well as interventions at the level of glazed openings, either by constructing new ones removing existing ones or installing ones with better thermal performance. Installation or replacement of more energy-efficient heating, cooling, ventilation and hot water preparation systems, application or reinforcement of insulation or water storage tanks, installation or replacement of high-efficiency shower systems, and performing preventive maintenance on technical systems are all actions that fall under the second category. The installation of more energy-efficient lighting systems (lamps and/or ballasts), as well as additional measures like the installation of presence detection sensors or the regulation of light flux control, are still measures that need to be taken into consideration. Finally, it is important to suggest improvement measures for the installation of renewable energy systems, promoting local energy production (primarily for self-consumption), and reducing the use of network energy to lower associated CO_2 emissions and building operating costs [31]. In addition to this broad variety of alternatives of energy-efficient measures, there are budgetary constraints associated with the investment.

An optimization model was developed based on the knapsack issue model to aid in decision-making, specifically in the selection of options to maximize energy efficiency. TRNSYS, a simulation tool that models the operation of the building's entire energy system and simulates thermal behaviour, has been integrated with the optimization model. During the winter and summer seasons, it is feasible to forecast energy requirements as well as heating and cooling power. To ensure that the TRNSYS software for simulating thermal processes accurately represents the building under consideration, the simulation parameters must be specified. The simulation parameters are specified and calibrated using current energy bills. This calibration mode adheres to ASHRAE standards and recommendations [32]. TRNSYS implements the building model to assess the thermal behaviour and energy requirements of the structure. A new simulation of the building's thermal behaviour and energy consumption is performed for each improvement measure or a combination of these. The computation of energy consumption is carried out based on the simulations supplied by the TRNSYS programme, and the savings realized by the adoption of the measure are evaluated.

In order to maximize energy savings and decrease the payback period computed as stipulated by the [30], the proposed optimization model selects a combination of retrofitting techniques while taking into account the financial availability for the investment.

4 The Mathematical Model

The established model is meant to assist in the selection of renovation projects that will best minimize annual energy consumption and shorten the payback period for the building to be renovated on a budget. A binary knapsack model is adapted, variables x_k taking the value of 1 if the kth retrofit action is carried out and 0 otherwise, are defined. The objective function will be

$$\text{Max} \sum_k S_k x_k \tag{1}$$

where S_k represents the utility score achieved with the implementation of retrofit action x_k.

The objective function in this chapter weights the two goals, one that consists of maximizing the annual energy savings, ES, and the other one consists of minimizing the payback period, PB. The values range of energy annual savings of the building for the k retrofit action, ES_k, is quite different from the values range of the payback period, PB_k. So a normalization of these values was made to aggregate both purposes in the same objective function. For this purpose, the following procedure was adopted,

$$\overline{ES_k} = \frac{ES_k}{ES^M};\quad ES^M = \max\{ES_k; \forall k\} \tag{2}$$

and

$$\overline{PB_k} = \frac{PB^m}{PB_k};\quad PB^m = \min\{PB_k; \forall k\} \tag{3}$$

where
ES_k represents the annual energy savings of kth type retrofit action.
PB_k represents the payback period of the kth retrofit action.

The two objectives, annual energy savings and the payback period, are normalized to a range of 0–1. The energy savings normalized, $\overline{ES_k}$, and the payback period normalized, $\overline{PB_k}$, are aggregated in a weight vector. So the objective function (1) can be rewritten as follows:

$$\text{Max} \sum_k \{\alpha_1 \overline{PB_k} + \alpha_2 \overline{ES_k}\} x_k \tag{4}$$

where

$$\overline{\alpha} = \left\{(\alpha_1, \alpha_2) : \alpha_i \geq 0 \wedge \sum_{i=1}^{2} \alpha_i = 1\right\} \tag{5}$$

These weights can be adjusted to produce various compromise results. First, the restrictions for the maximum investment amount were taken into consideration.

$$\sum_k x_k I_k \leq \text{Budget} \tag{6}$$

$$x_k \in \{0, 1\}, \forall k \tag{7}$$

where I_k represents the investment needed to implement kth retrofit action.

As previously indicated, there is a wide variety of retrofit tasks accessible, and this variety expands when different combinations are evaluated. The various options can be classified based on the sort of improvement desired. Consider the measure of improvement that examines heating equipment replacement; there are several scenarios for this measure because there are various types of equipment available, such as natural gas, biomass, heat pumps, and so on, each with a different model. Only one will, however, be installed in the building. To take into account the possibility of alternative materials or different models of equipment, retrofit operations are classified according to the intervention, such as construction components, equipment, lighting, renewable sources, and so on.

The set, A_i, represents the various retrofit action possibilities $\forall i = 1, ..N$, where N indicates the number of possible retrofit categories. A binary decision variable, $x_k^{A_i}$, takes the value 1 if the kth retrofit action of the A_i is carried out, else $x_k^{A_i} = 0$. The objective function and constraints (6) are updated to

$$\sum_i \sum_{k \in A_i} x_k^{A_i} I_k^{A_i} \leq \text{Budget} \tag{8}$$

A new restriction is added to the model to ensure that from each group, just one type of action is selected for a retrofit scenario.

$$\sum_{k \in A_i} x_k^{A_i} \leq 1, \forall i, i = 1, ..., N \tag{9}$$

5 Case Study

The applicability of the model was examined by increasing the energy efficiency of a utility building. The structure under analysis was built more than 50 years ago and is a senior housing complex on the north coast of Portugal. Currently, the building is not energy efficient, which leads to high energy consumption. Consequently, energy costs represent a considerable part of the Institution's annual budget, and the implementation of energy performance improvement measures that allow for the reduction of these costs is important. Furthermore, energy requalification of the building is required in order to acquire government cooperation agreements. Based on data (consumption bills) and simulation of energy consumption for the current building in TRNSYS, it appears that heating accounts for the majority of energy consumption in the structure, accounting for 71% of the total energy consumption. Lighting use, which accounts for 14% of the total energy consumption, is found to be the building's second highest energy consumer. In terms of energy consumption, it appears that natural gas demands correspond to 80% of the total energy consumption.

The requalification of this building aims to intervene to make it more energy efficient and decrease associated costs, as well as to improve the structure's thermal comfort. Because it is a building for the elderly, thermal comfort is critical, and it currently accounts for a major portion of energy consumption, and thus it is critical to outfit the building with technical equipment and more efficient practical thermal solutions. The promotion of energy efficiency is also looked at in a broader framework, with the objective of transforming the building into an almost zero energy building, or nZEB, by incorporating renewable energy sources. As a result, various improvement measures were discovered and analysed in line with the

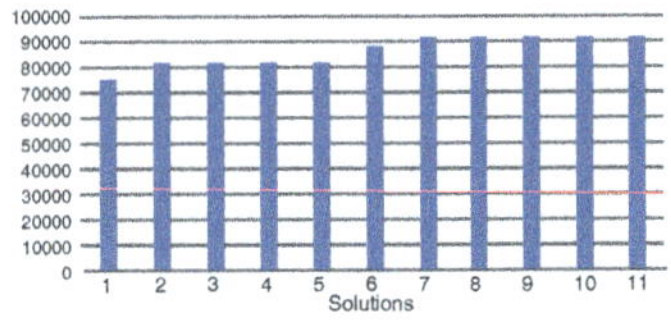

Fig. 1 Energy reduction and comparison of optimization issue solutions

Portuguese regulation, Despacho no. 7113/2015, of June 29, which establishes the set of energy efficiency measures. Building interventions should consider the following kinds of actions: outside wall improvements, exterior windows and doors, roofs and floors; efficient lighting; equipment and technical systems; and use of renewable energy sources.

The energy efficiency measures to be considered are divided into four categories based on Portuguese rules. Different measures can be considered and assessed in each of these groups, for example, different types of materials can be employed for constructive components, and alternative models can be evaluated for technical systems. It is difficult to decide which set of actions to carry out given the available budget. The requalification of the building entails selecting a set of actions that include interventions on the external walls, floor, windows, doors, and roof, as well as efficient lighting, technical equipment replacement and the installation of renewable energy sources while taking into account the characteristics and costs of the various materials/models.

The TRNSYS programme is used to simulate the building's behaviour, allowing the energy consumption of the building to be estimated. A simulation in TRNSYS is performed for each of the improvement measures under consideration, which models the thermal behaviour of the building with the adoption of the measure, allowing the computation of the building's energy costs. TRNSYS simulations were performed in this study for each of the independent measures as well as their combinations whenever there was an interaction between them. Building element improvements, for example, clash with the thermal behaviour of the building and the type of heating/cooling equipment required. The engagement in the construction components, on the other hand, has nothing to do with the hot water preparation requirements.

Currently, the management has only 250,000 euros for building renovations. The problem was solved with an Excel add-in solver, and several weight parameters of the objective function were examined, with varied values ranging from 0 to 1. Each bar in Fig. 1 illustrates a tried solution. The behaviour of the objective function was investigated by assigning different values to its parameters. The annual energy needs reduction associated with each solution is shown in Fig. 1, where each bar reflects the energy reduction of each solution in (Khw/year).

The graph in Fig. 2 indicates the payback period, in years, associated with each of the solutions: According to an analysis of the various alternatives, 6, 7, 8, and 9 are equivalent and provide the best compromise given the financial restrictions. This method entails altering the building's outside walls, switching to LED lighting from incandescent, replacing the heating boiler with a biomass boiler, and adding

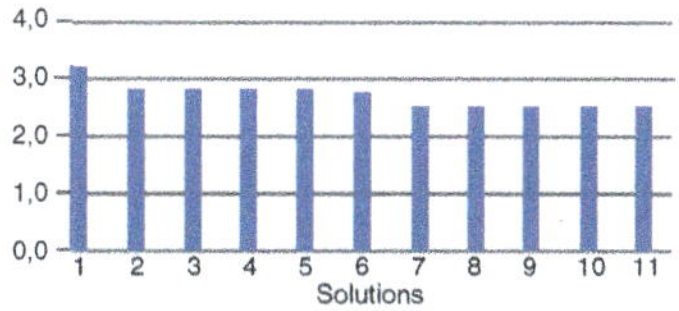

Fig. 2 Comparison of optimization problem solutions, payback period

photovoltaic panels. The adoption of this package of improvement measures enables the building to be categorized in energy class B and equates to an investment of 234,720 euros while adhering to the stipulated financial constraints.

6 Conclusions

Recently, new construction materials and technical equipment have appeared with the goal of improving building energy efficiency in order to better address human requirements and environmental challenges. As a result, given the variety and quantity of options available, determining the set of energy efficiency measures to be applied in a building is difficult. The provided approach model, a binary Knapsack model in conjunction with TRNSYS, an energy modelling software, replies to the challenge. Allows you to choose the collection of activities to be produced while keeping your budget in mind.

The model's solutions enable optimum energy savings with a short payback period. The proposed method was applied to a Portuguese royal service building, allowing the model's applicability and functionality to be tested.

Traditionally, in this type of intervention, few scenarios are evaluated. The ability to evaluate a large number of renovation actions and choose the best portfolio is the strength of the model when used in energy renovation projects. In the future, a sensitivity analysis should be carried out that takes into account the characteristics of the various renewal measures.

References

1. Global Status Report for Buildings and Construction, Global Alliance for Buildings and Construction (2021).
2. Sapnken, F. E., Hamed, M. M., Soldo, B., & Tamba, J. G. (2023). Modeling energy-efficient building loads using machine-learning algorithms for the design phase. *Energy and Buildings, 283*, 112807.
3. European Parliament and Council Directive 2002/91/EC of 16 of December of 2002 on the Energy Performance of Buildings.
4. European Parliament and Council Directive 2018/844 of 30 of May of 2018 amending Directive 2010/31/EU on the Energy Performance and Directive 2012/27/EU on Energy Efficiency.

5. European Parliament and Council Directive 2018/2002 of 11 of December of 2018 Amending Directive 2012/27/EU on Energy Efficiency.
6. Filippidou, F., & Navarro, J. P. J. (2019). *Achieving the cost-effective energy transformation of Europe's buildings*. Publications Office of the European Union, Luxembourg.
7. PNEC 2030. (2019). Plano Nacional Energia e Clima 2021–2030.
8. RNC 2050, Resolução do Conselho de Ministros n.º 107/2019, de 1 de julho de 2019, aprova o Roteiro para a Neutralidade Carbónica 2050.
9. Soares, N., Bastos, J., Pereira, L. D., Soares, A., Amaral, A. R., Asadi, E., Rodrigues, E., Lamas, F. B., Monteiro, H., Lopes, M. A. R., & Gaspar, A. R. (2017). A review on current advances in the energy and environmental performance of buildings towards a more sustainable built environment. *Renewable and Sustainable Energy Reviews, 77*, 845–860.
10. Wu, Z., Wang, B., & Xia, X. (2016). Large-scale building energy efficiency retrofit: Concept, model and control. *Energy, 109*, 456–465.
11. Vasconcelos, A. B, Cabaço, A., Pinheiro, M. D., & Manso, A. (2016). The impact of building orientation and discount rates on a Portuguese reference building refurbishment decision. *Energy Policy, 91*, 329–340.
12. Ascione, F., Bianco, N., Mauro, G. M., & Napolitano, D. F. (2019). Retrofit villas on Mediterranean coastlines: Pareto optimization with a view to energy-efficient and cost-effectiveness. *Applied Energy, 254*, 113705.
13. Economidou, M., Todeschi, V., Bertoldi, P., D'Agostino, D., Zangheri, P., & Castellazzi, L. (2020). Review of 50 years of EU energy efficiency policies for buildings. *Energy and Buildings, 225*, 110322.
14. Vasconcelos, A. B., Pinheiro, M. D., Manso, A., & Cabaço, A. (2016). EPBD cost-optimal methodology: Application to the thermal rehabilitation of the building envelope of a Portuguese residential reference building. *Energy and Buildings, 111*, 12–25.
15. Kaklauskas, A., Zavadskas, E. K., & Raslanas, S. (2005). Multivariant design and multiple criteria analysis of building refurbishments. *Energy and Buildings, 37*, 361–372.
16. Verbeeck, G., & Hens, H. (2005). Energy savings in retrofitted dwellings: economically viable. *Energy and Buildings, 37*, 747–754.
17. Peterson, S., & Svendsen, S. (2012). Method for component-based economical optimisation for use in design of new low–energy buildings. *Renewable Energy, 38*, 173–180.
18. Kumbaroğlu, G., & Madlener, R. (2012). Evaluation of economically optimal retrofit investment options for energy savings in buildings. *Energy and Buildings, 49*, 327–334.
19. Tadeu, S., Tadeu, A., Simões, N., Gonçalves, M., & Prado, R. (2018). A sensitivity analysis of a cost optimality study on the energy retrofit of a single-family reference building in Portugal. *Energy Efficiency, 11*, 1411–1432.
20. Commission Delegate Regulation 244/2012 of 16 January 2012 supplementing Directive 2010/31/EU of the European Parliament and of the Council on the energy performance of buildings by establishing a comparative methodology framework for calculating cost-optimal levels of minimum energy performance requirements for buildings and building elements.
21. Becchio, C., Dabbene, P., Fabrizio, E., Monetti,V., & Filippi, M. (2015). Cost optimality assessment of a single family house: Building and technical systems solutions for the nZEB target. *Energy Build, 90*, 173–187.
22. Jafari, A., & Valentin, V. (2018). Selection of optimization objectives for decision-making in building energy retrofits. *Building and Environment, 130*, 94–103.
23. Zalejska-Jonsson, A. (2014). Stated WTP and rational WTP: Willingness to pay for green apartments in Sweden. *Sustainable Cities and Societies, 13*, 46–56.
24. Bonifaci, P., & Copiello, S. (2015). Price premium for buildings energy efficiency: empirical findings from a hedonic model. *Valori e Valutazioni, 14*, 5–15.
25. Popescu, D., Bienert, S., Schützenhofer, C., & Boazu, R. (2012). Impact of energy efficiency measures on the economic value of buildings. *Applied Energy, 89*(1), 454–463.
26. Malatji, E. M., Zhang, J., & Xia, X. (2013). A multiple objective optimisation model for building energy efficiency investment decision. *Energy Build, 61*, 81–87.

27. Ascione, F., Bianco, N., Stasio, C. D., Mauro, G. M., & Vanoli, G. P. (2015). A new methodology for cost-optimal analysis by means of the multi-objective optimization of building energy performance. *Energy Build, 88*, 78–90.
28. Hashempour, N., Taherkhani, R., & Mahdikhani, M. (2020). Energy performance optimization of existing buildings: A literature review. *Sustainable Cities and Society, 54*, 101967.
29. Dingyuan, M., Xiaodong, L., Borong, L., Yimin, Z., & Siyu, Y. (2023). A dynamic intelligent building retrofit decision-making model in response to climate change. *Energy and Buildings, 284*, 112832.
30. Diário da República n.º 234/2013, 3º Suplemento, Série II de 2013-12-03 (pp. 88–88)
31. Diário da República n.º 179/2015, Série I de 2015-09-14 (pp. 7899–7922)
32. ASHRAE. Measurment of energy and demand savings. ASHRAE Guideline 14. US Standard 2002.

The Influence of Augmented Reality on the Consumer Purchasing Process

Federica Murmura, Laura Bravi, Giada Pierli, Gilberto Santos, and Fabio Musso

Abstract Augmented reality (AR) has recently gained the attention of both scholars and practitioners, thanks to its ability to provide captivating and immersive experiences that lead users to fully engage in the proposed content. Despite the increasing use of AR by companies and individuals, there are still few empirical studies investigating its influence on consumer behavior; therefore, the present study intends to analyze the different behaviors and attitudes of consumers during their purchasing process in front of AR technologies. To this end, an online questionnaire was submitted to individuals belonging to different generational cohorts, obtaining 337 responses. From the results, it emerged that most of the sample expressed a more than positive opinion regarding the help that AR technologies could give to the consumer when he/she chooses and buys a product. Understanding the characteristics of the product, having a unique experience with it during the purchasing process, and making choices with more awareness and fastness are some of the positive aspects that were identified by the respondents.

Keywords Augmented reality · Consumer behavior · Industry 4.0

1 Introduction

Nowadays, the enabling technologies of Industry 4.0 are not only introducing innovations at the industrial and economic level for companies but are completely transforming the attitudes and behavior at the sensory, emotional, and experiential

F. Murmura · L. Bravi (✉) · G. Pierli · F. Musso
Department of Economics, Society, Politics, Carlo Bo University of Urbino, Urbino, Italy
e-mail: federica.murmura@uniurb.it; laura.bravi@uniurb.it; g.pierli@campus.uniurb.it; fabio.musso@uniurb.it

G. Santos
School of Design, Campos do IPCA, Barcelos, Portugal
e-mail: gsantos@ipca.pt

J. Reis et al. (eds.), *Driving Quality Management and Sustainability in VUCA Environments*, Springer Proceedings in Business and Economics,
https://doi.org/10.1007/978-3-031-52723-4_5

level of individuals [1]. People are now daily users of digital devices, which have become integral parts of their behavior and decisions, especially in the purchasing process. Among these, augmented reality (AR) has recently gained the attention of both scholars and practitioners [2–4], thanks to its ability to provide captivating and immersive experiences that lead users to fully engage in the proposed content. Indeed, this technology allows individuals to experience an augmented world by superimposing virtual information on reality, incorporating sounds, visual representations, and other sensory stimuli within the same environment [5].

AR systems are emerging widely in most sectors [6]: gaming, visual arts, photography, film, interactive digital media, education, healthcare, and manufacturing. Due to the great capacity these technologies are showing in enhancing companies' ability to create value, the field of AR emerges as one of the most promising areas in the context of Industry 4.0[5]. The actual market for AR grew to $150 billion in 2021 [7]; in the period 2021–2028, it was estimated to reach approximately $340.16 billion [8]. Through the widespread use of smartphones and tablets, AR is also becoming increasingly accessible to consumers. Given this rapid growth and the presence of such digital devices, the interest of developers, retailers, and consumers has significantly increased [3]. In order to promote their products and/or services and to improve the user experience in real time, companies such as Sephora, L'Oreal, Nike, Adidas, Asos, Zara, IKEA, and Amazon have implemented a strategy of service augmentation, not focusing on the core product but on customer interaction [9, 10]. For instance, IKEA's mobile application allows users to place furniture items from the virtual environment in the real one, in front of their eyes.

Despite the increasing use of AR by companies and individuals, there are still few empirical studies investigating its influence on consumer behavior [2, 11]. Therefore, the present study intends to fill this gap by analyzing the different behaviors and attitudes of consumers during their purchasing process in front of AR technologies. To this end, an online questionnaire was submitted to individuals belonging to different generational cohorts, obtaining 337 responses.

In this perspective, the study contributes to enrich the literature on the topic, confirming the ability of AR technologies to positively influence purchasing behavior, improving the degree of consumer engagement and experience. Furthermore, it could offer interesting insights to companies in defining marketing strategies able to improve and enrich both the online and offline shopping experience of digital consumers using AR technology.

The paper is organized as follows. Section 2 outlines the theoretical framework concerning AR. Section 3 explains the research methodology used. Section 4 presents and discusses the results of the study. Finally, Sect. 5 provides conclusions, limitations, and directions for future research.

2 Literature Review

The introduction of new technologies by the Fourth Industrial Revolution has radically changed the needs of consumers, who are increasingly critical and selective about the market offerings. Through the opportunities related to the use of digital devices [12], individuals appear to be more aware and informed; they are able to gather more accurate information about the products or services to be purchased, to be more aware of the options on the market, to compare offers in terms of both quality and price, and to make purchasing decisions more quickly. Furthermore, the active participation of consumers in various online communication channels such as social media, web forums, blogs, and mobile apps significantly increases the level of interaction between peers as well as with companies and their brands [13, 14]. This higher degree of awareness and interaction has had inevitable effects on the behavior of consumers [15], who become less brand-loyal, individualistic in their choices, and increasingly attentive to the experiential and emotional components of the desired product or service. Nowadays, the consumer intends to satisfy needs that tap into the emotional and hedonistic sphere [9], not focusing exclusively on utilitarian and functional ones. In this perspective, individuals do not just demand the primary service or product they pay for but expect a valuable experience during purchase and demand to be enchanted [16]. As a consequence, it becomes essential to create purchasing situations that provide rewarding experiences in which the consumer plays an active role, becoming an integral part of the value creation process. The use of enabling technologies 4.0 has allowed companies to meet these new needs, significantly enriching the experience in the purchasing process. Such innovative tools enable users to interact with the company in an immersive environment [17], applying sensory and cognitive components to the experience. In a context mediated by technology, the consumer is thus able to immerse himself in an experience characterized by a high degree of interactivity and involvement [18], considerably enhancing the brand loyalty.

Among the many technological advances of Industry 4.0, AR is increasingly attracting the attention of scholars and practitioners [2–4], just because of its potential to engage consumers in a unique and vivid way [19]. The capacity of such technology to overlay the physical environment with virtual elements, including information and images, which can interact with the physical environment in real time, offers companies the possibility of generating authentic experiences throughout the entire consumer journey [3, 20]. In this sense, Rauschnabel et al. [4, p. 1141] define AR as "the strategic integration of AR experiences, alone or in combination with other media or brand-related cues, to achieve overarching marketing goals by creating value for the brand, its stakeholders, and societies at large, while considering ethical implications."

Specifically, AR has three main characteristics [21]: (i) combination of real and virtual elements, (ii) real-time interaction with the user, and (iii) the reproduction of content/elements in 3D. These attributes empower retailers to enhance the customer experience by creating virtual images and elements that users can combine with real objects. By dissolving the boundaries between the real and virtual worlds, such technology allows consumers to experience a new dimension, incorporating sound, visual elements, or other sensory stimuli, producing a fully immersive experience [5, 22]. Several studies state that, among the five senses, AR predominantly affects three of them: sight, touch, and sound [23, 24]. While sight and sound are senses that are easy to condition through augmented reality, AR technologies compensate for the feeling of touch by engaging the customer in an enjoyable shopping and consumption experience. Consumers have a physical in-store experience using AR tools in what is described as a "third realm," phygital, namely, individuals find themselves in physical environments that combine both the functional and emotional benefits of digital space and physical location [25]. However, it is important to point out that AR can be used both inside and outside the shop, creating an in-store (offline) experience directly within consumers' homes [26].

For a long time, the presence of large, bulky, and high-priced devices has been the primary barrier to the deployment and application of AR [27]. With the recent developments of smartphones, tablets, and other cutting-edge technologies, AR is becoming increasingly accessible to consumers [26], significantly changing the way they buy and consume. AR technologies seem to improve users' perceptions before, during, and after the purchase due to its appealing aspects, such as personalization, co-creation, user-friendliness and multitasking, time-saving, low effort, and playfulness of the virtual trial aspect [9, 28].

Today, consumers are constantly seeking instant gratification as well as a personalized, dynamic, and exciting experience. According to Kazmi et al. [2], AR is able to address this increasingly pressing need by creating a captivating and fascinating experience. By visually and contextually altering the reality of the physical world with the superimposition of virtual elements, such tool provides consumers with a unique combination of hedonistic and utilitarian/functional values, satisfaction, information, and learning opportunities [9]. This positively fosters the co-creation process [29], freeing potential buyers from the confines of the screen and leaving more time for imagination, exploration, and experimentation.

With the evident potential of AR and its increasing use by consumers [2, 3], it thus becomes crucial to investigate and understand its influence on the behavior of individuals during their purchasing process.

3 Research Methodology

The aim of the research was to develop an exploratory analysis [30], to analyze the different attitudes of consumers during their purchasing process in front of AR technologies.

The study was carried out by administering a questionnaire to a group of individuals belonging to different generational cohorts in Italy, using different channels (WhatsApp, email, Microsoft Teams). The data was collected from the beginning of October to mid-November 2022. A total of 337 individuals participated in the research.

The questionnaire was divided into three sections. In the first section, the sociodemographic data of the participants were collected. The second part of the analysis, on the other hand, relates to the purchasing process in general, focusing on the relationship between the consumer and technology. The data collected aims to investigate the types of technological tools owned by the respondents and the different activities carried out through these tools. The propensity to purchase both in the offline and online context of consumers was also analyzed. Finally, the last section focuses on the main topic of the research, i.e., understanding what consumers think about the use of AR during the purchasing process, whether it influences them or not in their decisions and if their experience is improved. The study adopts the American definition provided by the Pew Research Center for categorization of generational cohorts: Baby Boomers (1946–1964), Generation X (1965–1980), Generation Y (i.e., Millennials) (1981–1996), and Generation Z (born after 1996). The Silent Generation (or Matures) (1928–1945) was not taken into consideration for the study.

A descriptive analysis was developed to describe the profile of the consumers that participated to the survey. The questionnaire items have been evaluated considering a five-point Likert scale. To test the reliability of the items, Cronbach's alpha values were computed, considering only values greater than 0.60 as suggested by Nunnally and Bernstein [31]. Mean and standard deviation have been calculated to evaluate the items developed.

4 Results

4.1 Consumer Profile

As it can be seen from Table 1, most of the respondents are women (63.0%), 36.7% are men, and 0.3% preferred not to specify it. The 58.8% of the participants belong to the age group between 18 and 25 years (Generation Z), while 14.9% belong to the group between 26 and 41 years (Generation Y or Millennials) and 17.3% between 42 and 57 (Generation X). On the other hand, the 9.0% represents the Baby Boomers, respectively, from 58 to 76 years of age. As for the main education, the majority has a high school graduation (50.1%) followed by a bachelor degree (31.6%). Considering the respondents' employment, the majority are students (50.1%), followed by employees (28.1%), and they have a monthly wage of less than 500 euros (48.4%) or between 1001 and 1500 euros (17.6%).

Table 1 Sociodemographic feature of the respondents

		n	%
Gender	Male	123	36.7
	Female	211	63.0
	Prefer not to say	1	0.3
	Prefer not to say	1	0.3
Age	18–25 (Gen. Z)	197	58.8
	26–41 (Gen. Y/Millennials)	50	14.9
	42–57 (Gen. X)	58	17.3
	58–76 (Baby Boomers)	30	9.0
Education	Middle school graduation	12	3.6
	High school graduation	168	50.1
	Bachelor degree	106	31.6
	Master degree	37	11.0
	Postgraduate title	12	3.6
Employment	Student	168	50.1
	Working student	14	4.2
	Employee	94	28.1
	Freelance	21	6.3
	Worker	2	0.6
	Housewife	6	1.8
	Unemployed	5	1.5
	Retired	12	3.6
	Other	13	3.9
Monthly wage	Less than 500€	162	48.4
	501–1000€	52	15.5
	1001–1500€	59	17.6
	1501–2000€	33	9.9
	2001–2500€	17	5.1
	2501–3000€	4	1.2
	> 3000€	8	2.4

4.2 Consumer-Technology Relationship

The second section of the survey had the objective of collecting data and information regarding the relationship between the consumer and the various technological devices, especially at the time of the purchasing process. The most used devices are smartphones (97.0%), laptops (89.3%), modems for Internet access (68.0%), and smart TVs (65.3%). Apart from these, the desktop computer, game consoles, tablet, and smart watch have also experienced a high rate of usage. Subsequently, the respondents indicated the main social networks to which they are subscribed. The data confirms that Instagram (82.8%) and Facebook (79.5%) are the most used social networks. In addition to these, there is YouTube with a percentage of 55.8%, TikTok with 46.6%, and LinkedIn with 38.0% of individuals using it. Participants then indicated the activities they most frequently carry out using these tools (technologies and social networks). Activities such as sending/receiving instant messages (mean 4.71), taking photos and videos (4.35), searching for information on the web regarding new products and new releases (4.30), scrolling through news feeds on social networks, and sending/receiving emails (4.09) are among the most frequent ones from the sample investigated. As regards video calling (3.33), posting content on social networks (3.09), purchasing via e-commerce sites (3.05), and using video games available on the web (2.72), the participants considered these activities not to be vital to their daily lives.

Subsequently, respondents were asked how important they consider the fact of being able to proceed with the purchase of a product only after touching/seeing/trying it. This statement has yielded interesting results. The data confirm that subjects tend to prefer the purchase of a specific product only after having touched or seen it, with a percentage of 49.9%. However, it is important to underline that a significant part of the respondents (34%) stated that the action of touching and seeing the offer before buying is indifferent to them or minimally important for the purpose of the purchase. In particular, the 17.5% considers it indifferent to physically have the product before purchasing it, the 14.5% considers it irrelevant, and 2.4% does not take this aspect into consideration at all during the purchasing process.

Subsequently, the degree of agreement that the participants expressed regarding the advantages and disadvantages of online purchase was evaluated. Firstly, the sample unanimously believed that the main advantages were speed (4.15) and the possibility of comparing a vast range of offers (4.26), which is not always possible to do within the point of sale. In addition, the data confirmed the propensity of consumers to think that buying online costs less (3.72) and is less stressful (3.67). Regarding the idea that the online shopping process is complicated, the sample seems to disagree (2.08). These data seem to be in line with what Duarte et al. [28] state, namely, that among the reasons why consumers buy online, there are precisely the lower cost and less effort. Also, for the possibility of being able to compare multiple offers, the result confirms the study by Romano et al. [20], in which an entire analysis is dedicated to the possibility of technology to expand the set of considerations that customers have during the process of purchase decision.

4.3 Augmented Reality and Consumer Behavior

The third section of the questionnaire focused on the relationship between AR technologies and consumers and whether this innovation could affect their purchasing process. The results show that 53.1% of respondents know what AR is, 31.5% only partially, and 15.4% unaware of it. In addition, 57.6% said they have used or seen this type of technology at least once. Consequently, the analysis investigated in which contexts individuals found themselves using AR, especially as regards purchasing contexts. The two main contexts where AR was experienced to a greater extent by the participants concerned the video game sector and the real estate sector.

There is a reference, to the online sites of furniture and household items brands such as IKEA, Amazon, and so on (38.2%). Below there are experiences always lived within online sites of specific brands: in particular, with a percentage equal to 35.2%, the participants noticed the presence of these technologies in brand sites for personal care and makeup and 29.6% in sites specialized in the sale of clothing products. Only 5.1% of the interviewees said they had an experience through AR within the physical store, for example, interactive screens. As regards other contexts, online AR technologies were identified as relevant contexts for the research within eyewear sites (3.0%), jewelry sites (0.5%), and online sites for tourism-type activities and museums (0.5%). What was subsequently asked was how motivated they would have been to buy from a brand that uses interactive technologies such as AR for its promotional, marketing, and sales activities. For the sample, this aspect seems to be quite indifferent (3.45). Nonetheless, when asked how much they were inclined to try AR in a store if given the opportunity, 264 (78.8%) of 335 respondents said they would be inclined to have such an experience. This result can be somehow connected to the research carried out by Suh and Lee [32] in which it is underlined that the impact of this technology is greater when consumers try and experience these instead of just looking for information on the web.

The survey gave the opportunity to collect data regarding the perception that consumers have on the usefulness that AR technologies present in online and offline purchase contexts. The results showed that the subjects quite perceive the usefulness that AR innovations have when it comes to buying online (3.85) and are also of the opinion that they speed up the purchasing and decision-making process (3.87). The results, therefore, confirm some of the theories presented by Poushneh and Vasquez-Parraga [33] and Hilken et al. [9] when they delve into the phenomenon whereby AR technologies provide a high level of decision-making comfort in time of purchase decision. AR is perceived as valid and capable of improving the shopping experience within the physical store via screens and other interactive electronic devices (3.87). This theory is confirmed in the work of Poncin and Mimoun [34] who point out how "magic" mirrors and interactive screens inside the point of sale have a significant impact on the atmosphere of the store itself, on the shopping experiences, and obviously on consumer behavior. Table 2 shows the results collected regarding the evaluation that consumers make of this technology. First of all, it has been confirmed that AR allows to better understand the characteristics of a product even if a person is not physically in front of it (3.90). This confirms what has been highlighted by Kang et al. [24] and Garaus et al. [35] according to which these innovations in the field of marketing and the purchasing process convey more confidence to those who have to buy, and AR in contexts of purchase would decrease the overload of choices to which buyers are exposed, making it easier for them to choose the product. There is again confirmation that for many of the respondents, the use of these interactive technologies gives those who use them the opportunity to have a different and unique shopping experience (3.84), allowing the buyer to buy with greater awareness and faster (3.81). This last feature had also already been underlined in the study developed by Romano et al. [20], in which the authors demonstrated that AR can be used as a creative tool to attract and create greater awareness of products.

Table 2 Sociodemographic feature of the respondents

	All sample (100%; 335)		Gen. Z (58.8%; 197)		Gen. Y (14.9%; 50)		Gen. X (17.3%; 58)		Boomers (9.0%; 30)	
	Mean	SD	Mean	SD	Mean	SD	Mean	SD	Mean	SD
It allows me to use something new and enriches my digital knowledge	3.77	0.930	3.84	0.930	3.76	0.930	3.69	0.930	3.47	0.940
It allows me to understand product features	3.90	0.880	4.03	0.880	3.92	0.890	3.64	0.890	3.53	0.910
It allows me to buy faster and with more awareness (even online)	3.81	0.940	3.95	0.940	3.8	0.940	3.5	0.930	3.47	0.950
It allows me to have a different shopping experience	3.84	0.920	3.97	0.920	3.9	0.930	3.59	0.910	3.47	0.940
It allows me to increase satisfaction during and after the purchase	3.55	0.990	3.67	0.990	3.54	0.990	3.4	0.990	3.23	1.010
It would be useless; I don't know the technology enough to use it	2.31	1.140	2.13	1.140	2.36	1.140	2.84	1.150	2.33	1.140
It's the future so something everyone will have to get used to (even non-digital natives)	3.65	1.090	3.64	1.090	3.62	1.090	3.74	1.099	3.53	1.110
It's a waste of time; it would take me too long to figure out how it works	1.99	1.010	1.90	1.010	2.00	0.970	2.19	0.950	2.030	0.960
It allows companies to differentiate shopping experiences by making them more engaging	3.95	0.930	4.02	0.930	4.12	0.920	3.86	0.910	3.43	0.930

Consumers also believe that interactive technologies allow them to use something new that also enriches them in terms of digital knowledge (3.77) and are also of the opinion that the use of these new technological tools by companies helps them in creating different experiences from the competitors that can be "unique" in the eyes of the buyer (3.95). This therefore allows to confirm the "immersive" character that experiences through AR give, as underlined by Scholz and Smith [36] and Heller et al. [37, 38] as regards the degree of sensory stimulation and Yim et al. [19] as regards interactivity and playfulness. This last data regarding the experiential nature of these technologies is also in line with the opinions expressed by Alimamy and Nadeem [39] in their study, which highlights the importance in terms of creating new opportunities by companies especially regarding the creation of value together with the customer and the generation of authentic experiences with him.

It was deemed appropriate to divide opinions about AR technologies into generational classes. What the results show almost fully confirms what is observed by Wood [40]: those who belong to the age group between 18 and 25 or the Generation Z and those belonging to the group from 26 to 41 the so-called Generation Y are more inclined to the technological world. This can be highlighted by the data collected which demonstrate how the average opinion of these two generational classes is higher than that of Millennials and Boomers. Also with regard to the two questions concerning, respectively, the degree of agreement or disagreement on the uselessness of AR technologies given the lack of knowledge of the technology and the waste of time in using them, Generation Z has a value of 2.13 on average (therefore quite disagree) on the first and 1.90 on the second and Generation Y a mean of 2.36 and 2.00, while the Millennials with 2.84 and 2.19 and the Boomers with 2.33 and 2.03 do indeed disagree, but to a slightly lesser extent. This would also confirm what has been highlighted in the research of Ratchford et al. [41] where it is stated that elderly consumers perceive less the importance of these innovations in their purchasing process as they have to invest more in understanding how they work.

5 Conclusions and Limitations

The main objective of the research was to understand if and how AR technologies can influence consumer behavior during the purchasing process. Through the empirical analysis, it was possible to understand, through the sample of consumers, their attitudes, opinions, and behavior in the face of purchasing situations associated both with the use of technology in general and (and mainly) with the use of AR. The results confirmed that electronic devices such as smartphones, computers, tablets, and so on are now present in most homes of individuals of any age and gender. Although most of the sample prefers to shop in physical shops, some of the prepurchasing activities are carried out using technology, such as the search online for information about the product. Furthermore, it emerged a positive opinion regarding the help that AR technologies could give to the consumer when he/she chooses and buys a product. Understanding the characteristics of the product, having a unique experience with it during the purchasing process, and making choices with more awareness and faster are some of the benefits identified by the various generational courts analyzed. Therefore, it is certainly relevant for sellers to try to consider the possibility of using AR to improve the purchasing experience of their product/service. By doing this, they will be innovative and will achieve a competitive advantage from consumers' interest to try this technology and enjoy a more interactive shopping experience, bringing them closer to the brand to create solid and lasting relationships. The main limitation of the study derives from the sample of consumer considered. It would be relevant for future research to disseminate this research in Italy, to be able to obtain a larger sample of respondents and therefore more generalizable results and replicate the same in other European countries to understand the consumption trend at a European level.

References

1. Sima, V., Gheorghe, I. G., Subić, J., & Nancu, D. (2020). Influences of the industry 4.0 revolution on the human capital development and consumer behavior: A systematic review. *Sustainability, 12*(10), 4035.
2. Kazmi, S. H. A., Ahmed, R. R., Soomro, K. A., Hashem, E., Akhtar, H., & Parmar, V. (2021). Role of augmented reality in changing consumer behavior and decision making: Case of Pakistan. *Sustainability, 13*(24), 14064.
3. McLean, G., & Wilson, A. (2019). Shopping in the digital world: Examining customer engagement through augmented reality mobile applications. *Computers in Human Behavior, 101*, 210–224.
4. Rauschnabel, P. A., Babin, B. J., tom Dieck, M. C., Krey, N., & Jung, T. (2022). What is augmented reality marketing? Its definition, complexity, and future. *Journal of Business Research, 142*, 1140–1150.
5. Lavingia, K., & Tanwar, S. (2020). Augmented reality and industry 4.0. In A. Nayyar & A. Kumar (Eds.), *A roadmap to industry 4.0: Smart production, sharp business and sustainable development* (Advances in Science, Technology & Innovation) (pp. 143–155). Springer.
6. Uddin, S.-A., Hanna, G., Ross, L., Molina, C., Urakov, T., Johnson, P., Kim, T., & Drazin, D. (2021). Augmented reality in spinal surgery: Highlights from augmented reality lectures at the emerging technologies annual meetings. *Cureus, 13*(10).
7. ABI Research. The digital enterprise ecosystem: Augmented reality spotlight (2019). Available online: https://www.abiresearch.com/blogs/2019/10/01/digital-enterprise-ecosystem-augmented-reality-spotlight/. Accessed 27 Jan 2023.
8. Grand View Research Inc. The global augmented reality market size (2021). Available online: https://www.grandviewresearch.com/industry-analysis/augmented-reality-market. Accessed 27 Jan 2023.
9. Hilken, T., de Ruyter, K., Chylinski, M., Mahr, D., & Keeling, D. I. (2017). Augmenting the eye of the beholder: Exploring the strategic potential of augmented reality to enhance online service experiences. *Journal of the Academy of Marketing Science, 45*(6), 884–905.
10. Kumar, H. (2021). Augmented reality in online retailing: A systematic review and research agenda. *International Journal of Retail & Distribution Management, 50*(4), 537–559.
11. Riar, M., Xi, N., Korbel, J. J., Zarnekow, R., & Hamari, J. (2022). Using augmented reality for shopping: A framework for AR induced consumer behavior, literature review and future agenda. *Internet Research.*

12. İpek, İ. (2020). Understanding consumer behavior in technology-mediated spaces. In B. George & J. Paul (Eds.), *Digital transformation in business and society* (pp. 169–189). Palgrave Macmillan.
13. Qin, Y. S. (2020). Fostering brand–consumer interactions in social media: The role of social media uses and gratifications. *Journal of Research in Interactive Marketing, 14*(3), 337–354.
14. Cheung, M. L., Pires, G. D., Rosenberger, P. J., Leung, W. K., & Sharipudin, M. N. S. (2021). The role of consumer-consumer interaction and consumer-brand interaction in driving consumer-brand engagement and behavioral intentions. *Journal of Retailing and Consumer Services, 61*, 102574.
15. Roy, S. K., Gruner, R. L., & Guo, J. (2022). Exploring customer experience, commitment, and engagement behaviours. *Journal of Strategic Marketing, 30*(1), 45–68.
16. Ahmed, R. R., Streimikiene, D., Berchtold, G., Vveinhardt, J., Channar, Z. A., & Soomro, R. H. (2019). Effectiveness of online digital media advertising as a strategic tool for building brand sustainability: Evidence from FMCGs and services sectors of Pakistan. *Sustainability, 11*(12), 3436.
17. Rasool, A., Shah, F. A., & Islam, J. U. (2020). Customer engagement in the digital age: A review and research agenda. *Current Opinion in Psychology, 36*, 96–100.
18. Tom Dieck, M. C., & Han, D. I. D. (2022). The role of immersive technology in customer experience management. *Journal of Marketing Theory and Practice, 30*(1), 108–119.
19. Yim, M. Y.-C., Chu, S.-C., & Sauer, P. L. (2017). Is augmented reality technology an effective tool for Ecommerce? An interactivity and vividness perspective. *Journal of Interactive Marketing, 39*(1), 89–103.
20. Romano, B., Sands, S., & Pallant, J. I. (2021). Augmented reality and the customer journey: An exploratory study. *Australasian Marketing Journal, 29*(4), 354–363.
21. Azuma, R. T. (1997). A survey of augmented reality. *Presence: Teleoperators & Virtual Environments, 6*(4), 355–385.
22. Cao, M., Zheng, L., Jia, W., & Liu, X. (2020). Fast monocular visual odometry for augmented reality on smartphones. *IEEE Consumer Electronics Magazine*.
23. Wedel, M., Bigné, E., & Zhang, J. (2020). Virtual and augmented reality: Advancing research in consumer marketing. *International Journal of Research in Marketing, 37*(3), 443–465.
24. Kang, H. J., Shin, J., & Ponto, K. (2020). How 3D virtual reality stores can shape consumer purchase decisions: The roles of informativeness and playfulness. *Journal of Interactive Marketing, 49*, 70–85.
25. Batat, W. (2021). How augmented reality (AR) is transforming the restaurant sector: Investigating the impact of "Le Petit Chef" on customers' dining experiences. *Technological Forecasting and Social Change, 172*, 121013.
26. Rejeb, A., Rejeb, K., & Treiblmaier, H. (2021). How augmented reality impacts retail marketing: A state-of-the-art review from a consumer perspective. *Journal of Strategic Marketing, 31*(3), 718.
27. Rese, A., Baier, D., Geyer-Schulz, A., & Schreiber, S. (2017). How augmented reality apps are accepted by consumers: A comparative analysis using scales and opinions. *Technological Forecasting and Social Change, 124*, 306–319.
28. Duarte, P., Silva, S., & Ferreira, M. B. (2018). How convenient is it? Delivering online shopping convenience to enhance customer satisfaction and encourage e-WOM. *Journal of Retailing and Consumer Services, 44*, 161–169.
29. El-Jarn, H., & Southern, G. (2020). Can co-creation in extended reality technologies facilitate the design process? *Journal of Work-Applied Management, 12*(2), 191–205.
30. Malhotra, M. K., & Grover, V. (1998). An assessment of survey research in POM: From constructs to theory. *Journal of Operations Management, 16*(4), 407–425.
31. Nunnally, J. C., & Bernstein, I. H. (1994). *Psychometric theory* (3rd ed.). McGraw-Hill.
32. Suh, K. S., & Lee, Y. E. (2005). The effects of virtual reality on consumer learning: An empirical investigation. *Management Information Systems Quarterly, 29*, 673–697.

33. Poushneh, A., & Vasquez-Parraga, A. Z. (2017). Discernible impact of augmented reality on retail customer's experience, satisfaction and willingness to buy. *Journal of Retailing and Consumer Services, 34*, 229–234.
34. Poncin, I., & Mimoun, M. S. B. (2014). The impact of "e-atmospherics" on physical stores. *Journal of Retailing and Consumer Services, 21*(5), 851–859.
35. Garaus, M., Wagner, U., & Kummer, C. (2015). Cognitive fit, retail shopper confusion, and shopping value: Empirical investigation. *Journal of Business Research, 68*(5), 1003–1011.
36. Scholz, J., & Smith, A. N. (2016). Augmented reality: Designing immersive experiences that maximize consumer engagement. *Business Horizons, 59*(2), 149–161.
37. Heller, J., Chylinski, M., de Ruyter, K., Mahr, D., & Keeling, D. I. (2019a). Let me imagine that for you: Transforming the retail frontline through augmenting customer mental imagery ability. *Journal of Retailing, 95*(2), 94–114.
38. Heller, J., Chylinski, M., de Ruyter, K., Mahr, D., & Keeling, D. I. (2019b). Touching the untouchable: Exploring multi-sensory augmented reality in the context of online retailing. *Journal of Retailing, 95*(4), 219–234.
39. Alimamy, S., & Nadeem, W. (2021). Is this real? Cocreation of value through authentic experiential augmented reality: The mediating effect of perceived ethics and customer engagement. *Information Technology & People, 35*(2), 577–599.
40. Wood, S. L. (2002). Future fantasies: A social change perspective of retailing in the 21st century. *Journal of Retailing, 78*(1), 77–83.
41. Ratchford, B. T., Talukdar, D., & Lee, M. S. (2001). A model of consumer choice of the internet as an information source. *International Journal of Electronic Commerce, 5*(3), 7–21.

Adoption of Electronic Commerce by Portuguese Small and Medium-Sized Enterprises

José Carlos Ferreira Correia, Mário Caldeira Dias, and Dulce Matos Coelho

Abstract Since the beginning of the twenty-first century, small and medium-sized enterprises (SMEs) have been operating in the context of an economic crisis, with the added effect of the pandemic crisis, leading them to rethink how to act in the markets in which they operate. SMEs have assumed a relevant role in national economies, in terms of employment, production, and turnover, with the assimilation of new information and communication technologies (ICT) becoming increasingly important, with a view to adopting e-commerce. Given the fact that SMEs represent a considerable segment of the companies in Europe, European entities (European Commission and European Parliament) proceeded to redefine their concept and sought to enable them to enjoy the economic and social benefits provided by ICT, through the adoption of e-commerce, in order to ensure a more active and influential presence with their customers. When inquiring about this matter, we came across the fact that most of the existing studies, within the scope of the adoption of electronic commerce, refer to its application in the context of large companies, in nonagricultural activities, therefore the need to carry out scientific research on this issue, which allowed us to determine the factors that influence the adoption of electronic commerce on agricultural SMEs. In this study, the factors that stood out to the acceptance/adoption of electronic commerce by agricultural SMEs are as follows: perceived utility, risk perception/intent-adoption of e-commerce, perceived ease of use, and perception of social influence.

Keywords E-commerce · Adoption · Agricultural SMEs

J. C. F. Correia (✉)
Instituto Politécnico de Setúbal, Setúbal, Portugal

M. C. Dias
Universidade Lusíada, Lisboa, Portugal

D. M. Coelho
ESCE-IPS, Setúbal, Portugal

J. Reis et al. (eds.), *Driving Quality Management and Sustainability in VUCA Environments*, Springer Proceedings in Business and Economics,
https://doi.org/10.1007/978-3-031-52723-4_6

1 Introduction

The European Commission defines e-commerce in the Eurostat Statistics Glossary [1] as "the sale or purchase of goods or services, whether between businesses, households, individuals or private organizations, through electronic transactions conducted via the internet or other computer-mediated (online communication) networks. The term covers the ordering of goods and services which are sent over computer networks, but the payment and the ultimate delivery of the goods or service may be conducted either on- or off-line."

E-commerce is referred as one of the elements with most potential to change business models [2, 3]. The contribution of the Internet to change business models was fundamental, according to Laudon and Traver [4], because the Internet has enabled new forms of collaboration, communication, and innovation that were not possible before. According to these authors, the Internet has enabled new business models that leverage digital technologies to create, deliver, and capture value, e.g., connecting buyers and sellers more efficiently through e-marketplaces and digital platforms. The Internet has become an integral part of SME's business, presenting several advantages such as obtaining information of the products and services as well as information about existing and potential competitors [5].

Several authors name the benefits of e-commerce for entrepreneurs in SMEs, e.g., reach new markets, establish global network, and improve efficiency [6–9].

One of the most common modalities of e-commerce used by entrepreneurs in the agriculture sector is associated with e-marketplaces [10, 11]. These authors identified several benefits of e-marketplaces, such as increased access to markets, increased business opportunities, improved market intelligence (viz., providing SMEs with real-time price information), lower transaction costs, and improved market transparency and efficiency. Others (viz., Quayle [12]; Poon & Swatman [13]; Abell et al. [14]) have concluded that the tangible benefits derived from e-commerce, such as lower administrative and production costs, lower delivery time, and increased sales, are marginal in terms of direct gains [5].

Several studies identified, on the other hand, disadvantages of e-marketplaces, such as issues related to trust, security, and quality assurance [15, 16], increased competition and reduced bargaining power [17], product quality issues [18], job losses [19], and reduced flexibility of work practices [20, 21].

As the agriculture sector has shown growing readiness for digital reinvention, agricultural entities will be able to obtain added value from carrying out digital investments, which may be financed by community or national funds.

In order to understand how one can evaluate the determining factors in the adoption of ICT and e-commerce, it is important to bear in mind some of the existing theoretical models, which were used within the scope of the intention-adoption process of electronic commerce. We identified the Extension of the UTAUT Model [22]; the Extension of the Technology Acceptance Model – TAM [23]; the Model by

Table 1 Most relevant constructs to evaluate the determining factors in the adoption of ICT and e-commerce

Constructs	Description	Origin
Perceived ease of use (PEU)	"According to Davis [24], perceived ease of use refers to the degree to which an individual believes that using a given technology will be free of physical and mental effort" (Fernandes & Ramos, 2012, p. 6).	Gefen et al. (2003), Heidjen et al. (2003), Nakagawa (2008), Pavlou (2003) and Vijayasarathy (2004)
Perceived utility (PU)	"Perceived usefulness is defined by Karahanna, Straub and Chervany (1999) as a relative advantage, that is, the technology used is perceived as superior compared to traditional practice and optimizes the execution of tasks" (Fernandes & Ramos, 2012, p. 6)	Pavlou (2003), Gefen et al. (2003), Heidjen et al. (2003), Nakagawa (2008) and Vijayasarathy (2004)
Perception of business trust/perception of social influence (PBT/PSI)	"This construct represents the belief that the other party will behave in a socially responsible manner and, therefore, will comply with expectations, without taking advantage of the vulnerabilities of the person doing business with (Pavlou, 2003). The problem of trust in e-commerce is critical, since the purchase decision is based on the user's perception of the reliability aspects of the means of purchase and the e-commerce site" (Fernandes & Ramos, 2012, p. 6)	Gefen et al. (2003), Heidjen et al. (2003), Nakagawa (2008), Pavlou (2003) and Vijayasarathy (2004)
Risk perception (RP)	"According to Solomon [25], perceived risk is the belief that the purchase of a product or service will have negative consequences" (Fernandes & Ramos, 2012, p. 6)	Heidjen et al. (2003) and Pavlou (2003)
Intent-adoption of electronic commerce (IAEC)	"It refers to the degree to which an individual believes that people who are important to him think that a certain behavior should be performed (Fishbein & Ajzen [26], apud Venkatesh & Morris [27], p. 119). [. . .] is important for e-commerce because it also acts on online consumer behavior" (Fernandes & Ramos, 2012, p. 6)	Pavlou (2003)

Heidjen, Verhagen, and Creemers [28]; the Extension of the Technology Acceptance Model [29]; the Model by Bramal, Schoefer, and Mckechnie [30]; the Model by Vijayasarathy [31]; and the Model by Nakagawa [32].

From the analysis carried out on the models mentioned above, and based on the study of Fernandes and Ramos [33], we found that the most expressive constructs, given their importance and relevance, were those shown in Table 1.

2 Discussion

2.1 Proposed Model

To answer the research question, i.e., what are the factors that influence the intention/adoption of e-commerce in agricultural SMEs, the authors of the current paper used an extension of the technology acceptance model proposed by Pavlou [23]. Initially used to explain the acceptance of technology in the workplace, this model has been used by several researchers to explain the acceptance or rejection of the use or intention to use information systems, being one of the most influential models for understanding the adoption of technology [34].

Given the nature of this investigation, we sought, based on the model proposed by Pavlou [23], and adding some of the constructs used in research in the area of acceptance of technologies in e-commerce, to design a model that involved constructs that influence the intention-adoption of e-commerce by SMEs. Thus, the constructs PEU and PU, proposed by Davis [35] in TAM, social influence [32], trust [23], and perceived risk [23, 32], were used. Given the lack of information in the literature on the application of some constructs to the agricultural sector, and the need to obtain their perception by SMEs, we decided to insert the following constructs: perception of social influence (PSI), perception of business trust (PBT), risk perception (RP), and intent-adoption of electronic commerce (IAEC). The proposed model by Correia [36] is represented in Fig. 1.

In our research, five research hypotheses were considered, in order to understand the intention-adoption of e-commerce by SMEs in the agriculture sector.

- ***H1 Perceived ease of use has a positive effect on intent-e-commerce adoption.***

Pavlou [23] related the perceived ease of use with the transaction intention construct. The latter can be perceived as electronic commerce adoption intention, in the sense of confirming that workers believed that by using ICT they would be available to carry out other activities that do not require physical and mental efforts. In this context, the existence of a positive correlation between perceived ease of use and the intention of online behavior has been indicated in several studies [23, 28, 29, 31].

- ***H2 Perceived usefulness has a positive effect on intent-e-commerce adoption.***

Karahanna, Straub, and Chervany [37] considered the PU to be an advantage, as it aimed at optimizing the execution of tasks. In this context, several authors have mentioned the existence of a positive correlation between perceived usefulness and intention to purchase online [23, 29, 31].

- ***H3 Perception of social influence has a positive effect on intent-adoption of electronic commerce.***

According to Pavlou and Fygenson [38], when social influence is favorable, companies are more likely to adopt e-commerce. In this sense, it has been mentioned

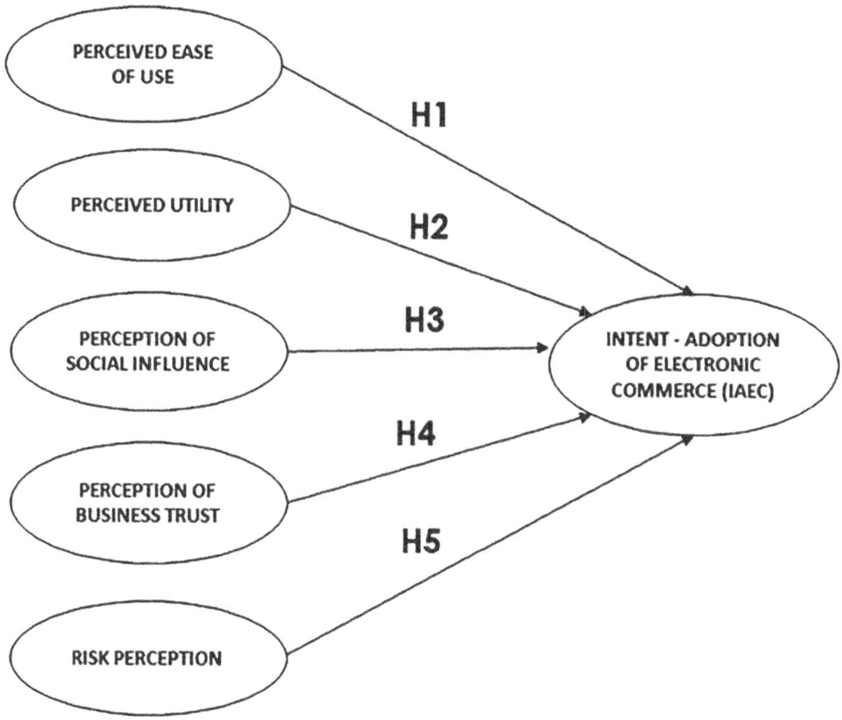

Fig. 1 Research model of e-commerce adoption by agricultural SMEs

that a positive correlation of social influence on online purchase intention can be observed [31, 32].

- ***H4 Perceived business trust has a positive effect on intent-e-commerce adoption.***

Pavlou [23] concluded that the trust construct is an important factor for electronic commerce, having highlighted that consumer confidence in Internet sales was fundamental for the growth of B2C electronic commerce. Gefen, Karahanna, and Straub [29] also highlighted that the level of trust was one of the factors to be taken into account when using this type of system. In this sense, it has been mentioned that a positive correlation of trust and online purchase intention can be observed [23, 28–30].

- ***H5 Perception of risk has a positive effect on intent-adoption of electronic commerce.***

Pavlou [23] introduced the perceived risk construct in his model, having associated it with e-commerce and online consumer behavior, due to the uncertainty existing in the e-commerce environment. For other authors [39], the perceived risk has been studied to understand the nonuse of the Internet as one of the purchasing channels made available by companies. In this context, a negative correlation of perceived risk in online purchase intention is pointed out [23, 28, 32].

2.2 Universe

The universe of this investigation consists of the entire business population that is included in an SME database (authors' own elaboration), covering entities related to the agriculture sector in Portugal. To determine the adequate size of a sample (n) for a finite population (N), we used the simple random sample method, which allows estimating the proportion of the population (p), once the confidence level (95%), the standardized normal (1.96), and precision level (D + −5%) have been defined. The minimum sample number for a finite population with N = 4279 was 353. We received 43 responses to the survey (response rate of 12.18%).

We used the method developed by Karl Popper [40], which consists of a hypothetical-deductive study, employing a quantitative approach of a descriptive and exploratory nature, comprising a set of procedures. We began with the choice of constructs based on the literature (in the omitted cases, we opted for the perception of the element under analysis), followed by the elaboration of a questionnaire, the realization of a pretest of the questionnaire, the collection of data, the description of the sample, and, finally, the analysis and discussion of results.

Given the lack of a standardized questionnaire, for the present investigation, we opted for the construction of a survey, in order to collect data on the constructs under analysis, with its validation going through several stages: I, elaboration; II, pretest; and III, data collection. The questionnaire was made available from September 14–30, 2021.

2.3 Statistical Treatment of Data

The data analysis was carried out through factor analysis, which intends to reveal latent variables, factors, or constructs, responsible for the behavior of certain observed variables, following a pattern resulting from an exploratory study or a certain theory [41]. The factor analysis comprised two types of analysis, exploratory factor analysis (EFA) and confirmatory factor analysis (CFA), through IBM SPSS application (Statistics Package for the Social Sciences). As for the interview data, we opted for a content analysis, as described by Sarmento [42].

All responses received by September 30, 2021, were validated and were considered valid. The following table contains the constructs of the initial model, as well as the constructs that were obtained resulting from the AFE (Table 2).

Table 2 Initial constructs versus constructs obtained from AFE

Constructs of the proposed initial model	Constructs obtained after AFE
PEU, PU, PSI, PBT, RP, IAEC	PU, RP/IAEC, PEU, PSI

2.4 Global Analysis of Results

Following the exploratory factor analysis, we proceeded to reduce the number of constructs in the initial model. A composite Likert scale from 1 to 5 was used in the questionnaire, where 1 corresponds to "Not at all relevant" and 5 to "Very relevant."

We found that the majority of participants, in terms of the mean of the constructs (where data from a normal distribution are concentrated), highlighted the PU construct with an average of 3.63, followed by the PEU construct with an average of 3.40, RP/IAEC construct with an average of 3.35, and, finally, PSI construct with an average of 3.10. Based on a frequency and percentage table of irrelevance, neutrality, and relevance of construct items, we concluded that participants also highlighted the PU construct (56.27%), followed by the PEU construct (47.65%) and the RP/IAEC construct (44.76%), as determining factors in the adoption of e-commerce. Given the value presented (32.55%), the inclusion of the PSI construct in the model should be considered.

In order to assess the adequacy of the sample, with regard to the correlation of the constructs, we carried out the Kaiser-Meyer-Olkin (KMO) and Bartlett's sphericity tests, with the values obtained in the case of the KMO test (0.771) inducing that the constructs determined in the exploratory factor analysis were adjusted to the data, testing their consistency. With regard to the Bartlett's sphericity test, as the p-value (sig) is equal to 0, less than the significance level = 5%, the adequacy of the EFA to the internal matrix was guaranteed, and, therefore, the data are considered adequate for the application of factorial analysis in the considered variables.

The commonalities, perceived as the proportion of common variance present in a variable, showed values greater than 0.60, revealing that the items were in accordance with the respective variables.

Regarding the total explained variance, we found that four components (constructs) were determined, in line with the inflection point of the curve that related the number of the factor (construct) and the respective eigenvalue, associated with the number 4, and, therefore, we could also induce that the four constructs were in accordance with the previously determined values.

The evaluation of the reliability of the exploratory factorial structure was carried out using the internal consistency coefficient, Cronbach's alpha, and Cronbach's alpha based on standardized items, obtaining the values shown in Table 3.

As previously mentioned, following the exploratory factor analysis, the initial model was reduced to four constructs, and these constructs were used to carry out the confirmatory factor analysis, in order to compare the results obtained in the EFA with the known theoretical hypothesis. We were thus able to confirm whether the

Table 3 Construct reliability statistics

Construct	Cronbach's alpha	Cronbach's alpha based on standardized items	Number of items	Evaluation of internal consistency[a]
PU	,962	,962	10	Excellent
RP/IAEC	,945	,946	8	Excellent
PEU	,932	,932	4	Excellent
PSI	,756	,758	2	Good

[a]In accordance with the scale referred by Sarmento [42]

Table 4 Factorial weight by item/constructs

Construct	Item	Factorial weight	Item	Factorial weight
PU	PEU1-Functionality	0,87	PSI2-Tec. resources	0,83
	PEU2-Responsiveness	0,86	PSI4-Expected results	0,83
	PU2-Productivity	**0,91**	PBT2-Growth	0,90
	PU3-Celerity	**0,91**	**RP2-Content**	**0,71**
	PU4-Safety	0,86	IAEC4-Future plan	0,80
RP/IAEC	PBT1-Competitiveness	0,86	RP4-Technologies	0,86
	PBT2-Growth	**0,62**	**IAEC1- Intention**	**0,94**
	RP1- Tasks	0,67	IAEC2-Strategy	0,84
	RP3-Human resources	0,87	IAEC3- App. Management	0,78
PEU	PEU3-Comprehensibility	0,90	**PU1-Performance**	**0,91**
	PEU4-Simplicity	0,88	**PBT4-Financing**	**0,82**

proposed model is fully determined or saturated, estimating only the relationships between the specified factors and variables.

From the confirmatory factor analysis, we found the following for three of the constructs, with the exception of the PSI construct (which had no relevant expression). In the PU construct, the items PU2-Productivity and PU3-Celerity contributed the most significantly, with a factorial weight of 0.91. The RP2-Content item, which contributed the least, presented a factorial weight of 0.71. Regarding the RP/IAEC construct, the item IAEC1-Intention contributed the most significantly, with a factorial weight of 0.94, instead of the item PBT2-growth, which contributed the least, with a factorial weight of 0.62. As for the PEU construct, the PU1-Performance item contributed most significantly, with a factorial weight of 0.91. On the contrary, we found that the Financing item was the least contributor to the PEU construct, with a factorial weight of 0.82 (Table 4).

To verify the suitability of the model to the data in the CFA, we used the assessment and adjustment measures mentioned in the studies by Hair et al. [43] and Byrne [44], namely, chi-square ($\chi 2$), comparative fit index (CFI), goodness of fit index (GFI), parsimony goodness of fit index (PGFI), normative fit index (NFI), adjusted goodness of fit index (AGFI), and root mean square error of approximation (RMSEA).

We concluded that in the PU construct, these values do not represent a high suitability, unlike the RP/IAEC and PEU constructs.

3 Conclusions

After carrying out the exploratory factor analysis, we found that there was a need to readjust the items/constructs, maintaining the initial research hypotheses, based on the new adequacy of the data. The confirmatory factor analysis was carried out considering the new constructs, and, from the analysis of the results obtained and meeting the initially established research hypotheses, we identified in the next table the hypotheses that were confirmed and those that were contested (Table 5).

We consider that the objective proposed in this investigation, to identify the factors that influence the adoption of electronic commerce, was achieved, following the analysis carried out on the data collected through the questionnaire survey, having identified the following factors/constructs: PU, PEU, PSI, and RP/IAEC.

We found that Portuguese agricultural SMEs consider the simplicity of applications to be important, understanding that they should enable users to understand how the apps can be used in certain tasks (PEU), as well as the installation of applications on different devices, reflecting the importance of productivity and speed of their use (PU), knowledge of the application by employees (PSI), and knowledge of applications, in order to contribute to the reduction of operating costs, adding to the need for applications to be competitive and technological (RP/IAEC).

This investigation allowed drawing relevant lessons for the literature associated with electronic commerce, identifying the elements that determined the intention-

Table 5 Verification of hypotheses according to EFA

Hypothesis	Construct	Highlighted items	Relevant items	Acceptance[b]	Cronbach's alpha	Conclusion
H1	PEU	PEU4-Simplicity	PEU4-Simplicity	47,65%	0,932	Confirmed
		PU1-Performance[a]				
H2	PU	PU2-Productivity	PU3-Celerity	56,27%	0,962	Confirmed
		PU3-Celerity				
		PU4-Safety				
		PBT2-Growth				
H3	PSIS	PSI3-Knowledge	PSI1-Control	32,55%	0,756	Confirmed
H4	PBT	–	–	–	–	–
H5	RP	IAEC1-Intention[a]	IAEC1-Intention	44,76%	0,945	Confirmed
H6		PBT1-Competitiv.				
		RP4-Technology				

[a]Items highlighted in CFA. [b]The results from average of the relevance in each construct

adoption in the case of Portuguese agricultural SMEs. It is also demonstrated that e-commerce is a necessity for agricultural SMEs, in order to become more competitive in the global market, using new working methods (PEU) and new technological solutions (PU), that is, SMEs must be receptive to innovation and available to assume a relationship of trust with customers (PSI). It should also be noted that the ability to implement changes is essential to achieve competitiveness, which is why it is important to invest in young people with some training in this area (RP), as we are facing an adaptation of current models to new times, with the aim of differentiation and improvement of indicators in terms of productivity and speed (IAEC).

From the initially proposed model, after conducting the interviews to five (5) persons, we found that the PBT construct was not considered relevant for agricultural SMEs. From the responses to the questionnaire, we also found that the RP and IAEC constructs are not equally relevant when analyzed separately.

According to the investigation, we confirmed that agricultural SMEs highlighted the constructs PU, PEU, RP/IAEC, PEU, and PSI, with 56,27%, 47,65%, 44,76%, and 32,55%, respectively.

SMEs highlighted, in the PU construct, productivity and speed, resulting from the use of applications allowing their users to achieve their goals in the shortest amount of time, adding the fact that applications can be installed on different devices. At the PEU, the simplicity of e-commerce applications was highlighted, enabling users to perform certain tasks. With regard to the constructs RP and IAEC, SMEs considered that it made sense to analyze them as if they were a single construct, with the intention of reducing operating costs and the need for applications to be competitive and technological. It should be noted, although of lesser importance, that the intention-adoption of e-commerce also depended on the PSI construct, resulting from the knowledge expressed by employees in handling this type of applications.

The changes resulting from the Internet and the web proved to be extremely important, in the sense of providing the reconfiguration of the value chain associated with the agriculture sector, reorienting agricultural SMEs in their relationship with customers, as well as in the establishment of cooperation networks between them. These changes include costs for SMEs, in order to maintain their value in a new type of market, whose differentiation is essential for their survival, which implies investment and development in new products/services, in order to maintain customers.

4 Limitations and Future Investigation

The limitations of this study relate to obtaining responses to surveys and, in part, to the lack of studies in this area. The small number of responses received (43) conditioned the methodologies and procedures to be applied and the respective testing of the validity of the constructs. A larger number of responses would have made it possible to complement the analysis, in order to consolidate the achievement of a more robust model.

For future investigations, we recommend increasing the sample and granting a longer period for sending responses, in order to work with the recommended universe of 353 responses, thus allowing a more complete confirmatory factor analysis, to ensure that the constructs mentioned in this investigation as factors that determine the adoption of e-commerce portray the observed reality. We also suggest the inclusion of other variables, in order to develop new concepts and obtain more expressive results.

It should be noted that this study could contribute to helping managers in the future, with regard to the factors that should deserve attention when considering the intention-adoption of electronic commerce in the companies where they work, in order to eliminate any adoption problems in the market in which they intend to operate.

References

1. Eurostat. (2019). E-commerce. In Eurostat statistics glossary. https://ec.europa.eu/eurostat/statistics-explained/index.php?title=Glossary:E-commerce
2. Lee, J., & Turban, E. (2001). A survey of electronic commerce research. *International Journal of Electronic Commerce, 5*(1), 7–22. https://doi.org/10.1080/10864415.2001.11044284
3. Turban, E., King, D., Lee, J., Liang, T.-P., & Turban, D. C. (2006). *Electronic commerce 2006: A managerial perspective*. Prentice Hall.
4. Laudon, K. C., & Traver, C. G. (2020). *E-commerce: Business, technology, society* (15th ed.). Pearson.
5. Mamede, J. H. P. S. (2008). Uma metodologia para a implementação de comércio e negócio electrónico em pequenas e médias empresas [Doctoral dissertation, Universidade do Minho]. RepositorioUM. https://hdl.handle.net/1822/9542
6. Bouncken, R. B., & Kraus, S. (2013). Innovation in knowledge-intensive industries: The double-edged sword of coopetition. *Journal of Business Research, 66*(10), 2060–2070. https://doi.org/10.1016/j.jbusres.2013.02.035
7. Chen, M. Y., Chen, J. S., & Tang, J. Y. (2010). Exploring the impact of e-commerce on organizational performance: A survey of small and medium enterprises in Taiwan. *Journal of Small Business Management, 48*(3), 332–355. https://doi.org/10.1111/j.1540-627X.2010.00297.x
8. Westland, J. C., & Clark, T. R. (2010). Global entrepreneurship and innovation through e-commerce. *Journal of World Business, 45*(4), 475–482.
9. Lu, J., & Beamish, P. W. (2004). International diversification and firm performance: The S-curve hypothesis. *Academy of Management Journal, 47*(4), 598–609. https://doi.org/10.5465/amj.2004.13680954
10. Ghazali, E., & Shamsudin, M. N. (2013). The use of e-commerce in agricultural marketing: A review of literature and empirical analysis. *Journal of Agribusiness Marketing, 6*(1), 48–67.
11. Durst, P., & Pauwels, L. (2005). E-commerce for agribusiness: Opportunities, challenges and concepts. *British Food Journal, 107*(5), 315–326. https://doi.org/10.1108/00070700510596836
12. Quayle, M. (2002). E-commerce: The challenge for UK SMEs in the twenty-first century. *International Journal of Operations and Production Management, 22*, 10.
13. Poon, S., & Swatman, P. (1997). Internet-based small business communication: Seven Australian cases. In B. F. Schmid, & D. Selz (Eds.), *EM - Electronic Commerce in Asia. EM - Electronic Markets*

14. Abell, W., & Limm, L. (1996). *Business use of the Internet in New Zealand: An exploratory study*. Lismore: AUSWeb, Southern Cross University.
15. He, Y., & Li, Y. (2018). E-marketplace: Advantages, challenges, and research directions. *Electronic Commerce Research and Applications, 28*, 67–73. https://doi.org/10.1016/j.elerap.2018.04.003
16. Ritchie, B., & Brindley, C. (2000). Disintermediation, disintegration and risk in the SME global supply chain. *Management Decision, 38*(10), 662–670. https://doi.org/10.1108/00251740010378309
17. Chen, Y., & Zhang, Y. (2018). An empirical study of the disadvantages of e-marketplace from the perspective of suppliers. *Journal of Business Research, 91*, 317–327. https://doi.org/10.1016/j.jbusres.2018.06.002
18. Tang, J., & Chen, Y. (2019). The dark side of e-marketplaces: An empirical investigation of online opportunism and product quality issues. *Information & Management, 56*(7), 103155. https://doi.org/10.1016/j.im.2019.103155
19. Lee, D. Y., Park, J., & Kim, M. (2021). Analysis of the negative effects of e-commerce on the retail industry. *Sustainability, 13*(12), 6598. https://doi.org/10.3390/su13126598
20. Lee, C.-S. (2001). An analytical framework for evaluating e-commerce business models and strategies. *Journal of Electronic Commerce Research, 2*(4), 178–190. https://doi.org/10.1108/10662240110402803
21. Sparkes, A., & Thomas, B. T. (2001). The use of the internet as a critical success factor for the marketing of Welsh agri-food SMEs in the twenty-first century. *British Food Journal, 103*(3), 177–186. https://doi.org/10.1108/00070700110395368
22. Benslimane, Y., Plaisent, M., & Bernard, P. (2003). Applying the task-technology fit model to WWW-based conceptualization and measurement. In *Proceedings of the 36th Annual Hawaii International Conference on System Sciences, 2003* (p. 10). IEEE. https://ieeexplore.ieee.org/stamp/stamp.jsp?tp=&arnumber=1174568
23. Pavlou, P. A. (2003). Consumer acceptance of electronic commerce: Integrating trust and risk with technology acceptance model. *International Journal of Electronic Commerce, 7*(3), 69–103. https://doi.org/10.2307/2514872
24. Davis, F. D. (1989). Perceived usefulness, perceived ease of use, and user acceptance of information technology. *MIS Quarterly, 13*(13), 319–340.
25. Solomon, M. R. (2002). *O comportamento do consumidor: comprando, possuindo e sendo*. Porto Alegre: Bookman.
26. Fishbein, M., & Ajzen, I. (1975). *Belief, attitude, intention and behavior: An introduction to theory and research*. Reading, MA: Addison-Wesley.
27. Venkatesh, V., & Morris, M. G. (2000). Why don't men ever stop to ask for directions? Gender, social influence, and their role in technology acceptance and usage behavior. *MIS Quarterly, 24*(1), 115–139. http://dx.doi.org/10.2307/3250981
28. van der Heidjen, H., Verhagen, T., & Creemers, M. (2003). Understanding online purchase intentions: Contributions from technology and trust perspectives. *European Journal of Information Systems, 12*(1), 41–48. https://doi.org/10.1057/palgrave.ejis.3000445
29. Gefen, D., Karahanna, E., & Straub, D. W. (2003). Inexperience and experience with online stores: The importance of TAM and trust. *IEEE Transactions on Engineering Management, 50*(3), 307–321. https://doi.org/10.1109/TEM.2003.817829
30. Bramall, C., Schoefer, K., & Mckechnie, S. (2004). The determinants and consequences of consumer trust in e-retailing: A conceptual framework. *Irish Marketing Review, 17*(1–2), 13–22.
31. Vijayasarathy, L. R. (2004). Predicting consumer intentions to use online shopping: The case for an augmented technology acceptance model. *Information and Management, 41*(6), 747–762. https://doi.org/10.1016/j.im.2003.08.011
32. Nakagawa, S. S. Y. (2008). A lealdade de consumidores nos ambientes de comércio online e offline [Doctoral dissertation, Faculdade de Economia, Administração e Contabilidade, Universidade de São Paulo]. [Online]. [s.n.]. https://doi.org/10.11606/T.12.2008.tde-12122008-154641

33. Fernandes, L. O., & Ramos, A. S. M. (2012). Intenção de compra online: aplicação de um modelo adaptado de aceitação da tecnologia para o comércio eletrônico. *Revista Eletrônica de Sistemas de Informação, 11*(2), 1–16. https://doi.org/10.5329/RESI.2012.1101006
34. Saraiva, C. M. L. (2012). *Determinantes do comportamento de compra online* [Master's thesis, Instituto Politécnico de Lisboa]. Lisboa, Portugal.
35. Davis, F. D. (1986). A technology acceptance model for empirically testing new end-user information systems: Theory and results [Doctoral thesis, Sloan School of Management, Massachusetts Institute of Technology].
36. Correia, J. C. F. (2022). Adoção do comércio eletrónico pelas PMEs agrícolas [Doctoral thesis, Faculdade de Ciências da Economia e da Empresa da Universidade Lusíada]. Repositório ULL. http://hdl.handle.net/11067/6660
37. Karahanna, E., Straub, D. W., & Chervany, N. L. (1999). Information technology adoption across time: A cross-sectional comparison of pre-adoption and pos-adoption beliefs. *MIS Quartely, 23*(2), 183–213. https://doi.org/10.2307/249751
38. Pavlou, P. A., & Fygenson, M. (2006). Understanding and predicting electronic commerce adoption: An extension of the theory of planned behavior. *MIS Quarterly, 30*(1), 115–143.
39. Pires, G., Stanton, J., & Eckford, A. (2004). Influences on the perceived risk of purchasing online. *Journal of Consumer Behaviour, 4*(2), 118–131. https://doi.org/10.1002/cb.160
40. Popper, K. R. (2013). *A lógica da pesquisa científica* (2nd ed.). Cultrix.
41. Hair Jr., J., Anderson, R., Tatham, R., & Black, W. (2005). Análise Multivariada de Dados. Bookman.
42. Sarmento, M. (2013). *Metodologia Científica para a elaboração, escrita e apresentação de teses*. Universidade Lusíada Editora.
43. Hair, J. F., Jr., Black, W. C., Babin, B. J., & Anderson, R. E. (2018). *Multivariate data analysis* (8th ed.). Prentice Hall.
44. Byrne, B. M. (2004). Testing for multigroup invariance using AMOS graphics: A road less travelled. *Structural Equation Modelling, 11*, 272–300. https://doi.org/10.1207/s15328007sem1102_8

When We Associate Innovation with Management and Apply It to National Parks: Case of Huascarán National Park, Peru

Ikrame Selkani

Abstract Managerial innovation is the invention and implementation of new management practices, processes, structures, or techniques compared to what is known in order to achieve the objectives of the organization better.

The purpose of this paper is to clarify via an example the link between the application of managerial innovation techniques and protected areas, that is to say, to identify an internal action plan that applies in protected areas. In other words, should the manager apply the old style of management more (give orders, orient, direct, get the last word, etc.) or attempt to bring to personnel into management, take into account their thoughts, and grow them as relevant human capital?

After the qualitative research: guide interview with national park manager, we understood that managerial innovation is being created to aid in the spontaneity of innovation and to more organically connect human relationships. To promote sharing and quality time, as was the case investigated in Huascarán National Park, Peru, we need to learn more about the collaborative component and cooperative modality.

The main questions of this paper are: within natural and protected areas such as parks, managerial innovation may well be applied; here, we will study the case of Huascarán National Park, Peru, and we will figure out how these areas are operated and the efficiency of managerial innovation.

Keywords Management · Direction · Innovation · Managerial innovation · Protected areas · National park

I. Selkani (✉)
University Santiago de Compostela, Santiago, Spain

Cesine Business School, Santander, Spain

J. Reis et al. (eds.), *Driving Quality Management and Sustainability in VUCA Environments*, Springer Proceedings in Business and Economics,
https://doi.org/10.1007/978-3-031-52723-4_7

1 Introduction

The management is a tool to assist in the planning of protected areas. They relate to the objective of land management. It's about referring to the objective of the protected area to know its management category. The managerial innovation is perceived as an evolution of the management techniques that constitutes in itself a real importance toward the companies that would finish with the old modes and begin with the innovation techniques respecting the well-being of the employees and reaching a better performance.

Innovation in management is viewed as an advancement of management strategies, which is in itself of real value to businesses that will end with the old modes and continue with innovation techniques that respect and achieve better results in terms of employee well-being [9]. We may cite some examples of procedures for managerial innovation, such as design thinking which is a process of innovation based on observation of uses and prototyping [7]. The technique helps to teach individuals new strategies in order to solve problems. Hackathon is a technology event with a date of accomplishment and a defined duration. They usually have a duration of only 24 h, but in some cases, programmers, designers, and project managers can stick together to create a new product or service for up to 1 week (Autissier et al., 2013).

One of the most famous techniques in the management innovation field is peer coaching, where two people from the same profession or with different skills exchange with each other on work situations they are experiencing and give a cross view according to the techniques of coaching. In turn, people become coachees and coaches with a results-oriented active listening posture.

Another technique would be the participative workshop which is a moment of exchange between people with an animation structured by a script. An example of a participatory workshop is what is called an icebreaker, which consists of breaking the ice on a topic and/or when the participants do not know each other. Sessions last between 1 and 3 h with a sequence of 1- to 45-min workshops [1].

Managerial innovation is the adoption of practice or method relative to new management; the organization puts into practice and develops during a certain period. In other words, the managerial innovation is the implementation by a company some conduct, methods that exist already with some differences on the application and which the organization did not practice before [8].

Natural protected areas could be defined as the following: "An area of land and/or sea especially dedicated to the protection and maintenance of biological diversity, and of natural and associated cultural resources, and managed through legal or other effective means" [12].

The main questions of this paper are: how are protected areas managed? Do protected areas use managerial innovation techniques? Here we will study the case of Huascarán National Park, Peru.

2 Classical Management Modes

Each manager has his or her own management style. The management style is determined by the manager's personality, the type of company, and the profile of the team members. There is no such thing as an ideal management style or a fundamentally bad management style.

Here, we will see the management mode according to Rensis Likert [14], an American psychologist, who defined four major management styles in the 1960s that are still relevant today: directive, persuasive, participative, and delegative.

2.1 The Four Most Common Management Profiles

Directive Manager
Also known as authoritarian management, it gives the manager complete control. In contrast to participative management, it is a very hierarchical vertical top-down approach to managing people. Managers clearly define the expected outcomes, and the team must carry them out. Directive managers do not expect their employees to question or provide feedback on the action plan.

Advantages include quick decision-making and increased productivity. This profile can be useful when the company is going through a difficult period and needs to be turned around.

Disadvantages include global rather than individual communication and a lack of humanity in relationships, which frequently leads to interpersonal issues. It is a management style that prioritizes results over relationships. Team members do not always understand decisions because they are not fully explained.

Participative Manager
This management mode is becoming more popular. It echoes the expectations of younger generations and provides flexibility and dialog, which are critical in an era of increasing remote work but are not permitted by directive management. It is a more democratic method of team management in which the manager cares, listens, and supports the team.

The *benefits* include a human and open management style that fosters each employee's responsibility and engagement. Employee motivation helps the company's competitiveness and creativity.

Disadvantages: because the management is attempting to find the best compromise, decision-making might be delayed, and there may be some disorder if the manager does not stay properly organized. One must also be cautious not to fully disregard performance.

Persuasive Manager
Here it takes a more paternalistic approach. He is more involved than the participatory manager and acts as a true mentor for his team. Persuasive managers motivate

their team members to succeed and attempt to persuade them with well-crafted arguments. They listen to their team members but always keep their goals in mind while making decisions.

Advantages: persuasive managers get right to the point, don't get bogged down in the minutiae, and assist their team members reach their goals. They understand how to bring their team together and create clear objectives. This combination of leadership and charity may inspire teams and foster loyalty.

Disadvantages: they are less dictatorial than a directive boss, yet they do not always listen to their people.

Delegative Manager

Here, the manager allows his team autonomy and freedom. Delegated managers are taking a back seat. They communicate knowledge and key objectives, and then they put their confidence in their team to achieve them. This form of management, sometimes known as "consultative management," is built on careful delegating.

The *benefits*: virtually complete autonomy, encouragement of initiative, and trust as the essential word! There is a strong sense of teamwork.

Disadvantages: employees may experience undue strain and stress as a result of the goals that must be met. Some people may be perplexed and may be afraid to ask inquiries.

2.2 Managerial Innovation

"Managerial innovation is the implementation of practices and processes, new management structures, which are significantly different from the standards usually used" [4]. The goal is to improve the quality of current organizational practices and develop the quality of the organization's performance.

Managerial innovation is perceived as an evolution of the management techniques. It finishes with the old management modes, respecting the well-being of the employees and reaching a better performance of the company. This one gives real importance to the collaborative work and the well-being of the people in order to have a better performance.

The managerial innovation is a set of micro action that helps to progress the practices and therefore the management principles.

The goal is to improve the quality of current organizational practices that significantly change the way management work is done. Simply put, managerial innovation changes the way managers do what they do" [11].

Author's Definition

"Managerial innovation is a set of practices, ideas, and/or programs, relating to the management and running of the organization. It's an association of information and means corresponding to the modification of old and existing concepts by new and improved ones for a better management and direction of the organization. It is a

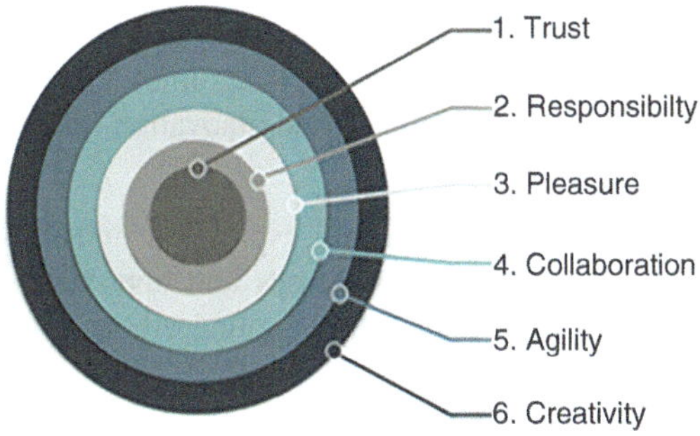

Fig. 1 Six pillars of managerial innovation. (Source: Boyer [5, 6])

mosaic of processes and concepts implementing new decisions for the practice of management" [15].

The Core Concept of Managerial Innovation

There are six pillars of managerial innovation, according to [5, 6]. To find the right balance between personal development and professional performance and to create new values for the company, employees, customers, and environment, there should be a display of the six pillars of managerial innovation (Fig. 1):

Trust

Trust is a very important element in building authentic relationships where everyone is convinced of the sincerity and respect of the agreed commitments. Avoid any feelings of betrayal or abandonment. Get a real and sincere commitment to the terms and conditions of collaboration agreed to by all parties. Manage with serenity and enthusiasm any uncertainties or difficulties encountered.

Responsibility

Empowering employees means a profound transformation of the processes of involvement, decision-making, freedom of action, and valuing successes and mistakes. A new way of working together that can be learned and appropriated regarding cultural and organizational specificities. To offer the possibility (to those who wish) of being a force for change at several levels (his job, his team, the company). Strengthen autonomy and encourage experimentation with new practices. Recognizing successes and learning from mistakes.

Pleasure

To develop the pleasure that each employee feels in carrying out the missions entrusted to them. Boost the conviviality and good atmosphere that prevails at all levels of the company.

Collaboration

Reinforce complementarily and the valorization of differences (personalities, professions, etc.). Valuing "collective intelligence," free ideas. Developing cooperation across functional and hierarchical boundaries.

Agility
Adapt constantly to the increasingly rapid and unpredictable changes in its environment. Reduce activity processing times (responsiveness). Anchor a "start-up" spirit that encourages boldness and innovation (proactivity).

Creativity
Allow and encourage the expression and experimentation of new ideas. Implementing innovation processes and approaches. Differentiate yourself from your competitors through differentiating offers.

3 Managerial Innovation Approaches

Managerial Innovation Matrix (Fig. 2)
In this part, we will go in-depth about the managerial innovation techniques according to Autissier et al. [3]. Managerial innovation means considering other modalities of cooperation and coordination between people in a constrained production system.

These techniques should be chosen prudently for a better understanding when working in teams inside an organization. These techniques should there good results in different big companies: Air France, Zapoos, Fiat Brazil... according to Autissier [3]. The following list of these techniques will help to comprehend on what they consist:

<table>
<tr><td rowspan="4">Delivery Development</td><td>Design thinking Co-design</td><td>Learning network</td><td>Incubator or Spin-Off</td></tr>
<tr><td>Hackathon</td><td>Participative workshop</td><td>Micro platform or the autonomous team</td></tr>
<tr><td>Tutoring, peer coaching and mentoring</td><td>The focus group and co-development</td><td>Disruptive experience and immersion</td></tr>
<tr><td colspan="3">Autonomy Development</td></tr>
</table>

Fig. 2 Managerial innovation matrix. (Source: adapted from [3])

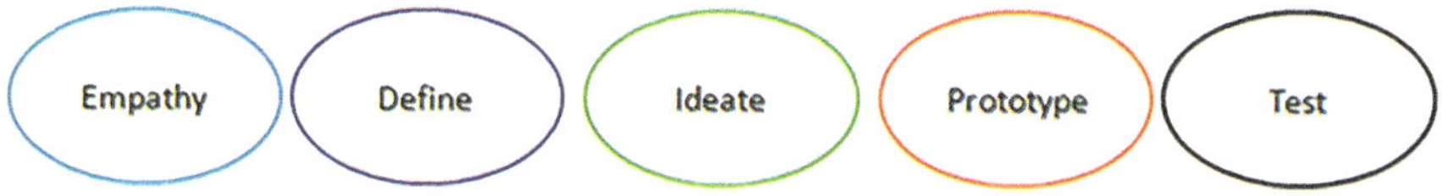

Fig. 3 Design thinking process. (Source: The Design-Thinking-micro process. Adapted from Stranford University 2016)

Design Thinking
Design thinking is a process of innovation based on observation of uses and prototyping [7]. Design thinking applies to product innovation but also to service innovation (Fig. 3).

Hackathon
Hackathon means hack and marathon for the development of new digital applications that are largely inspired by design thinking. For a fixed period of time (usually 2 days), teams of developers and users work together to produce a prototype from a given project.

A Learning Network
It is a hierarchical exchange device aimed at freeing speech.

The purpose of the exchanges is to deal with a subject that requires constructive participation.

Keywords the learning network is an opportunity to develop collective intelligence by putting people in situations of collective reflection and action [2]

Peer Coaching
Traditional coaching [13] has been supplemented in recent years by what is called peer coaching or cross coaching. Two people from the same profession or with different skills (especially in digital with digital natives – people who grew up in a digital environment) exchange with each other on work situations they are experiencing and give a cross view according to the techniques of coaching. In turn, people become coachees and coaches with a results-oriented active listening posture.

Spin-Off
Spin-off consists of creating a subsidiary and offering employees the opportunity to participate in it in an entrepreneurial and capital-intensive way. This makes it possible to carry out projects in a start-up logic without the administrative and political inertia of large companies.

Micro Platform or the Autonomous Team
The group has an obligation of results, and it solicits the company's hierarchy if there are problems of coordination and adequacy between the resources allocated and the results. The group is autonomous in its internal management. This mode of operation is intended for small groups for which the cost of coordination is low [16].

The Focus Group and Co-development
Focus group: the technique consists of having people meet in groups to exchange among themselves (with or without a facilitator) their practices on a subject with a view to changing these same practices.

A Participatory Workshop
It's a moment of exchange between people with an animation structured by a script.

An example of a participatory workshop is what is called an icebreaker, which consists of breaking the ice on a topic and/or when the participants do not know each other. Sessions last between 1 and 3 h with a sequence of one to four 45-min workshops (Autissier and Moutot, 2016).

Disruptive Experience and Immersion
Disruptive experience and immersion: visiting a local or a company for a few hours or a few days, discovering new ways of working and challenging, exchanging and sharing experience [3].

4 Case of Huascarán National Park, Peru

Huascarán National Park is a national park comprising most of the mountain range known as Cordillera Blanca (the highest tropical mountain range in the world), which is part of the central Andes in the Ancash region. With 3,400 km^2, operated by the Peruvian Network of Protected Natural Areas: SERNANP, it was designated a World Heritage Site by UNESCO in 1985.

4.1 Methodology Adopted

The qualitative study is a study that brings together a few participants but which allows us to go in-depth into the individual backgrounds of the respondent.

The method used in this study is the "semi-structured interview" which belongs to the distinctive data collection methods used in qualitative research [10]. In my study, it is important to develop the vision of the management inside of the protected areas. Throughout the manager answer, we will discover if the managerial innovation techniques are also used in a different place, here a national park, other than the big international companies.

4.2 Investigation Results and Findings

The respondent that answers the semi-structured interview is a man who is aged between 40 and 50 years old.

The First Part: About Management

He has been a manager for more than 5 years, with more than 20 people in his team.

For the third question, his team describes his style of management as direct management and participatory management, adding a little comment saying: "Although sometimes I go for all the options."

Concerning the fourth question, the stress of his team, he uses the option, "Assessment of the problem from different angles," and adding, "I develop companionship actions, humanize the relationship with colleagues," the manner of the motivation of his team. Representing the fifth question, he chooses "sending positive feedback," and he adds "In addition, group meetings to share a lunch, for example, I also look for incentives (orders, permits, training, etc.)."

The sixth question concerning the suppression of the manager position, the answer was no adding a comment: "it is necessary for someone to lead the group, under the indicated phases.... the manager's ability to analyze the progress of the team... .and other advantages it offers you."

For the seventh question asking about the understanding from the managerial innovation concept, this answer was: "Ability to promote new ideas or changes in the planning, management processes and especially in the results that arise. Surely this ability to manage with innovation will bring problems associated with the resistance of traditional management or planning."

The Second Part: Looking into Managerial Innovation Begins with

The eighth question, to this, the manager answered that the forms of managerial innovation that speaks the most are tutoring, peer coaching, and mentoring (Table 1).

The nineth question is: what method do you use? If not, which one would you like to practice? (Table 2).

For the tenth question, "Enter the type of innovative management you use for each situation" (Table 3):

The Third Part: Managerial Innovation in the Organization

In question 11, we presented four different definitions of the key concept: managerial innovation through some companies that use a special technique of managerial innovation. After presenting them, the manager answered and said that the one that

Table 1 Managerial innovation application in the Huascarán National Park, Peru

Design thinking, codesign	+
Learning networks	++
Tutoring, peer coaching, and mentoring	
Micro plateau or the autonomous team	+
Focus group and co-development	+
Participatory workshops	++
Disruptive experience and immersion	+++

Source: Author
+: used rarely
++: used sometimes
+++: used frequently

Table 2 Managerial innovation employed at the present

Design thinking, codesign	+
Learning networks	+
Tutoring, peer coaching, and mentoring	+
Focus group and co-development	+
Participatory workshops	++
Disruptive experience and immersion	++

Source: Author
+: used rarely
++: used sometimes
+++: used frequently

Table 3 Managerial innovation technique vs. situation

To solve problems	Participative workshops
Establish the management plan	Participative workshops
Preplanning	Participatory workshops and disruptive experience and immersion
Evaluation of data and identification of values	
Identification of pressures and threats	Participatory workshops
To alleviate differences between employees	Tutoring, peer coaching, and mentoring Y Micro stage or autonomous team
To develop teamwork	Design thinking, codesign
To establish the follow-up of the operations maintained	Focus group and co-development
Raising community awareness on issues of concern for the survival of the protected area	Disruptive experience and immersion Y Focus group and co-development

Source: Author

speaks most to his is **Poult Biscuits**, where the employees manage to get less control costs and more decision-making.

Concerning question 12, the manager was asked about his opinion about the performance in case of using the managerial innovation techniques.

The answer was yes, and he added "I consider that the hierarchy issue does not allow the development of new proposals for the human team, so it is necessary to remove or recover that knowledge, experience of the team group, in that logic also gives them more capacity to make decisions; the role of the manager with the accompaniment and support is key, gaining confidence in the team and the same toward his team."

To the 13th question, the name of the park is Huascarán National Park.

To the 14th question, the selection was for land protected area.

To the 15th question, the answer was 340,000 hectares or 3400 km^2.

To the 16th question, date of the creation, the answer was 1975.

Question 17th was answered affirmatively, that the protected area meets the IUCN definition.

To the 18th question, the answer was II: National Park.

Concerning the management category, question number 19, the answer was

Table 4 International conventions

Convention	Your country is a member country	Your country is not a member country	Registration year
Strategic Plan for Biodiversity 2011–2020 and the Aichi Targets	X		2010
The Convention on Biological Diversity	X		1993
Ngoya Protocol	X		2014
World Heritage	X		1981
Convention on International Trade in Endangered Species	X		1974
Convention on the Conservation of Migratory Species	X		1997
Ramsar and Wetlands Convention	X		1992

Source: Author

Table 5 Essential functions of Huascarán National Park

	1	2	3	4	5
Recreation and nature discovery					X
Public awareness of conservation issues				X	
Scientific research area for a better understanding of the animal and floral world				X	
Con preservation of cultural values and spiritual			X		
Places of education				x	

Source: Author
(1: not very important; 5: very important)

II. Ecosystem conservation and protection (e.g., national park).

Question 20, treating to international conventions, the answers were (Table 4).

Question 21, treat the important and essential functions of the protected area, the answers were (Table 5).

Question 22 was about the "World Heritage" label and the actions you think you will undertake to obtain this label, and the answer was "The Huascarán National Park is already a world heritage site; what has to continue working is to consolidate the site as a protected area of global importance, implement its master plan, and definitely change its management focus in the sense of improving its articulation to the territory in matters of management, planning, governance, and well-being for the local population."

Question 23 treated the subject of Green List and the plan within 5 years and if the protected area is planning to apply for inclusion on the IUCN Green List in order to benefit from this label, and the answer was "Yes, definitely, but it is a subject that requires time."

Concerning the 24th question treating the species in your protected area that are on the IUCN Red List, the answer was "yes, with more than five, and some examples were presented: Tremarctos ornatus; Vultur gryphus; puya de raimondi."

Talking about question 25, asking about the timing that protected areas benefit from ecotourism activities, the answer was: "All year, more between May and September."

The 26th question is about the numbers of ecotourist, and the answer was "More than 400,000 visitors (405,588)."

Having more details about ecotourism activities, question 27th, the answer was "scenic nature tourism, ecotourism expeditions, walking and camping, bird watching, rock and ice climbing, and mountain hiking."

Concerning question 28, the threats of the protected area, the answers were "biodiversity decline, tourist access, extraction (hunting, mining, logging, fishing), climate change, invasive species, overexploitation, inappropriate management, conflict with neighboring or associated communities."

For question 29, treating the pressures on biological diversity in the protected areas, the answer was "habitat loss and degradation, invasive alien species, climate change."

For the 30th question and last one, treating the actions taken for the protection of biodiversity, the answer was "We have a master plan for the protected area, which defines the strategies and actions to conserve biodiversity, addressing the main threats for a period of 5 years ... this is operationalized annually with the operational plans and the actions that are. They are developing are surveillance and control for threats such as hunting, mining, and ortho; in the same way, the ordering of the tourist activity, granting of rights, sensitization to the local population and now we are in articulation processes with the local authorities with the aim of strengthening local conservation and development."

5 Conclusion

The managerial innovation is created to help the innovation spontaneity and to link the peoples' connection more naturally. In this way, we need more about the collaborative dimension and cooperative modality to intense the sharing and quality time.

Here are some ideas that we propose for a better application of the technique of managerial innovation within protected areas: involving the other person in the decision-making, which means applying the adult/adult relationship between the manager and the coworkers, and giving autonomy to the team, establishing an environment of confidence, where listening, understanding, and non-judgment skill is valued and respected.

Concerning the ecotourism, it's very essential to minimize the negative impacts on nature and culture that can harm a destination, which can be done by these following ways: educate travelers about the importance of conservation.

Highlight the use of environmental and social studies, in addition to long-term monitoring programs, to assess and minimize impacts.

Management is an indispensable technique for any organization. When we talk about managerial innovation in protected areas, we refer to the set of techniques that allow staff to become autonomous and favor a scale of comfort and well-being, for better creativity alone or with a team.

Managerial innovation is developed to assist the spontaneity of innovation and to link the relationship between humans more naturally. In this way, to improve the sharing and quality time, which was the case studied in Huascarán National Park, Peru, we need more about the collaborative dimension and cooperative modality.

The empirical part comes in the form of a semi-structured interview: composed between closed and open questions and queries in order to study the point of view of a manager and find out what they think of managerial innovation and whether he is ready to apply it inside of the national park.

We could say finally that the managerial innovation is not just for companies but for all the institutions that remark the advantages of these techniques and the newness of those methods to better work and obtain improved performance.

References

1. Autissier, D., Johnson, K., & Moutot, J. M. (2015). *De la conduite du changement instrumentalisée au changement agile. Question (s) de management, 2*, 37–44.
2. Autissier, D., J. Johnson, K., & Moutot, J. M. (2016). L'innovation managériale: rupture ou évolution du management. *Question (s) de management, 2*, 25–33.
3. Autissier, D., Métais-Wiersch, E., & Peretti, J. M. (2019). *La boîte à outils de l'Innovation managériale*. Dunod.
4. Birkinshaw, J. M., & Mol, M. J. (2006). How management innovation happens. *MIT Sloan Management Review, 47*, 4.
5. Boyer, F. (2020a). 50 pratiques managériales innovantes – L'innovation managériale en action: S'inspirer et apprendre d'entreprises qui ont su concilier épanouissement individuel et performance collective. Éditions Eyrolles.
6. Boyer, F. (2020b). L'innovation managériale en action: 50 pratiques managériales innovantes. S'inspirer et apprendre des entreprises qui ont su concilier épanouissement individuel et performance collective. Eyrolles
7. Brown, T. (2008). Design thinking. *Harvard Business Review, 86*(6), 84.
8. Damanpour, F., & Aravind, D. (2012). Managerial innovation: Conceptions, processes, and antecedents. *Management and Organization Review, 8*(2), 423–454.
9. Dubouloz, S. (2014). Innovation organisationnelle et pratiques de mobilisation des RH. *Revue française de gestion, 1*, 59–85.
10. Eskola, J., & Suoranta, J. (2003). *Johdatus laadulliseen tutkimukseen. 6. painos. Vastapaino. Tampere. Finnish; Introduction to Qualitative Research* (6th ed.). Vastapaino.
11. Hamel, G. (2006). The why, what, and how of management innovation. *Harvard Business Review, 84*(2), 72.
12. IUCN. (1994). *Guidelines for protected area management categories*. IUCN.
13. Lenhardt, V., & Buratti, L. (2013). Découvrir le coaching-2e éd. InterEditions.
14. Likert, R. (1961). *New patterns of management*. McGraw-Hill.
15. Selkani, I. (2022). Managerial innovation: How to implement innovative management practices. *Alternatives Managériales Economiques, 4*(2), 540–561.
16. Womack, J. (2009). Constancy of purpose|| E-letter, LEI, lean.org, Feb. 11, 2009.

Adaptation of Portuguese Agricultural Small and Medium-Sized Enterprises to Digital Platforms

José Carlos Ferreira Correia, Mário Caldeira Dias, and Dulce Matos Coelho

Abstract The age of digital transformation is upon us, characterized by the demands of the markets in which companies operate, with a view of responding more efficiently and effectively to customer requests, where companies are encouraged to adopt new business models based on digital platforms. In the current context of uncertainty and risk, in which companies operate, it is important to question the contribution of digital platforms. This investigation stems from the need to validate the contribution of digital platforms in the process of adopting electronic commerce by small and medium-sized enterprises, considering that these platforms ended up being intrinsically linked to the business models of these companies. In this sense, the methodology employed included the realization of interviews, with the aim of validating the selection of constructs made from the existing e-commerce acceptance models in the literature. The present study focused on small and medium-sized enterprises in the agricultural sector in Portugal, with a view to the business sustainability of the primary sector in Portugal.

Keywords Business models · Digital platforms · Agricultural SMEs

1 Introduction

Over the centuries, companies responded positively to challenges, in what have been called revolutions, from the Industrial Revolution beginning in the eighteenth century to the revolutions associated with the business era (viz., the production

J. C. F. Correia (✉)
Instituto Politécnico de Setúbal, Setúbal, Portugal

M. C. Dias
Universidade Lusíada, Lisboa, Portugal

D. M. Coelho
ESCE-IPS, Setúbal, Portugal

J. Reis et al. (eds.), *Driving Quality Management and Sustainability in VUCA Environments*, Springer Proceedings in Business and Economics,
https://doi.org/10.1007/978-3-031-52723-4_8

era, the efficiency era, and the quality era, from the end of the nineteenth century until the 1970s) and the Information Revolution (from 1970 onward). The latter, characterized by increased access to information and the transformation of forms of communication and production, highlighted competitiveness and knowledge. Although technology plays an important role in these changes, some authors [1] stated that it is not the technology itself, but its use, that creates value. In an information economy, according to these authors, competitiveness is based on the ability of organizations to acquire and use information effectively, which highlights the importance of companies constantly adapting to new information and communication technologies (ICT).

Several scholars sustain that small and medium-sized enterprises (SMEs) need to be innovative and to adapt to new contexts of uncertainty, namely, through new business models, new forms of collaboration and access to information and other intangibles, in order to succeed. In the knowledge economy, there is a growing need of agility and adaptability for SMEs, where disruptive innovation gains importance and poses particular challenges to SMEs [2], and differentiation is crucial to create a unique value proposition and to face changed industry structures [3].

In this context, also characterized by greater technological dynamism in recent decades under the platform of globalization, by the redefinition of the concept of SMEs, by a global mandatory confinement in the face of the emergence of extreme phenomena (dictated by the COVID-19 pandemic), the right conditions were created for the virtualization of business models. These aim to expanding companies, not only in terms of new products/services but also in terms of time and space.

The fourth digital decade, resulting from the expansion of Internet phenomena and social networks on a global scale, resulted in the exponential growth of digital platforms associated with the adoption of electronic commerce by companies. Therefore, there has been an alteration of their way of interacting, not only among themselves, but also with customers, giving rise to the emergence of new business models.

A business model can be perceived as something resulting from the integration of three components, namely, value (associated with buyers/sellers), return (relationship with organizations), and logistics (how the supply chain can affect organizations) [4].

To determine the most appropriate business model for a company, several criteria may have to be used [5], such as functional integration and the degree of innovation [6], economic control and value integration [7], the type of relationships and the degree of externality [8], or the power of sellers and buyers [9]. Consequently, it can be said that business models determine how companies generate business that converts into revenue and results, increasing their profitability and sustainability, which is based on the strategy adopted by companies [10].

Currently, there are few areas without any propensity for digitization and digitalization. With regard to the agricultural sector, innovation has been considered the watchword, being a sector considered relevant for the survival of a country, which cannot remain stagnant in time, given the emergence of new challenges and opportunities.

Ransbotham and Kane [11] had found that companies increasingly see online communities as sources of valuable information, which leads them, for example, to inquire about product development ideas and their evaluation. This is, therefore, a fundamental aspect for SMEs to join online businesses, that is, the virtualization of businesses through the adoption of electronic commerce based on digital platforms.

In addition to simplifying the buying and selling process, electronic commerce also has a high potential for visualizing the development process of their products, encouraging companies to simplify their interaction processes with customers [12].

Some examples of digital platforms, within the scope of the agricultural sector in Portugal, are the following: mercachefe.pt and buyingportugal.pt (based on B2B business models available 24 h a day and 7 days a week), prodfarmer.pt (associated with operational management, inserted in the "Abrantes Cidade Inteligente" project, having as its mission the qualification, promotion, and dissemination of endogenous products), and smartfarmer.pt (with the objective of favoring direct marketing). This way, it can be said the business models developed by SMEs have come to depend on the ability of digital platforms to intervene, at the right time and in the right place.

2 Discussion

2.1 Methodological Considerations

This research begun with a quest for the factors that determine the adoption of e-commerce by SMEs in the agricultural sector, described in Correia [13]. Based on the existing e-commerce acceptance models, several constructs were identified, namely Perceived Ease of Use (PEU) and Perceived Utility (PU), proposed by Davis in Technology Acceptance Model [14], Perception of Social Influence (PIS), based on Nakagawa [15], Perception of Business Trust (PBT), based on Pavlou [16], Risk Perception (RP), proposed by Pavlou [16] and Nakagawa [15], and Intent to Adopt Electronic Commerce (IAEC).

Fernandes and Ramos [17] present a definition of each of these constructs as follows. Perceived Ease of Use refers to the degree to which an individual believes that using a given technology will be free of physical and mental effort [18]. Perceived Utility is comparable to a relative advantage, that is, the technology used is perceived as superior compared to traditional practice [19]. Perception of Business Trust/Perception of Social Influence represents the belief that the other party will comply with expectations, without taking advantage of the vulnerabilities of the person doing business with [16]. Perceived Risk is the belief that the purchase of a product or service will have negative consequences [20]. Intent-Adoption of Electronic Commerce refers to the degree to which an individual believes that people who are important to him think that a certain behavior should be performed (Fishbein & Ajzen, 1975, cit. in Venkatesh & Morris, [21]).

The proposed model by Correia [13] is graphically represented in Fig. 1.

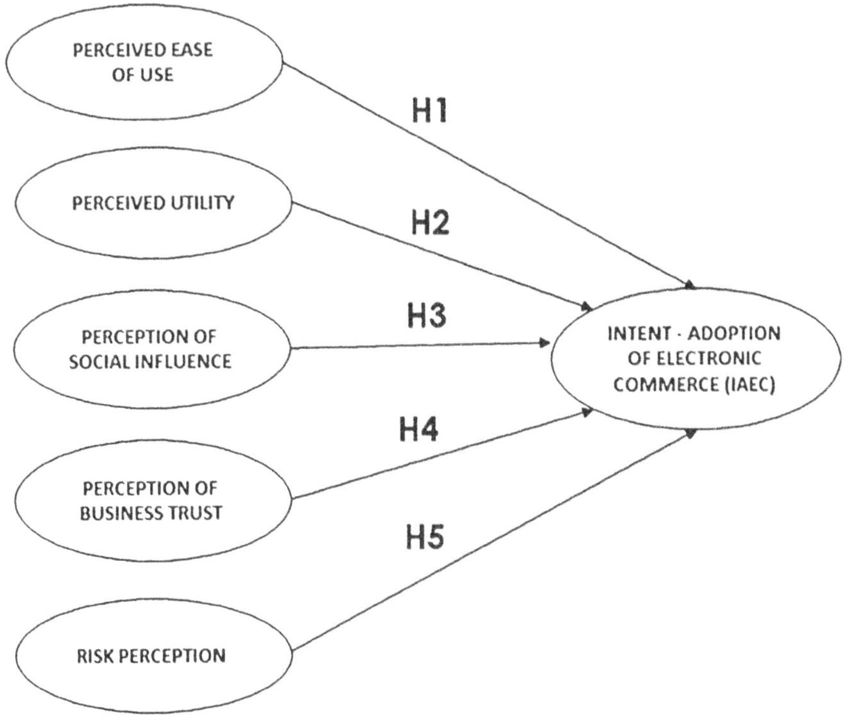

Fig. 1 Research model of e-commerce adoption by agricultural SMEs

Since there was no questionnaire that would allow us to understand the contribution of digital platforms in the acceptance/adoption of electronic commerce by agricultural SMEs, one had to be built from scratch, in order to collect data allusive to the constructs/factors present.

Bearing in mind that some of the mentioned constructs had not been applied to the agricultural sector, there was a need to obtain a perception of their applicability, from the perspective of various stakeholders, most of them associated with the agricultural sector.

To ensure issues related with validity and reliability, several interviews were conducted with key individuals related with agricultural SMEs, with the aim of validating the selection of constructs made from the literature, as well as to validate issues related to semantics and comprehensibility of the questions, among others. The methodology employed consisted of interviews conducted by digital means, i.e., by sending emails to key people both in academia and in agricultural companies, during August 2021.

The constructs in question derive from an exploratory factor analysis, namely, the rotational component matrix, whose values correspond to the factorial loads,

Table 1 Rotational component matrix

Comp. 1		Comp. 2		Comp. 3		Comp. 4	
PEU1	,904	RP3	,895	PBT4	,744	PSI1	,652
PU2	,883	IAEC1	,782	PEU4	,703	PSI3	,589
PU3	,859	IAEC2	,779	PEU3	,699		
PEU2	,852	IAEC3	,772	PU1	,649		
PBT2	,843	PBT3	,772				
PU4	,758	RP4	,728				
PSI2	,756	RP1	,707				
IAEC4	,697	PBT1	,658				
PSI4	,674						
RP2	,558						
N. Items	10		8	4		2	

that is, the correlation of each variable with each component (construct), from the performance of an orthogonal rotation.

Thus, a set of items belonging to each component (construct) were identified, based on the loads that presented the highest absolute value in that component (construct). The mentioned component matrix is presented in the next table (Table 1).

It can be seen that component 1 ("Perceived Usefulness" construct) was associated with ten items, component 2 ("Risk Perception/Intention-Adoption of Electronic Commerce" construct) was associated with eight items, component 3 ("Perceived Ease of Use" construct) was associated with four items, and component 4 ("Risk Perception" construct) was associated with two items.

Within the scope of the interviews, the following eight questions were asked:

1. What does the adoption of e-commerce by an agricultural SME mean to you?
2. What can distinguish an agricultural SME that has adopted e-commerce from another that has not?
3. What is possible to perceive in the adoption of the e-commerce by an agricultural SME?
4. How could the effectiveness of e-commerce adoption practices be quantified?
5. What are the most common characteristics when SME managers decide to adopt electronic commerce?
6. What are the types of organizations where e-commerce adoption by agricultural SMEs is more common?
7. What are the main aspects existing in agricultural SMEs that allow us to ensure that they have adopted e-commerce?
8. Indicate three other aspects that can induce the adoption of e-commerce by agricultural SMEs.

2.2 Universe

The application of the interview instrument was limited to a restricted group of people, having been considered fundamental in their choice the academic training (teachers with specialization in the areas of economics, electronic engineering, agricultural engineering) as well as the professional experience and qualification (engineers) of the intervenient. These were involved in generating value in the companies in which they developed their activity, some of them associated with the agricultural sector.

The following summary characterization of the interviewees can be presented:

- Interviewee A – performs professional functions in a state entity related to the primary sector
- Interviewee B – former leader of a state entity, as well as a professional association related to agriculture
- Interviewee C – belonged to the governing body of several state entities (related to agriculture and other areas), university professor
- Interviewee D – university professor
- Interviewee E – former leader of state entities, entrepreneur, and university professor

2.3 Global Analysis of Results

The content analysis of the interviews comprised several stages, namely, data collection, data aggregation, and interpretation. The opinions of the interviewees were taken into account, regarding the constructs taken from existing models in the literature on the process of intention-adoption of electronic commerce. As these models had not been tested in the agricultural sector, there was a need to obtain this information from the SMEs, that is, what perception they would have about the constructs. For example, the construct Risk is now considered Risk Perception, as it represents the perception that SMEs have about Risk. The conclusions drawn from the interviews according to the questions asked are presented below (Table 2).

The analyses carried out on the responses given by the participants were found to be in line with the SME survey, given that the constructs/factors reflected in the model by Correia [13] were revealed.

3 Conclusions

The study led us to conclude that, from the six initial constructs, four constructs were retained, namely, Perceived Utility, Perceived Ease of Use, Perception of Social Influence, and Risk Perception/Intention-Adoption of Electronic Commerce.

Table 2 Conclusions from the interviews carried out

Question	Answer
1. What does the adoption of e-commerce by an agricultural SME mean to you?	The adoption of e-commerce comprises one of the needs of agricultural SMEs, in order to be able to compete, enabling the use of new working methods (PEU) and new technological solutions (PU) in the search for new customers and guaranteeing the loyalty of their customers, with a view of achieving more ambitious goals and targets (RP)
2. What can distinguish an agricultural SME that has adopted e-commerce from another that has not?	Companies that adopt e-commerce are much more competitive, obtaining advantages (expected results) compared to those that have not adopted (PSI), resulting from the level of computer/technological literacy of managers/employees of these SMEs. The ability to implement changes is key to achieving competitiveness, which is why it is important to invest in young people with some training in this area (RP)
3. What is possible to perceive in the adoption of e-commerce by an agricultural SME?	We are facing the adaptation of current models to new times, with a view to differentiating and improving indicators in terms of productivity and speed (IAEC). The adoption of e-commerce has been absorbed by companies linked to large-scale distribution, isolating the consumer from the producer (RP)
4. How could the effectiveness of e-commerce adoption practices be quantified?	Given the lack of a framework that makes it possible to quantify effectiveness, it can be deduced that it can be measured through studies based on certain variables (production sales time, increase and diversification of customers), which allow understanding electronic commerce through ICT (IAEC), without which the intended benefits and competitive advantages for the company will not be obtained
5. What are the most common characteristics when SME managers decide to adopt electronic commerce?	Those responsible for managing SMEs want to make their companies more competitive in the global market (PBT), willing to adapt to new methodologies and innovative processes (PEU), using current technologies (PU), i.e., they are open to innovation and available to assume a relationship of trust with customers (PSI) It should also be noted that there are managers who want to maximize profits, sell production, and have other sales channels, including electronic commerce (IAEC), and sometimes there is a fear of making data available and carrying out online sales (RP), as it may be grounds for additional taxation
6. What are the types of organizations where e-commerce adoption by agricultural SMEs is more common?	SMEs that export their products, focusing on a more competitive aspect and directed toward electronic commerce (RP). It should be noted that the newly created SMEs have more up-to-date procedures (PEU and PU), whose production is for general consumption and not exclusively local, that is, they sell directly to the consumer; hence, they are more open to the use of electronic commerce (IAEC)

(continued)

Table 2 (continued)

Question	Answer
7. What are the main aspects existing in agricultural SMEs that allow us to ensure that they have adopted e-commerce?	SMEs have a growing success rate, due to the adoption of electronic commerce, resulting from the modernization of operating, production, dissemination, and marketing procedures (PEU and PU), with a view to improving performance and output (RP). In this context, it is important to keep in mind that it is necessary that the networks and access and dissemination devices are made available in a simple way (PEU and PU)
8. Indicate three other aspects that can induce the adoption of e-commerce by agricultural SMEs	It should be noted the ease of use of the applications by the user (PEU and PU), the creation of assistance networks/improvement of the distribution channel (RP), and the company's reputation (PSI)

Table 3 Constructs before and after EFA

Constructs of the initial model	Constructs obtained after EFA
Perceived Ease of Use	Perceived Utility
Perceived Utility	Risk Perception/Int-Adoption of E-Commerce
Perception of Social Influence	Perceived Ease of Use
Perception of Business Trust	Perception of Social Influence
Risk Perception	
Int-Adoption of E-Commerce	

Through an exploratory factor analysis (EFA), performed with IBM SPSS application (Statistics Package for the Social Sciences), the results shown in the table below are as follows (Table 3):

After conducting the interviews, through a content analysis, it was concluded that the constructs could be assessed through the following items.

"Perceived Usefulness" construct was associated with 10 items, namely, PEU1-Functionality, PEU2-Responsiveness, PU2-Productivity, PU3-Celerity, PU4-Security, PSI2-Technical Resources, PSI4-Expected Results, PBT2- Growth, RP2-Content, and IAEC4-Cost.

"Risk Perception/Intention-Adoption of Electronic Commerce" construct was associated with eight items, namely, PBT1-Competitiveness, PBT3-Competitive Market Pressure, RP1-Tasks, RP3-Human Resources, RP4-Technologies, IAEC1-Intention, IAEC2-Strategy, and IAEC3-Application Management.

"Perceived Ease of Use" construct was associated with four items, namely, PEU3-Comprehensibility, PEU4-Simplicity, PU1-Performance, and PBT-Financing.

"Perception of Social Influence" construct was associated with two items, namely, PSI1-Organizational Control and PSI3-Knowledge.

The objective proposed in this survey was considered to have been achieved. To identify the contribution of digital platforms to the acceptance/adoption of electronic commerce by agricultural SMEs, the following factors/constructs stood out: PU, productivity and speed in using platforms; PEU, the simplicity of applications; PSI, knowledge of the application by employees; and RP/IAEC, the reduction of operating costs.

From the analysis of the responses, it was found that the adoption of e-commerce responds to the needs of agricultural SMEs, enabling the use of new working methods (Perceived Ease of Use) and new technological solutions (Perceived Utility). This is related to the need to make companies more competitive in the global market, using current technologies (Perceived Utility), as well as being receptive to innovation and available to assume a relationship of trust with customers (Perception of Social Influence). Therefore, the ability to implement changes is essential to achieve competitiveness (Risk Perception), considering that it is necessary to adapt current models to new times, with a view to differentiating and improving indicators in terms of productivity and speed (Intention-Adoption of Electronic Commerce).

This investigation made it possible to draw conclusions, relevant to the literature, not only associated with electronic commerce but also with regard to the sustainability of the agricultural sector, which begins with the virtual interconnection of people with companies, for the acquisition of products, and results in the physical delivery of products.

4 Limitations and Future Investigation

The limitations of this study relate to the difficulty felt in finding key people with motivation and availability to respond in an appropriate time interval to the questions posed during the interview, especially because we were looking for people with different backgrounds who were somehow connected to the agricultural sectors.

For future investigations, the involvement of the heads of Portuguese agricultural associations/confederations is recommended, as well as entities related to the design of digital platforms, in order to establish the assumptions considered necessary so that future interconnections can result in the presentation of friendlier platforms for all players in the system.

It should be noted that this investigation could contribute to helping managers, at present, to take into account the use of digital platforms in their strategies, as a way of bringing companies closer to people, thus enabling an approximation of the rural world to the city space.

References

1. McGee, J., & Prusak, L. (1994). *Gerenciamento estratégico da informação*. Campus.
2. Christensen, C. M. (2015). Disruptive innovation: In need of better theory. *Harvard Business Review, 93*(12), 44–53.
3. Porter, M. E. (2001). Strategy and the internet. *Harvard Business Review, 79*(3), 63–78.

4. Hayes, J., & Finnegan, P. (2005). Assessing the potential of e-business models: Towards a framework for assisting decision-makers. *European Journal of Operational Research, 160*(2), 365–379. https://doi.org/10.1016/j.ejor.2003.07.013
5. Mamede, J. H. P. S. (2008). Uma metodologia para a implementação de comércio e negócio electrónico em pequenas e médias empresas [Doctoral dissertation, Universidade do Minho]. RepositorioUM. https://hdl.handle.net/1822/9542
6. TImmers, P. H. A. (1998). Business models for electronic markets. *International Journal of Electronic Markets [Em linha], 8*(2), 3–8. [Consult 31 ago. 2022]. Disponóvel em. https://doi.org/10.1080/10196789800000016
7. Tapscott, D., Ticoll, D., & Lowy, A. (1999). The rise of the business web. Business2.0 [Em linha]. [S.l. : s.n.]. p. 198–208. [Consult 31 ago. 2022]. Disponóvel em https://dl.acm.org/doi/fullHtml/10.1145/334425.334434
8. Amami, M., & Thévenot, J. (2000). L'internet marchand: caractérisation et positionnements stratégiques. *Systèmes d'Information et Management*, 5–40. https://doi.org/10.9876/sim.v5i1.74
9. Pigneur, Y. (1999). A framework for designing new information systems. In F. Bodart (Ed.), *The future of information systems – challenge and pitfalls* (pp. 61–102). Namur: FUNDP
10. Costa, T. J. S. G. (2010). A velocidade de internacionalização das empresas portuguesas via e-commerce – Estudo exploratório. Tese de Mestrado em Economia e Gestão Internacional, Faculdade de Economia do Porto.
11. Ransbotham, S., & Kane, G. C. (2011). Membership turnover and collaboration success in online communities: Explaining rises and falls from Grace in Wikipedia. *MIS Quarterly, 35*(3), 613–627.
12. Miller, E. (2001). Exchanges: More than marketplaces. *Computer-Aided Engineering, 20*(1), 54.
13. Correia, J. C. F. (2022). Adoção do comércio eletrónico pelas PMEs agrícolas [Doctoral thesis, Faculdade de Ciências da Economia e da Empresa da Universidade Lusíada]. Repositório ULL. http://hdl.handle.net/11067/6660
14. Davis, F. D. (1986). A technology acceptance model for empirically testing new end-user information systems: Theory and results [Doctoral thesis, Sloan School of Management, Massachusetts Institute of Technology].
15. Nakagawa, S. S. Y. (2008). A lealdade de consumidores nos ambientes de comércio online e offline [Doctoral dissertation, Faculdade de Economia, Administração e Contabilidade, Universidade de São Paulo]. [Online]. São Paulo, Brazil: [s.n.]. https://doi.org/10.11606/T.12.2008.tde-12122008-154641
16. Pavlou, P. A. (2003). Consumer acceptance of electronic commerce: Integrating trust and risk with technology acceptance model. *International Journal of Electronic Commerce (Spring 2003)*, 69–103. https://doi.org/10.2307/2514872
17. Fernandes, L. O., & Ramos, A. S. M. (2012). Intenção de compra online: Aplicação de um modelo adaptado de aceitação da tecnologia para o comércio eletrônico. *Revista Eletrônica de Sistemas de Informação, 11*(2), 1–16. https://doi.org/10.5329/RESI.2012.1101006
18. Davis, F. D. (1989). Perceived usefulness, perceived ease of use, and user acceptance of information technology. *MIS Quarterly,13*(3), 319–340. https://doi.org/10.2307/249008
19. Karahanna, E., Straub, D. W., & Chervany, N. L. (1999). Information technology adoption across time: A cross-sectional comparison of pre[1]adoption and pos-adoption beliefs. *MIS Quarterly, 23*(2), 183–213. https://doi.org/10.2307/249751
20. Solaymanni, S., Sohaili, K., & Yazdinejad, E. A. (2012). Adoption and use of e-commerce in SMEs. *Electronic Commerce Research*, 249–263. https://doi.org/10.1007/s10660-012-9096-6
21. Venkatesh, V., & Morris, M. G. (2000). Why don't men ever stop to ask for directions? Gender, social influence, and their role in technology acceptance and usage behavior. *MIS Quarterly, 24*(1), 115–139. https://doi.org/10.2307/3250981

Features of Cargo Capacity Finding for Vehicles to Increase the Sustainability Level of Supply Chains

Dmitriy Muzylyov, Natalya Shramenko, Christoph Hupfer, Justyna Trojanowska, and Piotr Trojanowski

Abstract This paper highlights the feature of determining cargo capacity for vehicle fleet using system assessment during perishable agricultural product supplying. The proposed approach can increase the sustainability level of the supply chain function, especially during agricultural cargo transportation over long distances. The unique approach is based on the fact that current delivery options were grouped for considered goods categories using the main technological criteria. This, in turn, made it possible to distinguish the average cargo capacity parameter for truck fleet used in transportation. Such logistics solution allows carriers to find a rational value of average truck tonnages for a particular type of supply chain. Firstly, rationalizing such a technological parameter as average load capacity allows for reducing unjustified operating costs that arise during perishable agricultural product transportation. The study determined quantitative values of average carrying capacities for vehicle fleets for each supply chain. This aspect is planned to be used for designing models that simultaneously consider the interests of carriers and customers of perishable agricultural products.

Keywords Sustainable supply chain · Agricultural perishable cargo · Transportation · Row of cargo capacity · Truck

---

D. Muzylyov
State Biotechnological University, Kharkiv, Ukraine
e-mail: murza_1@ukr.net

N. Shramenko (✉) · C. Hupfer
Baden-Württemberg Institute of Sustainable Mobility, Hochschule Karlsruhe University of Applied Sciences, Karlsruhe, Germany
e-mail: natalya.shramenko@h-ka.de; christoph.hupfer@h-ka.de

J. Trojanowska
Poznan University of Technology, Poznan, Poland
e-mail: justyna.trojanowska@put.poznan.pl

P. Trojanowski
West Pomeranian University of Technology in Szczecin, Szczecin, Poland
e-mail: piotr.trojanowski@zut.edu.pl

J. Reis et al. (eds.), *Driving Quality Management and Sustainability in VUCA Environments*, Springer Proceedings in Business and Economics,
https://doi.org/10.1007/978-3-031-52723-4_9

1 Introduction

In modern conditions, perishable agricultural cargoes (PAC) occupy a significant percentage of the overall structure in agricultural products' transportation. The delivery specifics of this goods category create some issues that reduce supply chain effectiveness between shippers and cargo recipients. This affects into the technical and operational parameters of trucks, the quality of customer service, and the economic components of the transportation process [1].

Technology specifics should be taken into account when solving the fundamental problems that arise during the delivery of perishable agricultural cargoes (PAC). At the same time, the specific character of this cargo category is determined by the following features list that arises directly during the PAC transportation [2]:

- Constant necessity to maintain a certain temperature range that ensures cargo preservation during transportation.
- The transport process must be carried out by a specialized fleet of vehicles (isothermal vans or refrigerators).
- The trucks' cargo capacities variation that carries out the PAC transportation process occurs in the range from 1.5 tons to 25 tons.
- The packaging resources must be structurally strong, ensure reliable cargo fixation, and also have no third-party odor.

Such aspects play a key role in the correct planning of supply chains for transporting perishable agricultural products, especially over long distances.

Therefore, the research aim is to formulate a unified criterion due to the average truckload capacity delivering perishable agricultural cargo in all groups of supply chains. It will help to define a rational option by the technological transportation aspect.

2 Literature Review

Researchers proposed various approaches based on modern logistics principles to partially solve problems [3, 4]. But despite this, almost all the presented mathematical models that are used to plan and organize the transportation process of perishable agricultural cargoes have the same significant drawback. We are talking about the unilateral nature of these approaches, because developed methods take into account only the interests of one of the parties: carriers or cargo customers.

More than 5% of fresh products are lost annually during transportation, according to experts [5]. It is this factor that is the main lever for improving agricultural products' transportation technology. That's why the company is obliged to take measures for PAC keeping, which are transported by specialized rolling stock, to avoid significant losses [6–8].

The approach to making management decisions to increase the synergistic effect when delivering goods through terminal systems is proposed in work [9]. This method is interesting from the viewpoint of efficiency increasing of organizing the transportation of mass categories of goods over long distances. However, it cannot be fully used to solve specific problems in the delivery of perishable agricultural goods [10, 11].

The review of linear and integrated methods for designing fresh fruit supply chains considers the full range of modern approaches to resolve issues of sustainability during functioning of perishable agricultural supply chains [12]. When assessing the sustainability of the supply chain, researchers [13], in the first place, pay attention to accounting of carbon emissions, operation costs, and freshness of perishable commodities. Additionally, scientifics [14] highlight attention to the fact that constant control over delivery process parameters is important to ensure sustainability in food supply chains. In the context of developing IoT era and Industry 4.0 and 5.0 conceptions, the use of intelligent approaches has become important. An example is research [15] about applying blockchain to design agricultural supply chains and to create food supply chains [16]. The use of fuzzy logic to optimize agricultural goods delivery also organically fits into the concept of smart solutions' implementation [17].

Analysis of studies aimed at optimizing the delivery process of goods shows that in order to effectively organize the delivery process of perishable agricultural cargoes, the selection of a transportation option, which should be carried out on the basis of an estimate of the transportation distance for each case, plays an important role. An example of such a technique is studies [18–20], which aimed at choosing a rational distance of perishable cargo delivery in international traffic. It is based on the determination of the equilibrium value of the transportation distance for alternative options: delivery by road in direct communication and piggyback rail delivery.

The authors [21] presented an efficient model for designing a truck fleet for transporting fruits but did not offer cargo capacity recommendations based on delivery options. Therefore, the approach does not provide a solution for organizing trunk transportation, allowing to select only a specific type of vehicle when delivering fruits over short distances.

It is advisable to use the methodology presented in the study [22] from the viewpoint of minimum costs for the perishable agricultural products' transportation, as well as considering the route planning based on time window delivery. But the conclusions made by the authors in this study are based only on the assessment of three delivery options, which cannot indicate the versatility of the model used. More interesting in this aspect is solving the routing problem for three-echelon perishable food supply applying robust approach [23]. Detailed issues of route designing are discussed in the overview, which is classified into single and multi-objective problems [24].

The correct choice of the type and carrying capacity of vehicles transporting agricultural products is one of the main tasks that carriers must solve when supplying PAC, both during the harvesting process [25] and during transportation to the doors of end consumers [26–28]. To increase the delivery efficiency of perishable agricultural goods and strike a balance between the carrier and PAC customers' interests, it's crucial to define the calculating method average cargo capacity of trucks for each SC. This approach will be useful for determining the optimal way of transporting perishable agricultural cargo.

Also, the importance of determining the average cargo capacity of trucks to ensure normal conditions for PAC delivery is evidenced by recent studies carried out in the works [29, 30] by scientists. In their studies, authors [29, 30] share the findings of experiments conducted on perishable agricultural goods storage, specifically bananas. The experiments tested various loading methods for containers with different weights, which could be utilized for transportation over both short and long distances. The methodology does not provide working mathematical tools, but it is a reasonable confirmation of the research topic's relevance.

An example of determining the loading parameters of a truck can be a methodological approach to select the optimal vehicle type based on the maximum performance of one truck per work shift during the day. This method was tested under conditions of PAC delivery in small-tonnage containers within the city limits [31].

According to the above, the main advantages of existing methods are depicted in Table 1.

It can be concluded based on the analysis presented that there is no basic methodology for determining the cargo capacity of truck fleet used for perishable agricultural goods transportation at this stage. Therefore, the study's purpose is to design a universal approach for determining the average cargo capacity of truck fleet during PAC transportation. The correct formation of the rolling stock fleet will minimize the cost of transporting perishable agricultural cargoes and fully satisfy the demand of customers for this type of transport service.

Table 1 Advantages and shortcomings of existing approaches for resolving issues of PAC delivery

Works	Advantages	Disadvantages
[3, 4]	Applied main logistics principles	Considered interests only one participant of transportation
[5–8]	Transportation executed by specialized rolling stock	Considered only the problem of cargo losses during transportation
[9–11]	Supply chain designed due to synergistic effect	Based on ergonomic and ecological aspects, without accounting for operational nuance
[12–20]	Modern approaches applied: Blockchain, fuzzy logic, neural network, etc. for supply chains in industry 4.0–5.0	The cargo truck capacity wasn't considered in the criteria
[21–30]	Different optimization aspects are considered	The proposed criteria are common
[31]	The main criteria for choosing SC option are the cargo capacity of the specific fleet	The approach was validated only for a fleet of low-tonnage containers, not for a vehicle fleet

Thus, if average cargo capacity of vehicle fleet is used as an input parameter for the integrated criterion for evaluating the transportation process, it becomes a determining factor in planning and organizing the PAC delivery in supply chains at a new quality level.

3 Research Methodology

Practical recommendations on the selection of refrigerated vehicle cargo capacity for PAC transportation are found on various portals of transport companies [32]. One can use a universal method to calculate the average cargo capacity of trucks that transport PAC. It involves combining practical knowledge from carriers with a methodology that evaluates the effectiveness of heuristic algorithms. This approach is detailed in the research paper [33].

An analysis of the theory and practice of the functioning of PAC delivery systems shows that a fairly wide variety of transport and technological schemes is currently being used. Therefore, it becomes necessary to summarize the schemes by identical characteristics in order to be able to determine the average value of the load capacity of trucks, which is advisable to use for each supply chain.

The preliminary analysis allowed to combine all existing types of PAC delivery technologies into five groups:

1. Group I. Direct supply chain option for perishable agricultural goods.
2. Group II. The variant of a supply chain of perishable agricultural cargoes (PAC) with a distribution center (DC) in the region of the sender (DC at the sender).
3. Group III. The variant of a supply chain of perishable agricultural cargoes (PAC) with a distribution center (DC) in the region of the recipient (DC at the recipient).
4. Group IV. The variant of a supply chain of perishable agricultural cargoes with a distribution center on a route on the middle between participants (where DC is not located at the sender or at the recipient).
5. Group V. The variant of a supply chain of PAC with a distribution center as in the region of the sender (DC at the sender) and in the region of the recipient (DC at the recipient).

3.1 *Direct Option of Supplying Perishable Cargoes*

It is advisable to use the results of previous studies [1] for a structural description of each of the five possible options for PAC delivery, as well as for mathematical formalization of determining delivery costs of perishable agricultural cargoes. This method provides an expert answer to understand how the rolling stock fleet changes depending on the unevenness of incoming and outgoing cargo flows from the warehouse system.

Fig. 1 The direct option of perishable cargoes supply chain

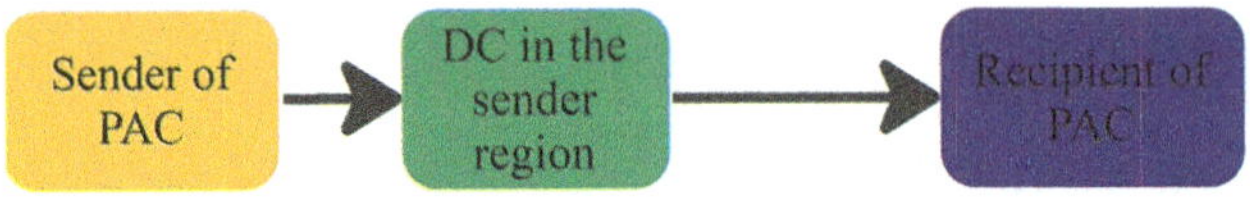

Fig. 2 The option of a supply chain of perishable agricultural cargoes (PAC) with a distribution center (DC) in the region of the sender (DC at the sender)

The most common option for delivering PAC is a scheme without the use of transit distribution centers or terminals (Fig. 1).

Considering the above, the average load capacity of truck fleets is proposed to be determined as follows for the most common technology of PAC delivery (direct option):

$$q_1^{\text{aver}} = \frac{\sum_{i=1}^{n} A_{\text{IntCity}_i} \cdot q_{\text{IntCity}_i}}{\sum_{i=1}^{n} A_{\text{IntCity}_i}}, \quad (1)$$

where q_{IntCity_i} is load capacity of the i-th type of vehicles during main transportation, tons; A_{IntCity_i} is quantity of i-type trucks on the main chain, units; q_1^{aver} is average load capacity of truck fleets for supply chains of Group I, tons; and n is number of truck type used for delivery PAC, unit.

3.2 *The PAC Supplying Through a Distributive Center in the Sender Region*

The following four schemes assume the use of a distributive center (DC). The second group is the DC option in the sender region (Fig. 2).

The average load capacity of the truck fleet can be determined by dependence on the second scheme:

$$q_2^{\text{aver}} = \frac{\sum_{i=1}^{n} A_{\text{DCSender}_i} \cdot q_{\text{DCSender}_i} + \sum_{i=1}^{n} A_{\text{IntCity}_i} \cdot q_{\text{IntCity}_i}}{\sum_{i=1}^{n} A_{\text{DCSender}_i} + \sum_{i=1}^{n} A_{\text{IntCity}_i}}, \quad (2)$$

Fig. 3 The option of a supply chain of perishable agricultural cargoes (PAC) with a distribution center (DC) in the region of the recipient (DC at the recipient)

Where q_{DCSender_i} is cargo capacity of the *i*-th type of vehicles when transporting PAC from the manufacturer to the DC in the shipper region, tons, and A_{DCSender_i} is quantity of *i*-th type trucks at the chain "Shipper-Distributive Center," units.

3.3 *The PAC Supplying Through a Distributive Center in the Recipient Region*

Also, such scheme is one of the common options for the delivery of agricultural products through the territory of Ukraine and other European countries. This type of supply chain is to mean that the distributed center is geographically located in a minimal distance from the PAC consumer (Fig. 3).

When forming the supply chain of perishable agricultural goods according to the scheme, when the distribution center is located in the region of the consignee, the average load capacity of the vehicle is determined by the following formula:

$$q_3^{\mathrm{aver}} = \frac{\sum_{i=1}^{n} A_{\mathrm{IntCity}_i} \cdot q_{\mathrm{IntCity}_i} + \sum_{i=1}^{n} A_{\mathrm{DCRecipient}_i} \cdot q_{\mathrm{DCRecipient}_i}}{\sum_{i=1}^{n} A_{\mathrm{IntCity}_i} + \sum_{i=1}^{n} A_{\mathrm{DCRecipient}_i}}, \tag{3}$$

where $q_{\mathrm{DCRecipient}_i}$ is load capacity of the *i*-th vehicle type when transporting PAC from the DC located in the customer's region, directly to the recipient of perishable agricultural cargo, tons, and $A_{\mathrm{DCRecipient}_i}$ is vehicle quantity of the *i*-th type in the "DC-Recipient" chain, units.

3.4 *The PAC Supplying Through a Central Distributive Center on the Route*

The fourth version of the scheme characterizes the transit transport status of Ukraine. In this case, transportation of PAC from one border to another can be carried out with the transshipment of cargo in the central part. A similar system is found in European Union countries. At the same time, the DC is geographically

Fig. 4 The variant of a supply chain of perishable agricultural cargoes with a distribution center on a route on the middle between participants (where DC is not located at the sender or at the recipient)

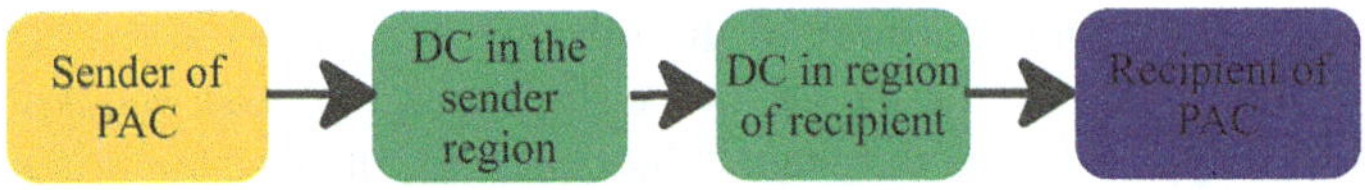

Fig. 5 The option of a supply chain of PAC with a distribution center as in the region of the sender (DC at the sender) and in the region of the recipient (DC at the recipient)

located in a third country, a transit hub on the route for delivery of cargo flows of perishable agricultural goods (Fig. 4).

The average load capacity of the vehicle is for the fourth transport and technological scheme:

$$q_4^{\text{aver}} = \frac{\sum_{i=1}^{n} A_{\text{IntCity1}_i} \cdot q_{\text{IntCity1}_i} + \sum_{i=1}^{n} A_{\text{IntCity2}_i} \cdot q_{\text{IntCity2}_i}}{\sum_{i=1}^{n} A_{\text{IntCity1}_i} + \sum_{i=1}^{n} A_{\text{IntCity2}_i}}, \tag{4}$$

Where q_{IntCity1_i}, q_{IntCity2_i} is load capacity of the i-th vehicle type during main transportation in chain 1 and 2, respectively, tons, and A_{IntCity1_i}, A_{IntCity2_i} is load capacity of the i-th vehicle type during main transportation in chain 1 and 2, respectively, tons.

3.5 *The PAC Supplying Through Two DC*

The fifth delivery option involves the presence of distribution centers, both in the region where the sender is located and in the region of the recipient (Fig. 5).

The fifth delivery option involves the presence of distribution centers, both in the region where the sender is located and in the region of the recipient (Fig. 5):

$$q_5^{\text{aver}} = \frac{\sum_{i=1}^{n} A_{\text{DCSender}_i} \cdot q_{\text{DCSender}_i} + \sum_{i=1}^{n} A_{\text{IntCity}_i} \cdot q_{\text{IntCity}_i} + \sum_{i=1}^{n} A_{\text{DCRecipient}_i} \cdot q_{\text{DCRecipient}_i}}{\sum_{i=1}^{n} A_{\text{DCSender}_i} + \sum_{i=1}^{n} A_{\text{IntCity}_i} + \sum_{i=1}^{n} A_{\text{DCRecipient}_i}}, \tag{5}$$

Due to Eqs. 1–5, the number of trucks required for transportation remains constant. It helps rationalize the load capacity of the rolling stock, which is essential for servicing long-distance routes efficiently. The load capacity has a significant impact on devising an effective method for delivering PAC and selecting the most appropriate supply chain option. We recommend calculating the average load capacity by predicting the transportation volumes of PAC.

4 Results and Discussion

It is necessary to form combinations of the fleet structure to determine the average value of the fleet load capacity.

It was proposed to conditionally divide the recommended truckload capacity for the transportation of this category of goods into four subgroups with an equal interval for further modeling. This approach will make it possible to better determine the numerical characteristics of the average load capacity indicator for each supply chain (Figs. 1–5).

It is proposed to calculate the average load capacity for four variants. At the same time, the primary generating element is the transportation chain and the presence of the DC, which determines the entire fleet of vehicles involved in the transportation of PAC both for main transportation and for transportation over short distances.

The values of the average load capacity of the vehicle fleet were obtained based on statistical information [34–38], which was collected for each of the five options for the supply chain of perishable agricultural goods across Ukraine territory over the past year. Combination's variants of vehicle fleet structure and results of calculation of average load capacity are given in Table 2.

Based on the formulas (1–5), we calculated the average weighted capacity of each SC's vehicle fleet (Table 2). These results helped define a criterion for selecting an appropriate load capacity for each SC. As an illustration, Group III has a recommended cargo capacity of 12 tons.

Table 2 Calculation results of average cargo capacity for truck fleets used on *i*-th supply chain

Subgroup of average truckload load capacity	Modeling rate, tons	Supply chain option of perishable agricultural cargoes				
		I	II	III	IV	V
Subgroup 1, q^{av1}	1.5–7.4	6.0	7.0	3.0	6.0	6.0
Subgroup 2, q^{av2}	7.4–13.3	10.0	9.0	9.0	12.0	13.0
Subgroup 3, q^{av3}	13.3–19.2	15.0	15.0	15.0	18.0	14.0
Subgroup 4, q^{av4}	19.2–25	23.0	25.0	21.0	24.0	24.0
Criteria for choosing rational SC, q^{av}_{total}, tons		13.5	14.0	12.0	15.0	14.3

Further calculations use the average load capacity, which is calculated as the arithmetic mean. The numerical value of the average load capacity is supposed to be used to graphically represent the scope of the relevant delivery scheme from the five alternative supply chains considered.

It is necessary to proceed from the degree of detail for the universality of the calculation of the average load capacity value, i.e., from the number of combinations allocated in a particular experiment according to the desired indicator:

$$q_{\text{total}}^{\text{av}} = \frac{\sum_{j=1}^{k} q^{\text{av}_k}}{k}, \tag{6}$$

Where q^{av_k} is average load capacity of the truck fleet in the k-th combination, tons, and k is quantity of subgroups of average cargo capacity, which is advisable to allocate for transportation of specific categories of PAC, units.

It should be noted that the average value of the load capacity of the vehicle fleet of PAC transportation will make it possible to more accurately determine the structure of the fleet according to this technical and operational parameter. This allows the best transportation technology to be chosen based on the minimum error value that is present in any calculations in technical systems operating with a large number of random factors.

5 Conclusions

Most of the approaches take into account only the interests of one of the parties: carriers or cargo customers that are used to plan and organize perishable agricultural cargoes' transportation. Therefore, it is recommended to use the average load capacity of the truck fleet transporting PAC. It has been established that this indicator can act as the main input parameter for designing models that allow researchers to choose the optimal delivery option taking into account the interests of the main subjects of the perishable agricultural goods supply chain.

Five dependencies are proposed to determine the average cargo capacity of the truck fleet that are used according to the i-th supply chain option. At the same time, the calculation of this indicator is carried out as a weighted average value by the number of truck brands having the appropriate load capacity both during main transportation and during input or output cargo flows to/from distributive centers, respectively.

Universal dependence is proposed for the calculation of quantity values of average load capacity of vehicle fleet. These values will determine the quantitative characteristics of this indicator when designing further models that take into account the interests of the main participants in the PAC delivery process. That's why the estimated average value of the fleet load capacity will become the initial information for creating models that allow a comprehensive assessment of perishable agricultural goods delivery.

The proposed approach has two managerial implications for carriers: selecting the appropriate vehicle fleet based on cargo capacity and determining a rational supply chain option due to the average cargo capacity of the trucks used.

References

1. Muzylyov, D., Shramenko, N., & Shramenko, V. (2020). Integrated business-criterion to choose a rational supply chain for perishable agricultural goods at automobile transportations. *International Journal of Business Performance Management, 21*(1/2), 166–183.
2. Kalinichenko, O. (2021). Improving the efficiency of delivery of perishable agricultural cargo in intercity transportation: Array. *Municipal Economy of Cities, 6*(166), 190–195.
3. Martínez-Val, J. P., & Ramos, A. (2021). Optimization of an industrial sector regulated by an international treaty. The case for transportation of perishable foodstuff. *Entropy, 23*, 109.
4. Negi, S., & Wood, L. C. (2019). Transportation lead time in perishable food value chains: An Indian perspective. *International Journal of Value Chain Management, 10*(4), 290–315.
5. Jeffrey K. B., Sargent, S. A., Brecht, P. E., Saenz, J., & Rodowick, L. (2019). Protecting perishable foods during transport by truck or rail, Preface, p. ix, HS1328, Horticultural Sciences Department, University of Florida/Institute of Food and Agricultural Sciences Extension, April.
6. Wu, C.-T., Huang, W.-H., Dy, K. B., Chang, C.-C., & Hsu, S.-H. (2022). Contribution of active controlled atmosphere (CA) technology to the value-chain of perishable fruits and to rural development: Case of atemoya in Taiwan. *Sustainability, 14*, 16013.
7. Kawasaki, N., Washio, T., Nakamura, K., & Nagahama, K.-i. (2023). Maintaining agricultural production by building local distribution systems in the Northern Area of Japan. *Land, 12*, 320.
8. Li, M., He, L., Yang, G., & Lian, Z. (2022). Profit-sharing contracts for fresh agricultural products supply chain considering spatio-temporal costs. *Sustainability, 14*, 2315.
9. Shramenko, N., & Shramenko, V. (2020). Simulation model of the process of delivering small consignments in international traffic through the terminal system. *CEUR Workshop Proceedings, 2711*, 443–454.
10. Pandey, N. (2022). Agile agriculture supply chain management (AASCM) for managing shifting consumer food preferences: Framework development using grounded theory approach. In R. S. Mor, S. S. Kamble, & K. S. Sangwan (Eds.), *Operations and supply chain management in the food industry* (Lecture notes in management and industrial engineering). Springer.
11. Bhat, R., & Jõudu, I. (2019). Chapter 2 – Emerging issues and challenges in agri-food supply chain. In R. Accorsi & R. Manzini (Eds.), *Sustainable food supply chains* (pp. 23–37). Academic.
12. Nguyen, T.-D., Nguyen-Quang, T., Venkatadri, U., Diallo, C., & Adams, M. (2021). Mathematical programming models for fresh fruit supply chain optimization: A review of the literature and emerging trends. *AgriEngineering, 3*, 519–541.
13. Liu, A., Zhu, Q., Xu, L., Lu, Q., & Fan, Y. (2021). Sustainable supply chain management for perishable products in emerging markets: An integrated location-inventory-routing model. *Transportation Research Part E: Logistics and Transportation Review, 150*, 102319.
14. Modisa, T. J. (2022). Sustainability in food supply chains: Case of Southeast Asia. In R. S. Mor, S. S. Kamble, & K. S. Sangwan (Eds.), *Operations and supply chain management in the food industry* (Lecture notes in management and industrial engineering). Springer.
15. Sharma, R., Shishodia, A., & Kamble, S. S. (2022). Blockchain technology for enhancing sustainability in agricultural supply chains. In R. S. Mor, S. S. Kamble, & K. S. Sangwan (Eds.), *Operations and SCM in the food industry* (Lecture notes in management and industrial engineering). Springer.

16. Burgess, P., Sunmola, F., & Wertheim-Heck, S. (2022). Blockchain enabled quality management in short food supply chains. *Procedia Computer Science, 200*, 904–913.
17. Kieu, P. T., Nguyen, V. T., Nguyen, V. T., & Ho, T. P. (2021). A Spherical Fuzzy Analytic Hierarchy Process (SF-AHP) and Combined Compromise Solution (CoCoSo) algorithm in distribution center location selection: A case study in agricultural supply chain. *Axioms, 10*, 53.
18. Ait Hou, M., Grazia, C., & Malorgio, G. (2015). Food safety standards and international supply chain organization: A case study of the Moroccan fruit and vegetable exports. *Food Control, 55*, 190–199.
19. Volkov, V., Taran, I., Volkova, T., Pavlenko, O., & Berezhnaja, N. (2020). Determining the efficient management system for a specialized transport Enterprise. *Naukovyi Visnyk Natsionalnoho Hirnychoho Universytetu, 4*, 185–191.
20. Liu, E., Zhang, W., Zheng, H., Qian, Y., Li, W., Zhou, N., & Feng, H. (2021). Fractal calculus for refrigerated transportation of perishable foods: Energy consumption and energy saving. *Thermal Science, 25*(2B), 1255–1260.
21. Barman, K., Ahmad, M. S., & Siddiqui, M. W. (2015). Factors affecting the quality of fruits and vegetables: Recent understandings. In M. W. Siddiqui (Ed.), *Postharvest biology and technology of horticultural crops: Principles and practices for quality maintenance* (pp. 1–50). Apple Academic Press.
22. Komijan, R. A., & Delavari, D. (2017). Vehicle routing and scheduling problem for a multi-period, multi-perishable product system with time window: A case study. *International Journal of Production Management and Engineering, 5*(2), 45–53.
23. Hashemi-Amiri, O., Ghorbani, F., & Ji, R. (2023). Integrated supplier selection, scheduling, and routing problem for perishable product supply chain: A distributionally robust approach. *Computers & Industrial Engineering, 175*, 108845.
24. Utama, D. M., Dewi, S. K., Wahid, A., & Santoso, I. (2020). The vehicle routing problem for perishable goods: A systematic review. *Computer Science*. Taylor and Francis Online, 1816148.
25. Giallombardo, G., Mirabelli, G., & Solina, V. (2021). An integrated model for the harvest, storage, and distribution of perishable crops. *Applied Sciences, 11*, 6855.
26. Jouzdani, J., & Govindan, K. (2021). On the sustainable perishable food supply chain network design: A dairy products case to achieve sustainable development goals. *Journal of Cleaner Production, 278*, 123060.
27. Qiu, F., Hu, Q., & Xu, B. (2020). Fresh agricultural products supply chain coordination and volume loss reduction based on strategic consumer. *International Journal of Environmental Research and Public Health, 17*, 7915.
28. Muzylyov, D., & Shramenko, N. (2020). Mathematical model of reverse loading advisability for trucks considering idle times. In I. Karabegović (Ed.), *New technologies, development and application III. NT 2020* (Lecture notes in networks and systems) (Vol. 128, pp. 612–620). Springer.
29. Lloyd, C., Jedermann, R., & Lang, W. (2016). Airflow behavior under different loading schemes and its correspondence to temperature in perishables transported in refrigerated containers. Dynamics in logistics. Lecture notes in logistics. Proceedings of the 5th international conference LDIC, Springer, Cham: Bremen, Germany, 481–490.
30. Snowdon, A. L. (2010). Carriage of bananas (musa spp.) in refrigerated ships and containers: Preshipment and shipboard factors influencing cargo out-turn condition. ISHS Acta Horticulturae 879: IV International Symposium on Banana: International Conference on Banana and Plantain in Africa: Harnessing International Partnerships to Increase Research Impact [Electronic resource], 375–383.
31. Ronzoni, M., Accorsi, R., Guidani, B., & Manzini, R. (2022). Economic and environmental optimization of packaging containers choice in food catering supply chain. *Transportation Research Procedia, 67*, 163–171.
32. Refrigerator transportations across Ukraine/Transportica [Electronic resource]. – Mode of access: https://ua.transportica.com/transportation/refrigerator/

33. Lezama-León, E., Lezama-León, M., Ruiz-Vanoye, J. A., Díaz-Parra, O., Barrera-Cámara, R. A., & Fuentes-Penna, A. (2019). Review of heuristic algorithms applied to the solution problem of land transportation of perishable products. *International Journal of Combinatorial Optimization Problems and Informatics, 10*(1), 1–5.
34. Novikova, A., Redzyuk, A., Shum, I., Sabanska, O., & Yashchenko, T. (2021). GAP analysis of road transport statistics of Ukraine. *Economics, Finance and Management Review, 3*, 39–47. https://doi.org/10.36690/2674-5208-2021-3-39
35. Pavlenko, O., Muzylyov, D., Shramenko, N., Cagáňová, D., & Ivanov, V. (2023). Mathematical modeling as a tool for selecting a rational logistical route in multimodal transport systems. In D. Cagáňová & N. Horňáková (Eds.), *Industry 4.0 challenges in smart cities. EAI/springer innovations in communication and computing* (pp. 23–37). Springer.
36. Anufriyeva, T., Matsiuk, V., Shramenko, N., Ilchenko, N., Pryimuk, O., & Lebid, V. (2023). Construction of a simulation model for the transportation of perishable goods along variable routes. *Eastern-European Journal of Enterprise Technologies,* 2(4 (122), 42–51. https://doi.org/10.15587/1729-4061.2023.277948
37. State Statistics Service of Ukraine: official site. [Electronic resource]. – Mode of access: http://www.ukrstat.gov.ua/
38. Patrol police: official site. [Electronic resource]. – Mode of access: http://patrol.police.gov.ua/statystyka

Blockchain in a Maintenance Database

Suzana Lampreia, Teresa Morgado, Helena Navas, Valter Vairinhos, Victor Lobo, and João Alves

Abstract Nowadays, data management is part of the maintenance management. Organizations are now gathering a diverse range of data, including information from sensors and operator logs, among others. To structure a database may be a challenge when organizing data collected from sensors on a ship, and the difficulty increases when maintenance managers apply algorithms trying to give early results from data that will support decision-making. A blockchain structure is not unbreakable but is safer than other known structures. It will be exposed how a blockchain structure applied to data collected from equipment may enhance the database integrity and security. A state-of-the-art on blockchain and maintenance is made to expose its advantages in the maintenance management, and then it will be proposed a data structure for a maintenance ship system and present some directions for further studies. Also, the Portuguese Navy maintenance organization for surface ships and the advantages and disadvantages of blockchain in it are presented.

Keywords Blockchain · Database · Maintenance · Security

S. Lampreia (✉) · V. Vairinhos · V. Lobo
Portuguese Naval Academy & CINAV, Almada, Portugal

T. Morgado
Instituto Superior de Engenharia de Lisboa, Lisbon, Portugal

LASI – Intelligent Systems Associate Laboratory, Universidade do Minho, Portugal

H. Navas
UNIDEMI, NOVA School of Science and Technology, Universidade NOVA de Lisboa, Caparica, Portugal

LASI – Intelligent Systems Associate Laboratory, Universidade do Minho, Portugal

J. Alves
Instituto Superior de Engenharia de Lisboa, Lisbon, Portugal

J. Reis et al. (eds.), *Driving Quality Management and Sustainability in VUCA Environments*, Springer Proceedings in Business and Economics,
https://doi.org/10.1007/978-3-031-52723-4_10

1 Introduction

1.1 Database as Part of Organization Management

Nowadays, data storage and management are part of organization's data management. The strategies developed for database proper functioning can be various. Organizations along the years have adapted to the type of collected information but also to the risk of database hack or misuse.

The type of collected data can vary from data from sensors, captured video and images, manual registration, and others. To structure a database with all this information is complex, difficult, and a challenge. When it is allied the collected data with an implanted planned maintenance system, the complexity increases. If the maintenance system is based in data, that data is treated with mathematical algorithm, and the complexity is even higher.

There are various types of database structure, but in this study, we focused on blockchain structure.

Blockchain is a system originally developed for financial purposes that is now being adapted for integration into other information systems existing within a network. Access to blockchain information is granted through a variety of user certifications. While the structure is not impervious, it is more secure than other known architectures.

The blockchain concept has been gaining "terrain" in the databases type. This is an important concept that in our opinion can be revealed by the quantity and quality articles we can find in 68uukt6kutquick research in Google Scholar, Fig. 1, and legend at Table 1.

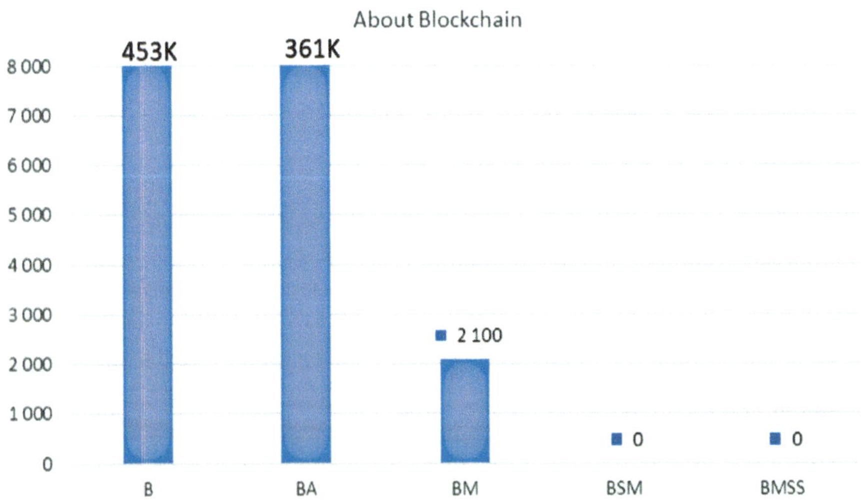

Fig. 1 Research about blockchain

Table 1 Research abbreviation

Abbreviation	Name
B	Blockchain
BA	Blockchain applications
BM	Blockchain "in maintenance"
BSM	Blockchain "in ships maintenance"
BMSS	Blockchain "in military ships maintenance"

In Google Scholar research, in the military area, we couldn't find, yet, any articles or maintenance in military ships, but there are publications in other relevant types of network publication.

One characteristic that made blockchain attractive and secure to apply to databases is the fact of the distributed cloud storage [1].

There are three types of blockchain: "the public blockchain is decentralized, consortium blockchain is partially centralized, and private blockchain is fully centralized as it is controlled by a single group" [2, 3].

In the present study, we conducted a concise overview of the state-of-the-art in blockchain technology, followed by an exploration of some of its applications. Next, we introduced the maintenance database used for surface ships in the Portuguese Navy, and subsequently, we presented a proposal for restructuring the existing database using blockchain technology. Finally, we will conclude the study and discuss potential avenues for further research and development.

2 What About Blockchain?

2.1 Blockchain Applications

A blockchain is a cryptographic technology that functions like a database comprising various interconnected entries referred to as blocks. These blocks can store data in a decentralized organizational structure [4] (Green, et al., 2020).

The blockchain applied to financial matters uses multiple stages of validation on the network; these stages are represented by nodes in Fig. 2 [5]. Although malfunction occurred in the network usual, the system maintains its operationality [5].

To enhance data security and flexibilization, many companies are decentralizing the information along a common network instead of a centralized information way [4]. In the network, the responsibility is shared by the various users and not from a centralized authority [4].

Blockchain technology (BT) by Petkovic et al. (2019) was proposed to guaranty secure network information change and "data storage exchange between Marine Autonomous Surface Ships and shore control system."

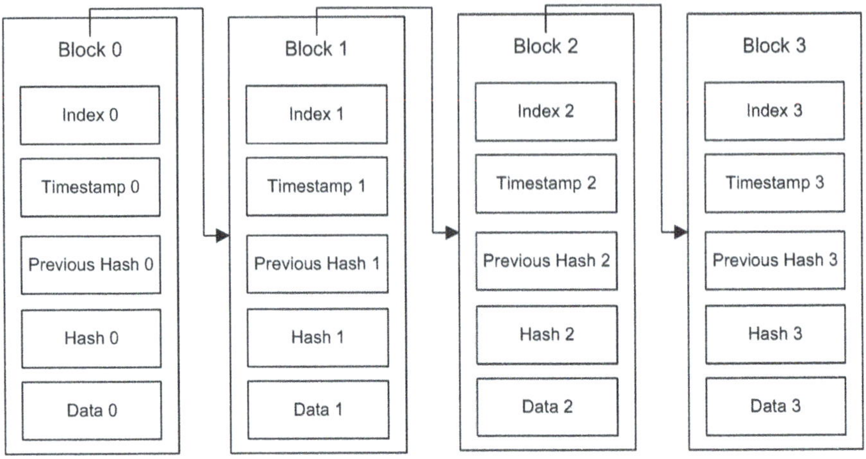

Fig. 2 Blockchain structure. (Source: Adapted from [5])

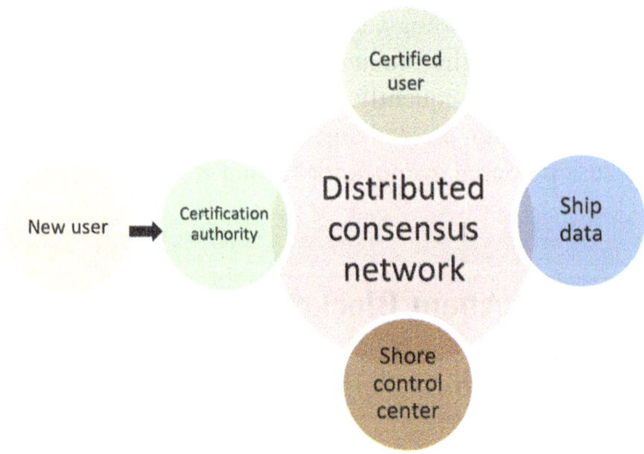

Fig. 3 Example of application of blockchain in maritime area. (Adapted from Petkovic et al., 2019)

A blockchain concept in data from ships allows a secure network with connections between users from the ship, the ship itself, and shore control center, for that the user interaction with the database requires certified access granted by those responsible from database [6]. Who access the network must continuously certify their access. This procedure decreases the cyber threats to the systems (Fig. 3) [6] (Petkovic et al., 2019).

In the past, blockchain was considered an unbreakable security system, but now, in public blockchain networks, it has already been compromised [7, 8].

The blockchain structure is now being applied in new concept, with "benefit of ensuring that war fighters maintain high confidence in the authenticity and security of the data they get from DOD[1] networks" [9].

[1] DOD – Department of Defense

"Chinese military publications have consistently proclaimed that blockchain technology will provide people's Liberal Army (PLA) with an edge in intelligence, weapons lifecycle, personnel management, and information warfare" contributing for military supply chains [10].

The concept of blockchain, owing to its inherent trustworthiness, could potentially be applied across a range of domains within Aerospace and Defense, including "operations management," "logistics and supply chain management," among others [11].

The critical factors in determining when blockchain technology might apply to military or not may be if the process involves collaboration or not and what type of correlation (e.g., interrelated or sequential), or if there is climate of distrust between involved ones, and if decentralization is aligned with applicable command directions and normative [12].

"Blockchain technology presents a new chapter of capabilities for cyber, operations, and even acquisitions" [13].

For the US, China, and Russia militaries, the determinant for blockchain implementation is the network security and military logistics [14].

A private blockchain applicable in a maintenance system, with accreditation levels, with nodes corresponding to identify users may be a decentralized system with the same structure of original blockchains [15]. In a military context, where data is typically confidential, blockchain technology appears highly suitable, as it continues to be a more secure system compared to other concepts [15].

Usually, the Military Command has centralized information, so in case of a hack event, in wartime, or in peacetime, confidential information may be vulnerable being available or even loss of information [16]. If a blockchain database is implemented, some risks, like access and uncertified information share, may be mitigated, by having data layers or network layers that will enhance data reliability and credibility [16].

2.2 *Blockchain Advantages and Disadvantages*

In financial applications, the data collection and arrangement in database is fluid as the transactions are made, because of the certification steps to carry out the transactions [17].

Blockchain structure data has now been applied in other concepts, which do not finance.

Blockchain main advantages are

- Decentralized system.
- Multiply clearance chain to carry out an objective, which enhances security in the actions [18].

- People trust more in database's structure like blockchain than others; may be used in faster … "bank payment transactions," decentralized systems which are conducted to maintain the original data [17].
- Hierarchical responsibility [19].
- Immutable information [19].
- Transparency—all participants can view the operations made [19].
- Cyberattacks are possible, but most of it are not possible to intervene in blockchain operation.

Blockchain main disadvantages are [19]

- Energy consumption to maintain real-time information and access.
- If a node fails, and redundancy does not exist, it may mean blockchain system failure.
- The balance between nodes and the costs for the users.

If blockchain is applied in the maintenance management, it will probably enhance the database security and the maintenance interventions and action.

3 Maintenance Database in Portuguese Navy

3.1 Current Database in Portuguese Navy Maintenance Management

Maintenance in the Portuguese Navy has three levels. The first level is carried out by the ship's crew, the second level is carried out by a shore based technical support workshop, and the third level is under the responsibility of the Navy Technical Management (NTMC) that defines where third level work is carried out by the Navy main support shipyard (AA—Arsenal do Alfeite) or provided by outsourcing.

The maintenance data management system implemented on surface ships is called Data Collection and Treatment System (in Portuguese, *Sistema de Recolha e Tratamento de Dados* (SRTD)) [20].

The SRTD is a subsystem of the Portuguese Navy Maintenance Management System (SGM—Sistema de Gestão da Manutenção). The other subsystem is the Planned Maintenance System (SMP—Sistema de Manutenção Planeada) which will not be explored in the present study [20]. The SGM in turn is part of the Portuguese Navy logistic system [20].

The today software for SRTD is called SICALN (an ORACLE language software, which represents the Navy Logistic Integrated System software for surface ships) and supports the maintenance information from SMP and some technical manuals about the equipment from surface ships. And its work has a maintenance registration and maintenance intervention requests from ships.

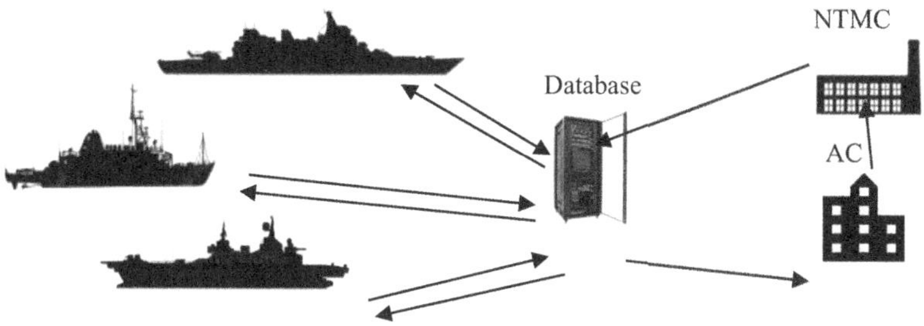

Fig. 4 Adapted SRTD model [20]

When maintenance need occurs, the ship reports the maintenance if it is a first stage one or makes a request for maintenance at SICALN to the administrative command (AC) of the ships responsible for processing the request and forwarding it to the second stage workshop or sending it to the third stage, that is, to NTMC to define the work and send it to the shipyard.

The servers from SICALN are not physical in NTMC; this system feeds the integrated logistic system (ILS). The SICALN has network connection between NTMC, the organism also responsible for software management, between ships and the AC. These organisms have online actualized maintenance information if it is connected to the network. But if not, they use another channel for maintenance occurrences and request.

Fig. 4 is an adaptation of SRTD [20]. The connections between ships and database mean ships sending processed information or requests in three types of registering data models. For example, the DSM58 is where ships report or request maintenance support; the DSM59 is where ships report monthly the equipment hours of functioning, oil, and marine gas oil consumption; and the DSM60 is where the technical services or departments on board request (internally the ship) or report consumption of spare parts for a maintenance action [20]. From the database, the next procedure in the database is responsibility of AC, and if the work is considered to be on the third level, the NTMC takes action.

The SICALN database should be reconfigured and fully integrated, to make data available in a dynamic way showing, for example, multilayer pages of equipment's state graphs, anomaly tendency, available spare parts, available ships, and performance to all interested Navy organization.

Although an integrated logistic system enhances the decision-making process, it brings other cybersecurity preoccupations, but with proper cybersafety or with blockchain implementation, an eventual network attack can be minimized or mitigated.

We saw the SRTD in a certain type of maintenance works in a decentralized way, although the Technical Management can always check the maintenance interventions from the ships.

4 How May a Blockchain Database Be Implemented in Ship Maintenance?

The main advantage of blockchain in maintenance may be to introduce faster decision maintenance procedures in high complex and technologic machines [21].

The implementation of a blockchain database in the Navy maintenance system may represent some challenges and risks, mostly because it means that data from more than 20 years, from the still operational ships, must be imported, and some procedures should be adapted along the various and related to maintenance organizations in the Navy.

But a structure like blockchain may also represent an enhancement in the maintenance database integrity and security.

The implementation of a blockchain database in the Navy's maintenance system may pose certain challenges and risks. This is primarily due to the necessity of importing data spanning over 20 years from ships that are still in operation, as well as the need to adapt various procedures across the maintenance organizations within the Navy. However, a blockchain-based structure can also bring about improvements in the integrity and security of the maintenance database.

In Fig. 5, it is proposed a data flux structured for a maintenance of few systems from a ship (not specifying them) that should be collected in an autonomous way, but with installed control of malware system between the various steps of data processing. All the equipment and systems of a ship made part of its maintenance plan. So, the various sensors from various equipment collect the data to an AI system to distribute data for a compilation of each equipment, and then the data is treated accordingly the specifications and the respective applied algorithm considering clusters for a previous analysis, modified control charts, and neural networks, and

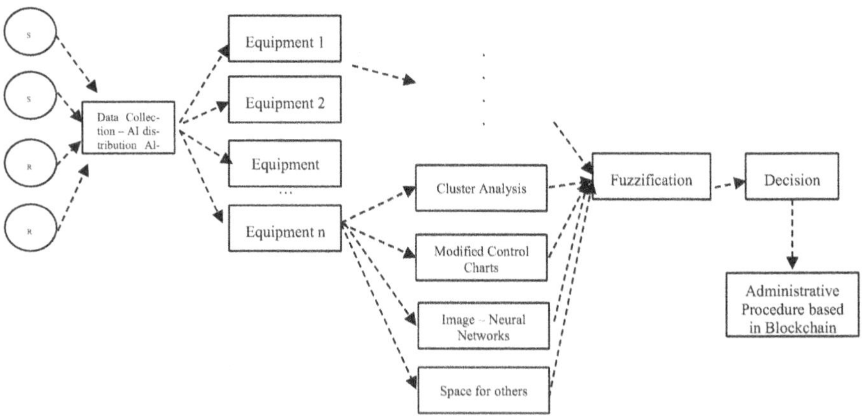

Fig. 5 Data flux

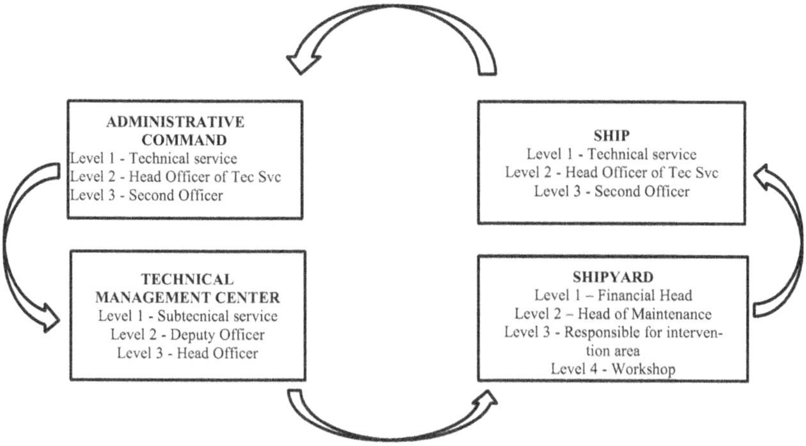

Fig. 6 Organizations involved in maintenance administrative procedure

many others have introduced in the system and appliable. Then a fuzzification of the data is made and then the decision. The data analysis may be considered the ship responsible, but also the AC and NTMC should evaluate the data and carry studies between similar ships and equipment.

In an intervention of the third level, four main organizations participate, the ship, the administrative command, the Navy Technical Management Center, and the shipyard or workshop. Between this organizations and in the own organizations, specifically validation access exists (Fig. 6).

The advantages of blockchain in the Portuguese Navy may be the decentralized information. The SICALN is a private database and secured by military firewalls, so it is probably more unbreakable than others, and introducing the blockchain concept safety may be enhanced. However, if data centers and computers are not actualized, it may affect the system performance, because the data is increased every day, and many users may be trying to access the blockchain platform at the same time.

When analyzing the main disadvantages of blockchain in relation to the maintenance database for the Portuguese Navy, we can identify a few key aspects. Initially, there may be challenges in striking the right balance between nodes and addressing user costs. However, given that we are observing a finite organizational universe, these costs are unlikely to significantly escalate over time.

Energy consumption will persist, but it will also remain limited due to the finite number of users. In the event of a node failure, there is no cause for concern, as the Navy has alternative communication processes (or redundancy) in place and can initiate a maintenance action accordingly.

5 Conclusions

Blockchain is a database structure that initially was only applied in the financial areas. But by the state-of-the-art, it was possible to conclude that blockchain is now being applied in other areas in various organizations and may have the potential to be applied in various areas, such as in defense and maritime areas.

The actual maintenance organization on the Portuguese Navy was present, describing the three levels of maintenance, the involved organizations on the maintenance process, and presents the SRTD that is used in the surface ships for maintenance administrative procedures between the respective responsible in the process.

It was present the data flux in the blockchain database with its diverse steps.

Finally, it was exposed some advantages and disadvantages by implementing the blockchain in the Portuguese Navy.

It was concluded that blockchain structure may be applied in a restructuration of the actual maintenance database in Portuguese Navy.

For future studies, it is proposed to experiment automatically data arrangement in the blockchain maintenance database and explore the algorithms that may be applied to data in blockchain maintenance database to support decision-making on maintenance interventions.

Acknowledgments The authors from UNIDEMI acknowledge the Portuguese Fundação para a Ciência e a Tecnologia (FCT – MCTES) for its financial support via the project UIDB/00667/2020 and UIDP/00667/2020 (UNIDEMI).

References

1. Sri, P., & Bhaskari, L. (2018). A study on blockchain technology. *International Journal of Engineering & Technology, 418–421.*
2. Zheng, Z., Xie, S., Dai, H., Chen, X., & Wang, H. (2017). An overview of blockchain technology: Architecture, consensus, and future trends. *2017 IEEE 6th International Congress on Big Data*, 557–564.
3. Buterin, V. *On public and private blockchains.* https://blog.ethereum.org/2015/08/07/on-public-and-private-blockchains/. Last accessed 2022/05/12.
4. Green, E., Carr, E., Winebrake, J., & Corbett, J. (2020). *Blockchain technology and maritime shipping: A primer.* U.S. Maritime Administration.
5. Angin, P. (2020). Blockchain-based data security in military autonomous systems. European. *Journal of Science and Technology*, 362–368.
6. Petkovic, M., Mihanovic, V., & Vujovic, I. (2019). Blockchain security of autonomous maritime transport. *Journal of Applied Engineering Science, 3*(613), 333–337.
7. Orcutt, M. *Once hailed as unhackable, blockchains are now getting hacked.* MIT Technology. https://www.technologyreview.com/2019/02/19/239592/once-hailed-as-unhackable-blockchains-are-now-getting-hacked./. Last accessed 2022/05/09.
8. Waldman, A., 2022. *Cryptocurrency cyber attacks on the rise as industry expands.* https://www.techtarget.com/searchsecurity/feature/Cryptocurrency-cyber-attacks-on-the-rise-as-industry-expands. Last accessed 2022/05/09.

9. Alcazar, V. (2017). Data you can trust. *Air & Space Power Journal*, (Air Power), 91–101.
10. Logan, T., & Lebryk, T. *America and its military need a blockchain strategy*. https://www.c4isrnet.com/opinion/2021/04/05/america-and-its-military-need-a-blockchain-strategy/. Last accessed 2022/05/09.
11. Ahmad, R., et al. (2021). Blockchain for aerospace and defense: Opportunities and open research challenges. *Computers & Industrial Engineering Journal, 151*, 106982.
12. McAbee, A., Tummala, M., & McEachen, J. (2019). *Military intelligence applications for Blockchain technology*. Hawaii.
13. Korsak, D., & Fuqua, E. (2021). Decrypting Bitcoin and Blockchain. The JAG Reporter, Issue Military Justice and Discipline, 1–11.
14. Lilly, B., & Lilly, S. (2021). Weaponising Blockchain – Military applications of Blockchain technology in the US, China and Russia. *The Rusi Journal, 166*(3), 46–56.
15. Mohril, R., Solanki, B., Lad, B., & Kulkarni, M. (2021). Blockchain enabled maintenance management framework for military equipment. *Transactions on Engineering Management, Issue Engineering Management*, 1–14.
16. Zhu, Y., Zhang, X., Ju, Z., & Wang, C. (2020). A study of blockchain technology development and military application prospects. *Journal of Physics*.: Conference Series, *1507*, 052018.
17. Chowdhury, M., et al. (2020). *X.: Blockchain versus database: A critical analysis*. s.l., IEEE.
18. Rehmani, M. (2021). Blockchain technology and database management system. In T. i. T. Engineering (Ed.), *Blockchain systems and communication networks: From concepts to implementation* (pp. 15–22). Springer.
19. Strebko, J., & Romanovs, A. (2018). *The advantages and disadvantages of the Blockchain technology*. IEEE.
20. Marinha: Data Collection and Processing System Manual (Manual do Sistema de Recolha e Tratamento de Dados (SRTD)) (ILMANT512). Direção de Navios. (1984).
21. Singh, M., Rathi, R., & Goyat, R. (2018). Maintenance in industrial sector based on emerging Blockchain technology. *Journal of Emerging Technologies and Innovative Research*, 431–439.

Developing a TRIZ-Lean Methodology in Portuguese Navy Complex Systems

João Alves, Teresa Morgado, Suzana Lampreia, Helena Navas, António Abreu, Ana Dias, and João Calado

Abstract To maximize safety and the useful life of the equipment while reducing costs, new philosophies and tools must be used to replace the old maintenance habits. A TRIZ-Lean approach, applied to maintenance management, promises to reduce non-value activities maintenance and minimize energy and resource consumption by giving innovative solutions. This paper aims to develop a methodology that uses TRIZ and Lean tools to implement a new maintenance management solution in complex naval systems. The equipment studied in this work was the central air system on naval ships. This system is responsible for pressuring

J. Alves · A. Dias
Polytechnic Institute of Lisbon, Lisbon, Portugal

T. Morgado (✉)
Polytechnic Institute of Lisbon, Lisbon, Portugal

Navy Research Center – Portuguese Naval Academy, Almada, Portugal

Research and Development Unit for Mechanical and Industrial Engineering – NOVA University of Lisbon, Caparica, Portugal

Intelligent Systems Associate Laboratory, Guimarães, Portugal
e-mail: teresa.morgado@isel.pt

S. Lampreia
Navy Research Center – Portuguese Naval Academy, Almada, Portugal

H. Navas
Research and Development Unit for Mechanical and Industrial Engineering – NOVA University of Lisbon, Caparica, Portugal

Intelligent Systems Associate Laboratory, Guimarães, Portugal

A. Abreu
Polytechnic Institute of Lisbon, Lisbon, Portugal

Centre of Technology and Systems of UNINOVA, Caparica, Portugal

J. Calado
Polytechnic Institute of Lisbon, Lisbon, Portugal

Institute of Mechanical Engineering, Lisbon, Portugal

J. Reis et al. (eds.), *Driving Quality Management and Sustainability in VUCA Environments*, Springer Proceedings in Business and Economics,
https://doi.org/10.1007/978-3-031-52723-4_11

atmospheric air, allowing many ship functions to be performed. In conclusion, the TRIZ-Lean methodology developed showed the potential to improve efficiency, reduce waste, and increase productivity, allowing an easy understanding of the influence of the parameters in the value chain of the Portuguese Navy's ship system.

Keywords TRIZ-LEAN · Navy systems · Maintenance management

1 Introduction

Effective maintenance management is critical for the reliable and safe operation of industrial equipment, particularly concerning naval ships where equipment failure can have severe consequences for the entire system. One clear example is the central air system on naval ships, represented in Fig. 1. This system is responsible for pressuring atmospheric air, which allows many ship functions to be performed. Some examples are the engine starting, where the main engine is typically started with pressurized air, which drives the piston, the control of the ship systems through the management of pneumatic valves, the feed of the boiler combustion, and the soot blowing of the exhaust boiler.

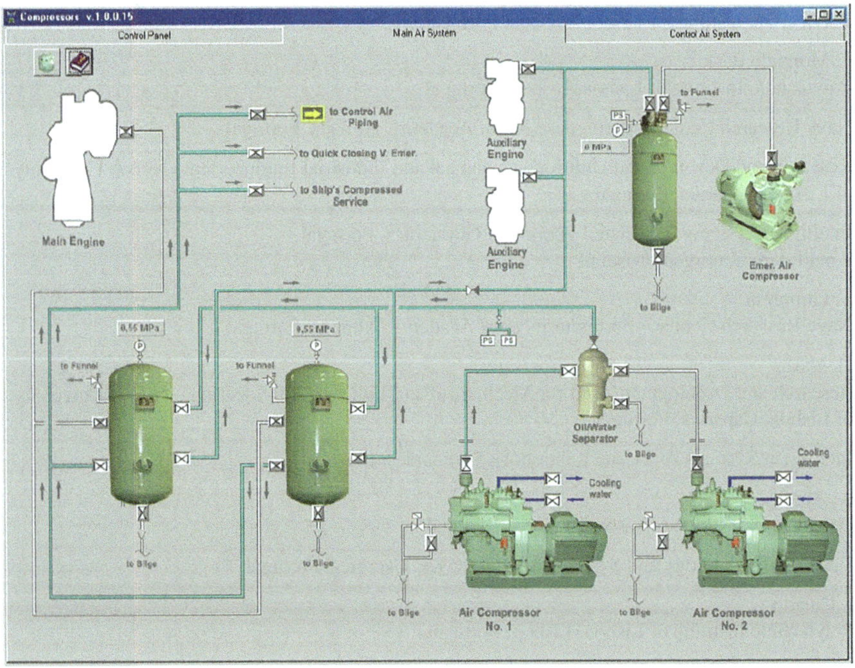

Fig. 1 Simulation plant from the air system

The theory of inventive problem-solving (TRIZ) is a systematic approach that contributes to break mental inertia through innovation. It focuses on identifying and solving problems by using inventive principles and patterns [1–4].

Lean, on the other hand, is a philosophy and a set of tools for continuous improvement that aims to eliminate waste and improve efficiency [5].

Therefore, to improve maintenance management, a variety of tools and methodologies have been developed and implemented such as TRIZ and Lean [6, 7].

This article presents a case study using a combination between TRIZ and Lean for maintenance management of the air system on a naval ship in the Portuguese Navy. It describes the methodology used for implementing these tools. Finally, it discusses any unexpected results or limitations of the implementation and recommendations for future research and practice in this area.

2 Literature Review

2.1 TRIZ Tools

The TRIZ was first developed by Genrich Altshuller [1]. Inspired by the innovative work of inventors such as Thomas Edison and Henry Ford and believing that anyone could be an inventor or create creatively, Altshuller studied patents and patent applications in various fields to identify problem-solving patterns. As a result, Altshuller concluded that not all inventions had the same inventive value and could be divided into five levels, as represented in Table 1 [2–4].

Table 1 Levels of invention

Level	Characterization	Description
First level	Simple Solutions	Personal information solutions are around 32% of all patents.
Second level	Improvements	The solution introduces a little progress and improvement in the overall system. These solutions are 45 % of all patents.
Third level	Innovation inside of the studied area	A solution that introduces progress on the system, notoriously improving the system. For example, Cordless phones. This level includes 18% of all patents.
Fourth Level	Innovation outside of the studied area	The solution changes the main principle of the function of the current system. For example, Jet Engine: This level includes 4% of all patents.
Fifth Level	Extraordinary	Solutions that were found based on new knowledge. Usually, this kind of solution is the initiator of a new system. For example, the Laser technology. This level includes 1% of all patents.

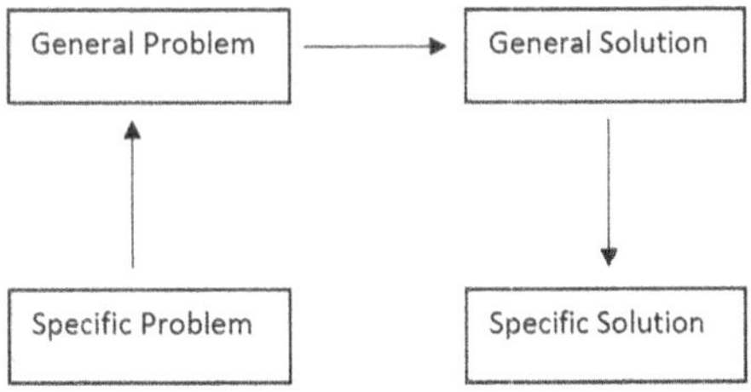

Fig. 2 General methodology from problem-solving using TRIZ

Usually, TRIZ is used in levels 3 and 4 of innovation and is based on a set of analytic tools which allow the transformation of a specific problem, into a specific solution, by generalizing both [3]. Figure 2 presents the general TRIZ methodology applied in the context of this work.

One of the tools that is worth to mention is the ideality matrix, based on the principle of ideality, which focuses on the ideal state of a system, meaning that all parts should be in their best performance possible and implying that all systems can be enchanted to a near-ideal state [3]. This enchantment can be measured mathematically, and it can be described by Eq. 1 [6]:

$$\text{Ideality} = \frac{\sum \text{Benefits}}{\left(\sum \text{costs} + \sum \text{Waste products}\right)} \tag{1}$$

where benefits are useful functions determined by the system and the cost and waste products are the system's output.

Therefore, according to TRIZ, the ideal state of a system is one where it fully meets all the necessary functions (benefits) while having no negative impact on the environment or other systems (cost and waste) [2].

To build the ideality matrix, the most critical parameters from the given application must be defined and then tested in their relationship with each other. If there is no relationship, the space which characterized their conjunction must be empty. However, a "+" and a "-" must be added if the relationship is beneficial or harmful.

Another important tool in TRIZ methodology is the matrix of contradictions which is based on the existence of contradictions in every technical system, and they can be classified into two different types [3]: technical contradictions, which happen when the improvement of one aspect of the system comes at the cost of another, and physical contradictions, in contrast, which appear when a physical variable brings both benefits and harm [7]. Altshuller build a matrix based in 39 engineering parameters or product attributes that engineers usually try to improve and 40 principles of invention that can be used to overcome system conflicts and contradictions between the parameters [1].

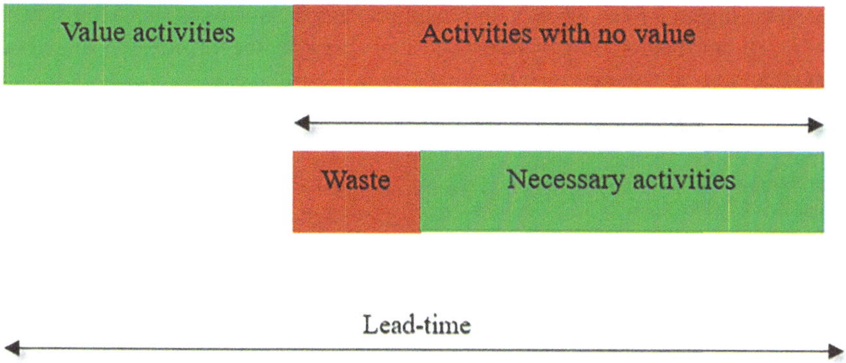

Fig. 3 Conceptualization of activities in any industry

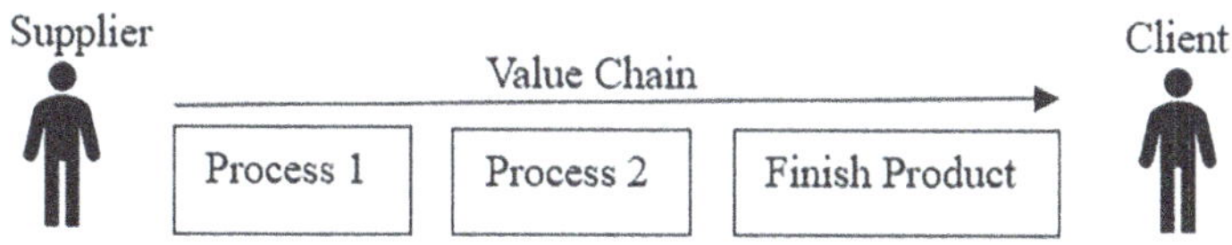

Fig. 4 Generical illustration of a value chain

2.2 *Lean Tools*

Lean is a philosophy and set of tools for process improvement that has been widely adopted in industries ranging from manufacturing to healthcare to finance [9–12]. The origins of Lean can be traced back to the Toyota construction company in Japan, and Taiichi Ohno invented it in 1940 [7]. At its core, Lean pretends to establish a continuous improvement process, using various tools that help increase the time spent on value activities while reducing waste. In Fig. 3, it can be seen how these activities are related to what concerns lead time.

For the aims of this study, the Lean tool used is value map streaming (VMS), which enables the visualization of the material and the information flow through all the desired processes between a customer and a supplier [8], as can be seen in Fig. 4.

This tool, as a specific solution, can be divided into four steps, as proposed by Sheth et al. [8]:

1. Calculate Takt Time given by Eq. 2:

$$\text{TAKT Time} = \frac{\text{Daily Available time}}{\text{Daily Demand}} \tag{2}$$

2. Map the process flow: where the value chain between the process should be drawn.

3. Map the material flow: where value chain between the materials should be drawn.
4. Map the information flow: where value chain between the information should be drawn.

2.3 TRIZ-Lean Methodologies

TRIZ and Lean combination is revealed to be a powerful tool in various industries, guiding them toward sustainable development [13–16]. While TRIZ offers a structured framework for innovation, Lean systematically identifies and eliminates process waste. Moreover, both methodologies are aligned since they promote continuous improvement effort and learning.

Some authors have been applying a combination of these two tools. For example, Morgado et al. [6] proposed using tools from TRIZ, such as the ideality matrix, to identify the most harmful parameters in the process of an agri-food sector and the matrix of contradiction to find a solution. At the same time, Lean aimed to transform the general solution into a specific one.

David et al. [9] reveal another path for combining these tools; in their study case related to a metalworking company with nonexistent Lean and TRIZ levels, the authors used the 5S tool from Lean to identify the problems related to the processes. They used the TRIZ ideality matrix to evaluate the evolution of the company. A similar approach was taken by Pombo et al. [14], who used Lean philosophy to identify waste in the material management of the warehouse and applied a contradiction matrix to find innovative solutions to overcome the problem.

A different approach was taken by Godina et al. [9] in the management of a ham production line. The authors used substance-field analysis, a method from TRIZ, to characterize the production process and identify opportunities for improvement and used 5S from Lean to implement a solution. Likewise, David et al. [9] and Godina et al. [15] used the ideality matrix to evaluate the improvement by increasing the level of ideality.

Nurcahyo et al. [16], in their research, expose how lean tools, such as value stream mapping, can be used to identify waste in the production of rear motorcycle arms and how contradiction matrix can help to overcome inefficiencies, waste, and bottlenecks from the process. To evaluate the improvement from implementing these combining tools, the authors used Six Sigma, which allowed them to identify a reduction in the variation of the process.

3 Study Case

The methodology proposed in this work for maintenance management of the air system from Portuguese Navy ships combines TRIZ and Lean to improve the ship's air system's overall performance.

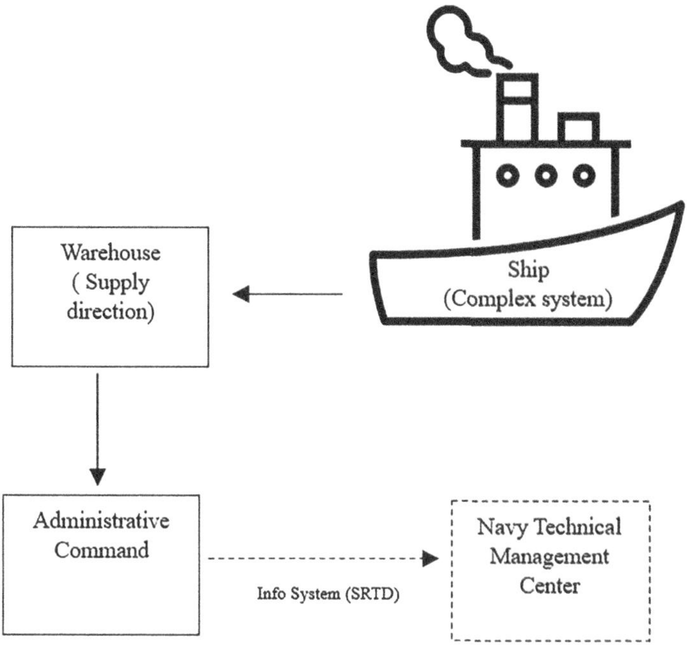

Fig. 5 VMS maintenance first level of the Portuguese Navy

Figures 5, 6, and 7 represent the value map chain (VMC) from the three existing maintenance levels in the Portuguese Navy, including the information communication by traced arrows [17, 18].

The proposed method starts with an ideality matrix, as shown in Table 2. This matrix was produced based on brainstorming meetings, where it was defined which parameters would be the most important for the marine system maintenance management and what would be objective to increase or decrease the parameters. For a more straightforward and clear understanding, each parameter was marked with "↑" and "↓" representing the need for an increment or reduction, respectively. The parameters present in Table 2 are defined and discussed as follows:

1. *Manutability:* this parameter represents how easy and fast it is to replace a component from a system. This parameter has a positive relationship between reliability, intrinsic availability, and operational availability since it is easier to repair and maintain when issues arise. To improve this parameter, more spare parts would be needed while the maintenance cost would necessarily increase as well.
2. *Reliability:* this parameter is described as risk of breakdowns or failures that could impact the system's performance. Improving this parameter would increase intrinsic availability and operation availability since more reliable systems can function more hours without fail. Still, reliability also negatively

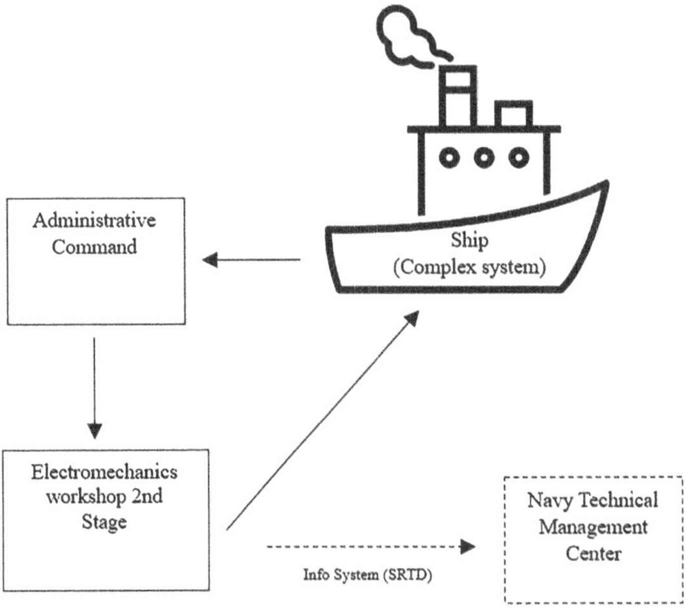

Fig. 6 VMS maintenance second level of the Portuguese Navy

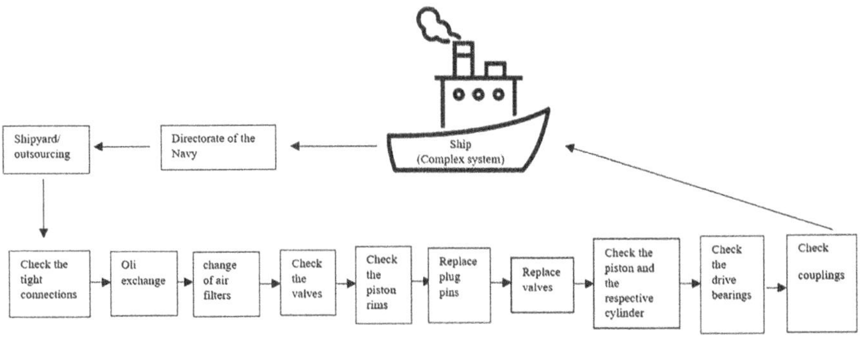

Fig. 7 VMS maintenance third level of the Portuguese Navy

Table 2 Ideality matrix

Parameters:	1	2	3	4	5	6	7	8	9
1-Manutability ↑		+	+		+	-			-
2-Reliability ↑			+	-	+	-			-
3-Intrinsic availability ↑				-		-			-
4- Availability of human resources ↑		+	+		+				-
5-Operation Availability ↑			+	-					-
6- Spare parts availability ↑	+	+	+		+			+	-
7- Inspection ↑	+	+	+		+	-		+	-
8- Redundancy ↑	+	+	+		+				-
9-Maintenance Costs ↓	-	-	-	-	-	-		-	

influences other parameters since more parts and human resources will be needed to expand it, which comes with a higher cost in the maintenance process and has a negative effect.

3. *Intrinsic availability:* this parameter is related with the ability of the system to remain available and operational without any external intervention. The rise of it would come with the need of more human resources and parts, accompanied by increased maintenance costs.
4. *Availability of human resources:* the increasing number of available workers positively affects system's reliability since there is a need of more workforce to maintain the air system and the other systems efficiently. Jointly, intrinsic availability and operational availability also observe some improvement, although maintenance costs also increase since more workers represent a higher cost.
5. *Operation availability*: can be described for Eq. 3, where available productive time is the total time available for functioning, subtracting the scheduled downtime. Meanwhile, time used in the production is related to subtracting the available productive time for the nonscheduled downtime. Therefore, the increment in this parameter, in the Portuguese Navy context, will positively increase system's availability. However, it would require the use of more workers, with a higher cost for maintenance.
6. *Spare parts availability*: Portuguese Navy needs more parts for the maintenance of the ship's systems; this parameter positively affects the manutability of the system, since more parts available will reduce the time related to finding the part. It also shows positive effects on the system's reliability since there is an increment of redundancies and an increase in intrinsic and operational availability. Like the other mentioned parameters, increasing spare part availability will also increase the maintenance costs.
7. *Inspections:* the increasing number of inspections would positively influence all parameters, apart from availability of human resources, which has no influence, since it would define the bare minimum for a safer operation, and therefore the reliability, redundancy, and intrinsic and operation availability would benefit from it. However, maintenance cost would necessarily increase, with a decrease in the spare parts to ensure that everything can pass the inspection.
8. *Redundancy:* this parameter is related to the number of existing redundancies in the systems, and the increment of it would promote an increase of manutability, intrinsic availability, and operation availability since the system would be capable of working, for instances, by the possibility of one air compressor being unavailable. In what comes maintenance cost, it is expected to have a negative effect since more resources are needed to improve redundancy.
9. *Maintenance costs:* the reduction of maintenance costs would have a negative effect in all parameters, except in what comes to inspections, since the reduction of this costs would not show a difference in the number of inspections.

$$\text{Operation availability} = \frac{\text{Available productive time}}{\text{Time used in production}} \tag{3}$$

4 Discussion and Conclusions

This study represents a study case of application of a TRIZ-Lean solution to improve the maintenance management of the air system from Portuguese Navy. The proposed methodology helped to identify which parameter was the most harmful for the maintenance management in the Portuguese Navy context. Maintenance cost showed the most negative effect with all the other parameters, apart from inspection, revealing the need of innovative solutions to overcome the existing problem of disinvestment in the Portuguese Navy. This TRIZ-Lean approach also presents a solution to identifying the value chain by using the Lean tool, VMS, enabling the visualization of the waste and all types of activities which directly influence each parameter chosen in the ideality matrix. These two tools represent a systematic approach to clarify where management solutions must be applied.

This new methodology, applied to maritime applications, represents a further development in the spread of this powerful tools since they can simplify understanding the influence of the parameters in the value chain. However, more studies must be applied since the value stream mapping was based on a specific complex system, the air system, and from the Portuguese Navy Maintenance Service Manual not from direct observation of the value chain. Therefore, applying all four steps proposed took time. Nevertheless, it was developed a general overview of the maintenance process since the objective of this work was to present a methodology.

The developed TRIZ-Lean methodology has the potential to improve efficiency, reduce waste, and increase productivity, giving an easy understanding of the parameters' influence in value chain.

The next step in the proposed methodology is the use of a value map streaming of the maintenance management from Portuguese Navy as a generic tool to identify the value chain in the maintenance process and identify where should the measures be applied to improve the ideality from each parameter, established in the ideality matrix (Table 2).

Acknowledgments The authors acknowledge the Polytechnic Institute of Lisbon for its financial support through the grant IPL/2022/MEVSIsMan_ISEL.

References

1. Altshuller, G.. (1999). *The innovation algorithm: TRIZ, systematic innovation and technical creativity*.
2. Ilevbare, I. M., Probert, D., & Phaal, R. (2013). A review of TRIZ, and its benefits and challenges in practice. *Technovation, 33*, 30–37.
3. Navas, H. V. G. (2013). TRIZ: Design problem solving with systematic innovation. In *Advances in Industrial Design Engineering*. InTech.
4. Ekmekci, I., & Nebati, E. E. (2019). Triz methodology and applications. *Procedia Computer Science, Elsevier B.V*, 303–315.
5. Sehested, C., & Sonnenberg, H. (2011). Introduction. In *Lean innovation*. Springer.

6. Morgado, T. L. M., Sandiães, J., & Navas, H. V. G. (2019). TRIZ and lean philosophies applied together in management activities. *Quality Innovation Prosperity, 23*, 90–102.
7. Navas, H., & Cruz-Machado, V. (2014). Application of lean and TRIZ methodologies in the maintenance service of a food company. In *2014 International Conference on Systematic Innovation*.
8. Sheth, P. P., Deshpande, V. A., & Kardani, H. R. (2014). Value stream mapping: A case study of automotive industry. *IJRET: International Journal of Research in Engineering and Technology*, 310–314.
9. David, A., Morgado, T., Sartal, A., & Navas, H. (2019). TRIZ methodologies for improving LEAN production processes. In *Proceedings of the 1st Conference on Quality Innovation and Sustainability—ICQIS2019* (pp. 164–171).
10. Castro, C., Pereira, T., Sá, J. C., & Santos, G. (2020). Logistics reorganization and Management of the ambulatory pharmacy of a local health unit in Portugal. In *Evaluation and program planning* (Vol. 80, pp. 1–10). Elsevier.
11. Jimenez, G., Santos, G., Sá, J. C., Ricardo, S., Pulido, J., Pizarro, A., & Hernandez, H. (2019). Improvement of productivity and quality in the value chain through lean manufacturing—A case study. In *Procedia manufacturing* (Vol. 41, pp. 882–889). Elsevier.
12. Morgado, T., Pinto, A., Navas, H., & Lampreia, S. (2022). Development of a computerized maintenance management model of a laboratory testing service enterprise. In J. Machado, F. Soares, J. Trojanowska, & E. Ottaviano (Eds.), *Innovations in Mechanical Engineering. ICIENG 2021* (Lecture notes in mechanical engineering). Springer.
13. Alves, A. C., Sousa, P., & Navas, H. (2019). Lean and TRIZ: From the problems to creative and sustainable solutions. In *Proceedings of the 6th European Lean Educator Conference* (pp. 103–116).
14. Pombo, I. M. C., Navas, H., & Cruz-Machado, V. (2015). TRIZ and lean methodologies in an industry of air handling and ventilation. In *The 2015 international conference on systematic innovation*.
15. Godina, R., Carvalho, H., Rodrigo, P., & Navas, H. (2019). Towards continuous improvement by using a lean-TRIZ approach. In *Proceedings of the 6th European Lean Educator Conference* (pp. 169–178).
16. Nurcahyo, R., Apdillah, R., & Yadrifil. (2017). Application of lean six sigma and theory of inventive problem solving to reduce waste and improve the quality motorcycle rear arm. *Journal of Asian Scientific Research, 7*(5), 176–186.
17. Lampreia, S., Lobo, V., & Vairinhos, V. (2021). *Maritime cybersecurity risk assessment: The ship maintenance databases' case study*. ISMS.
18. Marinha.: Direção do serviço de manutenção, Portugal, 1984.

Intrinsic Availability of a Critical Productive Equipment

Teresa Morgado, Gonçalo Alves, António Abreu, and Ana Dias

Abstract The current context of the world economy represents a significant challenge for industries. The prevailing process of globalization, constant technological evolution, and increased market competitiveness are forcing companies to act and adapt. Continuous improvement actions are needed to reduce waste; increase productivity, quality, and delivery levels; and simultaneously reduce the costs associated with production processes. The main objective of this research was to increase the Intrinsic Availability of assembly line production equipment to achieve and guarantee a world-class goal of 90%. This work was developed in partnership with a company dedicated to manufacturing and assembling metallic structures for the aeronautic industry. The methodology applied in this research is based on TPM and Lean tools. Positive results were achieved since, at the end of the method used, it was possible to reach 91.80% Intrinsic Availability of the equipment, exceeding the objective of 90%, and an average OEE value of 66.7% compared to the initial 58.1%, associated with a 170.9% increase in MTBF and a 48.4% reduction in the total downtime of the machine, compared to the initial situation.

Keywords TPM · Lean · Intrinsic Availability

T. Morgado (✉)
Polytechnic Institute of Lisbon, Lisbon, Portugal

Navy Research Center – Portuguese Naval Academy, Almada, Portugal

Research and Development Unit for Mechanical and Industrial Engineering – NOVA University of Lisbon, Caparica, Portugal

Intelligent Systems Associate Laboratory, Guimarães, Portugal
e-mail: teresa.morgado@isel.pt

G. Alves · A. Dias
Polytechnic Institute of Lisbon, Lisbon, Portugal

A. Abreu
Polytechnic Institute of Lisbon, Lisbon, Portugal

Centre of Technology and Systems of UNINOVA, Caparica, Portugal

J. Reis et al. (eds.), *Driving Quality Management and Sustainability in VUCA Environments*, Springer Proceedings in Business and Economics,
https://doi.org/10.1007/978-3-031-52723-4_12

1 Introduction

Organizations pursue one or more of the following objectives: profit maximization, meeting deadlines, a specific level of service or product quality, cost minimization, safety and environmental protection, and employee development. Of course, all these objectives are impacted to some degree by Maintenance. Therefore, Maintenance's targets must be aligned with the organization's goals. Maintenance's primary responsibility is to provide a service that enables an organization to achieve its objectives [1–4].

The methodology applied during this work was Action-Research (see Fig. 1). This methodology is characterized by the fact that it is an essentially practical and applied research methodology, which is governed by the need to solve real problems and where the action to improve the process is based on a previously carried out investigation. Action-Research covers any process that follows an improvement cycle in a practical context through research. This cycle comprises four stages: defining the problem or diagnosis, planning the actions, implementing the actions or making decisions, and finally, monitoring and evaluating the results [5, 6].

This work was developed in partnership with an aeronautic company. This company currently has the TPM methodology [3], [7, 8] implemented in the critical equipment of the production lines.

The principal objective of this project is to increase the Intrinsic Availability of Critical Productive Equipment on the assembly line of the steel structures factory to reach and remain at the world-class target of 90% through the application of the TPM methodology and the use of Lean tools and consequently improve the Mean Time Between Failures (MTBF), Mean Time to Repair (MTTR), Intrinsic Availability, and Overall Equipment Efficiency (OEE) indicators [9, 10].

1.1 *State-of-the-Art*

According to BS EN 13306:2017 [11], Maintenance activities can be classified in two ways: considering the intrinsic reliability characteristics of the equipment

Fig. 1 Action-Research cycle

or from a management and planning point of view. According to the inherent reliability characteristics, activities can be divided into three types: Improvement Maintenance, Preventive Maintenance, and Corrective Maintenance. From the point of view of activity management, these can be divided into two distinct types: Planned Maintenance and Unplanned Maintenance.

Over the past decades, Maintenance has played a fundamental role in the sustainability of organizations [3, 8, 12–19]. In terms of costs, Maintenance represents, after energy expenses, the most significant part of a company's budget and can represent a large part of the value added to a product [4, 20].

Philosophies such as Just-In-Time (JIT) and Lean Manufacturing require a continuous flow of production and increasingly efficient planning of operations [21, 22]. Equipment is expected to have high reliability and availability for this to be possible. A well-developed, solidly structured, and properly planned Maintenance strategy can help reduce costs, increase productivity, and improve quality and safety in production processes [16, 23].

Total Productive Maintenance (TPM) aims to optimize productivity, improve equipment efficiency, and eliminate failures by training operators and team leaders to proactively participate in daily maintenance tasks, such as cleaning, inspection, and equipment lubrication. Therefore, TPM methodologies are used to achieve the organization's strategic goals to reduce costs. For these reasons, it is widely used in many industries, and its implementation has shown relevant results [3, 8, 24].

It is imperative to have well-formulated Maintenance based on the organization's strategy and production to develop a structured approach to measuring the performance of the maintenance function [3, 6].

Performance measurement is fundamental to Maintenance Management. Well-defined key performance indicators (KPIs) make visualizing the gaps between current and desired performance possible [9, 10]. Maintenance performance must be measured so that it can be controlled and monitored to take appropriate and timely action to minimize and mitigate safety risks, meet social responsibilities, and increase the effectiveness and efficiency of assets, creating long-term value and increasing the economic viability of organizations [10, 23, 25].

2 Experimental Work

Based on the Action-Research methodology, the theoretical framework was first analyzed to improve knowledge of the subject covered in this work. At the same time, an analysis was made of the initial situation of the assembly area of the metal structures factory and its equipment to choose the equipment where this project would be carried out. Therefore, various factors and indicators were analyzed, such as the criticality of the equipment, the Overall Equipment Efficiency (OEE), the number of breakdowns, the Mean Time Between Failure (MTBF), and the Mean Time to Repair (MTTR), during the period from January to July 2021. This analysis used Maintenance Management and TPM Management tools/software.

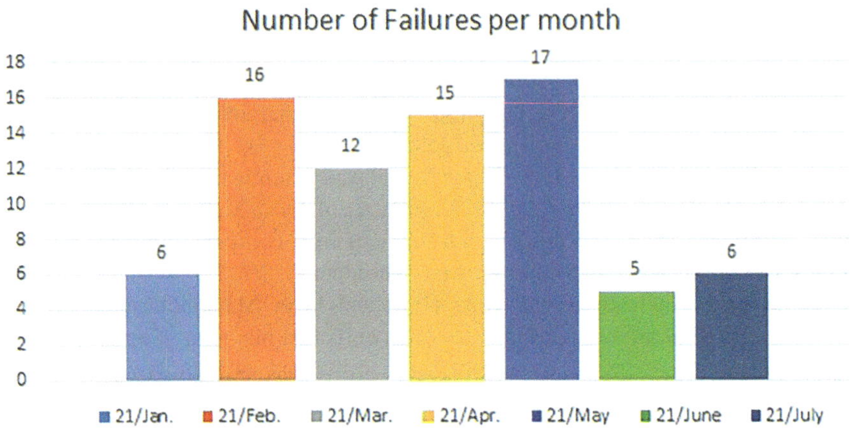

Fig. 2 The number of failures per month from January 2021 until July 2021

Initially, the number of failures and machine downtime was surveyed each month from January to July 2021 to understand their trend. Figures 1 and 2 show graphs of the number of failures and machine downtime compared to the available production time by month. During this period, 77 corrective maintenance operations were registered (Fig. 1), totaling 237 hours and 48 minutes of machine downtime out of approximately 1259 hours of available production time for the same period, equivalent to 18.9% of the total available production time (Fig. 2). During this period, the average MTBF was 14 hours and 32 minutes, and the average MTTR was 4 hours. The average equipment availability was 80%.

All the corrective work orders were analyzed, and the information regarding the equipment systems where the breakdowns occurred was corrected. After correcting all this information, a Pareto diagram was generated to understand which systems were responsible for the most significant number of breakdowns and thus filter and direct root cause analysis. Figure 3 shows the Pareto diagram of the number of failures per system. The analysis of Fig. 4 shows that most of the failures and machine downtime are caused by problems with V-Side Tools and U-Side Systems.

2.1 Root Cause Analysis and Implement Actions

Through the analysis of Table 1, the main root causes of the identified problems are the lack of condition control, dirty components of the V-Side Tools, and the lack of regular cleaning of the V-Anvil tools. In addition, the absence of a preventive maintenance plan for the V-Anvil tools significantly impacted availability due to numerous corrective shutdowns and product quality, as marks were frequently found on products.

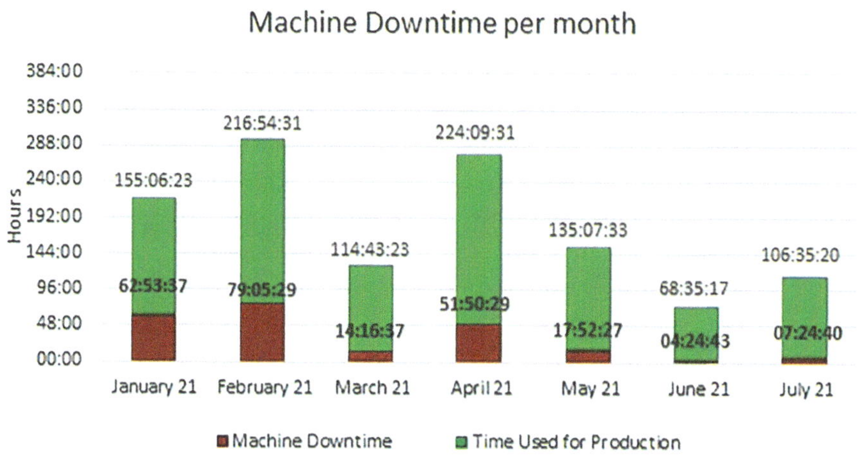

Fig. 3 The monthly machine downtime from January 2021 until July 2021

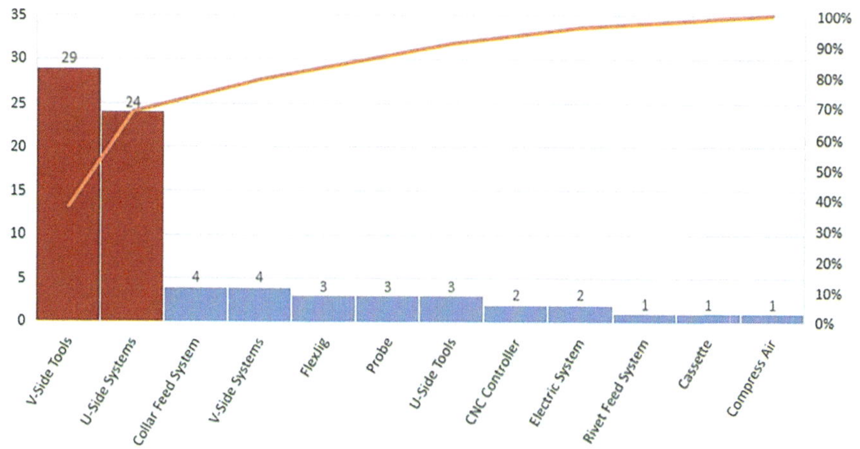

Fig. 4 Number of failures per system

Table 2 shows that the main root causes of the identified problems are the lack of cleaning and tuning of the fastener injection system, the lack of preventive routines in the camera system, and the lack of condition control of the sealant application system.

After identifying the root causes of the recurring problems, several improvement actions were developed and implemented. These actions essentially consisted of creating a preventive plan for the V-Side Tools; reviewing the preventive Maintenance and autonomous maintenance plans (cleaning and inspection); creating simple procedures for training operators and Maintenance technicians; and occasional improvement of some systems and tools of the equipment.

Table 1 Recurrent problems in V-Side Tools and root causes

Component	Recurrent problem	Number of work orders	Root causes
V-side tools	Incorrect swaging of collars	4	Dirt/lack of cleaning tools Lack of preventive maintenance and condition control Lack of knowledge of the operators Wear of components
	Stuck/broken components	5	
	Emergency stop during collar swaging	3	
	Marks on the product	6	
	Scratched/marked/crushed collars	3	
	Sensor malfunction	6	

Table 2 Recurrent problems in the U-Side Systems and root causes

U-side systems	Recurrent problem	Number of work orders	Root causes
Fastener injector	The machine does not pick up fasteners	8	Dirt/lack of cleanliness Wear of components Lack of preventive maintenance and condition control
	Alarms	5	
Camera	Alarms	4	
Sealant	Struck components and alarms	5	

3 Results

The number of failures and the respective machine downtime, the Mean Time Between Failures, the Mean Time to Repair, the Intrinsic Availability, and the Global Efficiency of the Equipment studied and the results are discussed in this section. The chosen period for analyzing the evolution of indicators begins in August 2021, corresponding to the first month after the initial situation analysis and the start of the implementation of improvement actions, and ends in May 2022. This choice was made for two reasons: first, because the MTBF and MTTR indicators are calculated over 3 months, presenting evolution with a slower behavior, and second, because during the last 2 months of 2021 and January 2022, a very low production cadence was observed, contributing little to the 3-month Net Scheduled Time.

Figures 5 and 6 show the graphs of the number of failures per month and the machine downtime relative to the available monthly production time from August 2021 to May 2022, respectively. During this period, 74 corrective maintenance

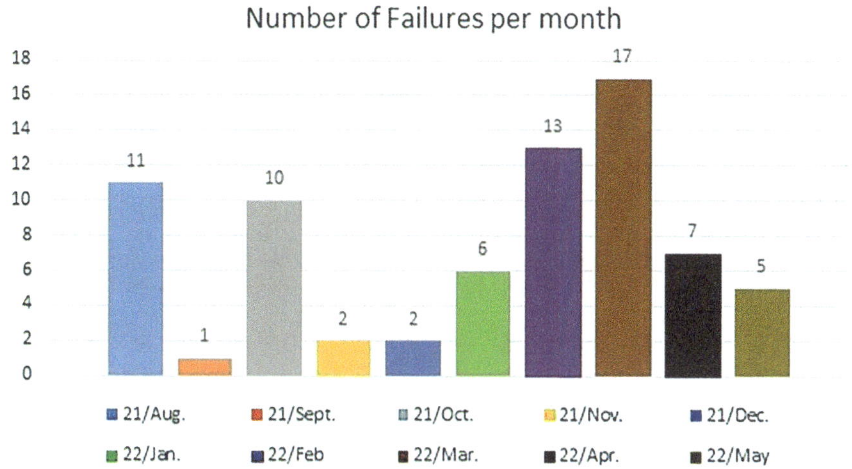

Fig. 5 Number of failures per month from August 2021 until May 2022

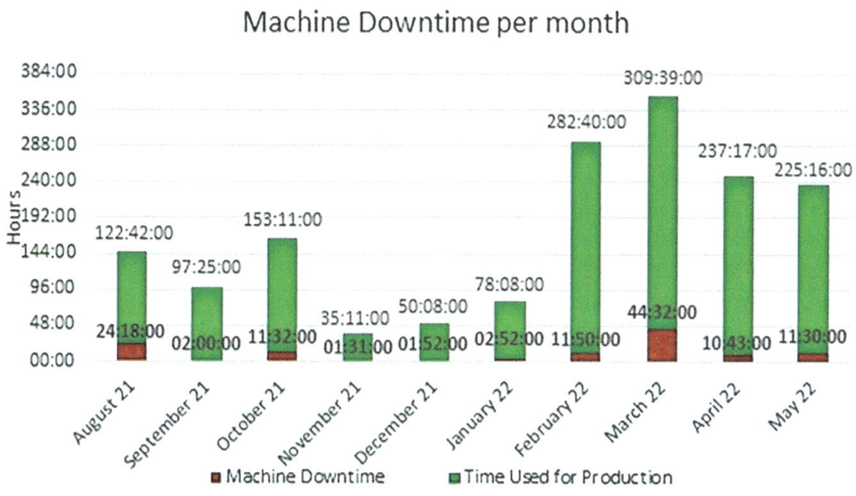

Fig. 6 Machine downtime per month from August 2021 until May 2022

events were recorded, totaling 122 hours and 40 minutes of machine downtime and approximately 1714 hours of available production time. Therefore, it is possible to verify a reduction of approximately 48.42% compared to the total machine downtime identified in the initial situation from January to July 2021.

In Fig. 7, the evolution of the MTBF and MTTR during the period from January 2021 to May 2022 is shown. Through the analysis of the chart, it is possible to verify the reversal of the trend starting in July 2021, the period in which the analysis of failures, the TPM monitoring, and the implementation of improvement actions

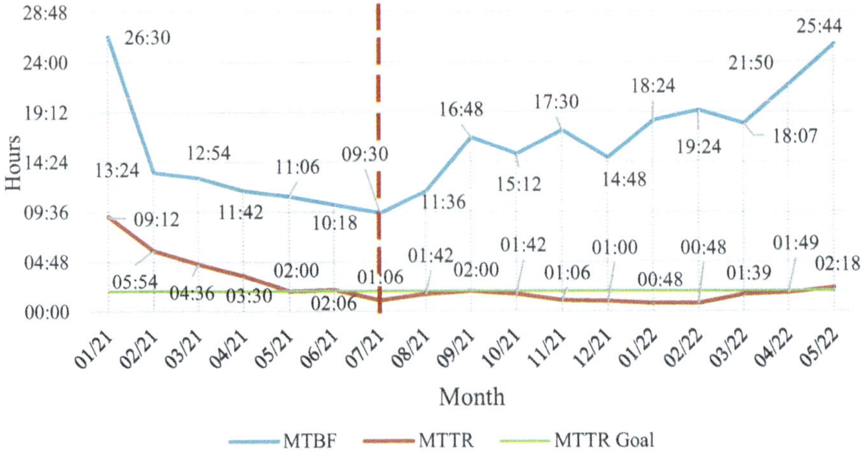

Fig. 7 Evolution of MTBF and MTTR

began. In the last month of the case study, the MTBF was 25 h and 44 min, representing an increase of 170.9% compared to the 9 h and 30 min of MTBF observed in the initial situation. The number of failure reduction justified the MTBF evolution and, consequently, the machine downtime and high monthly production times. On the other hand, the values of MTTR remained constant and always with values below the target of 2 h.

Figure 8 shows the evolution of the Intrinsic Availability of the equipment from January 2021 to May 2022. As expected, it is possible to verify an increase in the equipment's availability due to low MTTR during the case study. The world-class goal of 90% was achieved, and an availability value of 96.04% was reached in February. The average availability during the case study's development was 92.19%. Thus, the improvement actions implemented impacted Intrinsic Availability by reducing the number of corrective Maintenance and the time the machine was stopped. As for the evolution of OEE, it is possible to verify a positive trend throughout the case study's development. In November and December 2021, it was noted that, despite the high percentage of Intrinsic Availability, the OEE values were relatively low due to losses such as product analysis, program validation, waiting period for process solutions/information, and lack of material. It is important to note that this type of loss was not analyzed in this case study, and the implemented actions did not aim to reduce these losses. In March 2022, the low % OEE value of 60.9% can be justified by many unplanned stops due to a lack of operators, materials, programs, and corrective Maintenance. These unplanned stops directly affect the operational availability indicator used in calculating OEE. Despite the OEE average for the period from August 2021 to May 2022 being 66.7% compared to 58.1% for the period from January to July 2021, there is still a need to positively evolve this indicator so that the company's target of 67% is reached and, later, a world-class OEE of 85% is achieved.

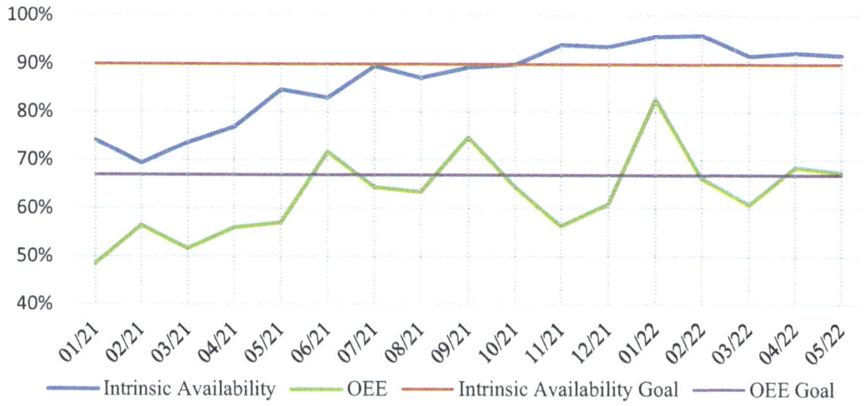

Fig. 8 Evolution of Intrinsic Availability and OEE

In addition to the quantitative results presented, results that are not measurable were also obtained, namely, the development of the operators' autonomy and technical capacities; cooperation and involvement among all areas involved in TPM; improving communication between operation and Maintenance; and increasing operators' motivation and sense of responsibility for the equipment they operate.

4 Discussion

The work carried out in this study case consisted of analyzing failures to improve the availability indicator of critical equipment. However, although the company had the necessary tools for correctly implementing the TPM methodology, its practice needed to be carried out as intended due to the lack of presence and participation of all parties involved, particularly the Planned Maintenance pillar, and the absence of critical analysis of failures. In an initial investigation, the equipment under study presented MTBF, MTTR, and OEE values below the established targets, caused by the high number of failures and machine downtime. Therefore, a failure study was conducted from January to July 2021 to reverse this situation. Subsequently, from August 2021 to May 2022, improvement actions were developed and implemented in the equipment systems that significantly impacted the equipment's availability. These actions consisted mainly of reviewing the preventive and autonomous maintenance plans, creating procedures for training operators and Maintenance technicians, and improving specific systems and tools of the equipment.

5 Conclusions

From the results obtained, it is concluded that during the development period of the case study, the implementation of the improvement actions contributed to a significant decrease in the failure rate compared to the initial period and a 48.4% reduction in total machine downtime compared to the initially analyzed period. Consequently, the MTTR fell to values below the 2-hour target, the MTBF increased 170.9% compared to the initial value, and the intrinsic availability increased to values above the 90% target, achieving the primary goal of this case study.

Concerning OEE, it was concluded that although the average OEE increased by 8.6% (from 58.1% in the initial period to 66.7% at the end of the study), the approach and actions taken did not solve the other types of losses associated with operational availability, such as lack of material, lack of operator, and lack of program. Thus, the increase in Intrinsic Availability related to the equipment did not allow for such a direct gain that would be reflected in the OEE indicator since it is only a part of operational availability. There is still much room for improvement in this regard. One of the principal limitations identified was the need for more motivation and resistance to change on the part of the operators, as they were accustomed to a particular work method. Moreover, they often considered that their participation in maintenance tasks was irrelevant and was seen as additional work and responsibilities. However, the continuous monitoring and active participation in TPM meetings throughout this case study strengthened the relationship between operators and maintenance technicians. It motivated operators to be more autonomous and develop a sense of equipment ownership. It is concluded that the whole organization should participate and commit to continuous improvement to achieve world-class values of equipment and production line efficiency.

Acknowledgments The authors acknowledge the Polytechnic Institute of Lisbon for its financial support through the grant IPL/2022/MEVSIsMan_ISEL.

References

1. Bousdekis, A., Apostolou, D., & Mentzas, G. (2020). Predictive maintenance in the 4th industrial revolution: Benefits, business opportunities, and managerial implications. *IEEE Engineering Management Review. IEEE, 48*(1), 57–62.
2. Costa, R., Barbosa, M., Silva, F., Sá, J. C., Ferreira, L. P., & Pinto, B. (2022). *Improving procedures for production and maintenance control towards industry 4.0 implementation. Innovations in mechanical engineering II* (pp. 58–67). Springer.
3. Jain, A., Bhatti, R., & Singh, H. (2014). Total productive maintenance (TPM) implementation practice: A literature review and directions. *International Journal of Lean Six Sigma, 5*(3), 293–323.
4. Moreira, A., Silva, F., Correia, A., Pereira, T., Ferreira, L., & Almeida, F. (2018). Cost reduction and quality improvements in the printing industry. *Procedia Manufacturing, 17*, 623–630.

5. Loorbach, D., Bakel, J.C. van, Whiteman, G., Rotmans, J.: Business strategies for transitions towards sustainable systems. Business Strategy and the Environment, 19, 133–146 (2010).
6. Bevilacqua, M., & Braglia, M. (2000). Analytic hierarchy process applied to maintenance strategy selection. *Reliability Engineering and System Safety, 70*(1), 71–83.
7. Abhishek, J., Bhatti, R., & Singh, H. (2014). Total productive maintenance (TPM) implementation practice: A literature review and directions. *International Journal of Lean Six Sigma, 5*(3), 293–323.
8. Kaur, M., Singh, K., & Ahuja, I. S. (2013). An evaluation of the synergic implementation of TQM and TPM paradigms on business performance. *International Journal of Productivity and Performance Management, 62*(1), 66–84.
9. European Committee for Standardization: 15341. (2007). *Maintenance – Maintenance key performance indicators* (p. 28). CEN.
10. Muchiri, P., Liliane, P., Ludo, G., & Harry, M. (2011). Development of maintenance function performance measurement framework and indicators. *International Journal of Production Economics, 131*(1), 295–302.
11. British Standards Institution: BS EN 13306:2017 – Maintenance – Maintenance Terminology. 94 (2018).
12. Pinto, G., Silva, F., Campilho, R., Casais, R., Fernandes, A., & Baptista, A. (2019). Continuous improvement in maintenance: A case study in the automotive industry involving lean tools. *Procedia Manufacturing, 38*, 1582–1591.
13. Guariente, P., Antoniolli, I., Ferreira, L., Pereira, T., & Silva, F. (2017). Implementing autonomous maintenance in an automotive components manufacturer. *Procedia Manufacturing, 13*, 1128–1134.
14. Reis, M., Godina, R., Pimentel, C., Silva, F., & Matias, J. (2019). A TPM strategy implementation in an automotive production line through loss reduction. *Procedia Manufacturing, 38*, 908–915.
15. Ribeiro, I., Godina, R., Pimentel, C., Silva, F., & Matias, J. (2019). Implementing TPM supported by 5S to improve the availability of an automotive production line. *Procedia Manufacturing, 38*, 1574–1581.
16. Santos, T., Silva, F., Ramos, S., Campilho, R., & Ferreira, L. (2019). Asset priority setting for maintenance Management in the Food Industry. *Procedia Manufacturing, 38*(2019), 1623–1633.
17. Dias, A., Antunes, R., Abreu, A., Anes, V., Navas, H., Morgado, T., & Calado, J. (2022). Utilization of the arena simulation software and lean improvements in the management of metal surface treatment processes. *Procedia Computer Science, 204*, 140–147.
18. Lampreia, S., Morgado, T., Navas, H., Cabrita, R., & Requeijo, J. (2021). An approach to ship equipment maintenance management. In J. Machado, F. Soares, J. Trojanowska, & E. Ottaviano (Eds.), *Innovations in mechanical engineering. ICIENG* (Lecture notes in mechanical engineering) (p. 2022). Springer.
19. Morgado, T., Pinto, A., Navas, H., & Lampreia, S. (2021). Development of a computerized maintenance management model of a laboratory testing service enterprise. In J. Machado, F. Soares, J. Trojanowska, & E. Ottaviano (Eds.), *Innovations in mechanical engineering. ICIENG* (Lecture notes in mechanical engineering) (p. 2022). Springer.
20. Horenbeek, A., Bey-Temsamani, A., Vandenplas, S., Pintelon, L., & Deketelaere, B. (2011). Prognostics for optimal maintenance: Maintenance cost versus product quality optimization for industrial cases. In J. Lee, J. Ni, J. Sarangapani, & J. Mathew (Eds.), *Engineering asset management* (Lecture notes in mechanical engineering) (p. 2014). Springer.
21. Ferreira, C., Sá, J., Ferreira, L., Lopes, M., Pereira, T., & Silva, F. (2019). iLeanDMAIC—A methodology for implementing the lean tools. *Procedia Manufacturing, 41*, 1095–1102.
22. Neves, P., Silva, F., Ferreira, L., Pereira, T., Gouveia, A., & Pimentel, C. (2018). Implementing lean tools in the manufacturing process of trimmings products. *Procedia Manufacturing, 17*, 696–704.
23. Smith, R., & Hawkins, B. (2004). *Lean maintenance: Reduce costs, improve quality, and increase market share*. Elsevier Butterworth–Heinemann.

24. Levitt, J. (2010). *TPM reloaded – Total productive maintenance*. Industrial Press.
25. Singh, J., & Singh, H. (2020). Justification of TPM pillars for enhancing the performance of manufacturing industry of northern India. *International Journal of Productivity and Performance Management, 69*(1), 109–133.

Study of the Time Windows Range Impact on Vehicle Routing Problem

Pedro Afonso Oliveira, Maria Teresa Pereira, Marisa Oliveira, Filipe R. Ramos, and Fernanda Amélia Ferreira

Abstract The purpose of this chapter is to study and explain the impact of the length of the time windows in the vehicle routing problem with time windows (VRPTW). A case study was explored with the aim of understanding the relationship between the use of time windows with different intervals and the cost of charges, client satisfaction, and travel time. The solution presented studies the influence that time windows have on delivery costs. In it, it is possible to verify that the number and capacity of needed vehicles to the service vary, as well as the number of served clients in a route, according to their schedule and demand, and the size of the time windows in which deliveries are made. As a result, the time window negotiation is important and related to the operational services with a huge impact on the global business.

Keywords VRP · VRPTW · Cost optimization · Route optimization

P. A. Oliveira
School of Engineering of Porto (ISEP) - Polytechnic of Porto, Porto, Portugal

M. T. Pereira (✉)
School of Engineering of Porto (ISEP) - Polytechnic of Porto, Porto, Portugal

Associate Laboratory for Energy, Transports and Aerospace (LAETA-INEGI), Porto, Portugal
e-mail: mtp@isep.ipp.pt

M. Oliveira
School of Engineering of Porto (ISEP) - Polytechnic of Porto, Porto, Portugal

Associate Laboratory for Energy, Transports and Aerospace (LAETA-INEGI), Porto, Portugal

F. R. Ramos
CEAUL-Centro de Estatística e Aplicações, Faculdade de Ciências, Universidade de Lisboa, Lisboa, Portugal

F. A. Ferreira
UNIAG, School of Hospitality and Tourism of Polytechnic Institute of Porto, Porto, Portugal

J. Reis et al. (eds.), *Driving Quality Management and Sustainability in VUCA Environments*, Springer Proceedings in Business and Economics,
https://doi.org/10.1007/978-3-031-52723-4_13

1 Introduction

Due to globalization, the transportation sector is the one that has the most impact on the logistic cost, adding the technological evolution that allows dynamic operational support and the use of optimization algorithms. One of the issues with a high impact on transport service computation and routing is the range of time windows negotiated or imposed by the clients. Currently, available technology and real-time decisions have given a special focus on dynamic routing, allowing gains on vehicle routing problems (VRPs). The VRP consists in finding the lowest cost set of routes for a set of vehicles in order to satisfy the total demand of a geographically dispersed set of clients in a network [1]. Over time, extensions and variants to the VRP appeared. Among these is the Vehicle Routing Problem with Time Windows (VRPTW), which consists in applying and complying with imposed time windows by suppliers/clients based on the concept of VRP. This chapter explored the impact of the time window range on services' costs as a contribution to understanding the importance of time windows negotiation in transportation services. For this, a brief contextualization of the theme and a literature review was done. A VRPTW model was implemented, and a set of instances with different time window ranges for low, medium, and high demand from a set of clients was considered to simulate real routing problems context. The rest of the chapter is organized as follows: In the second section, a literature review on the VRP and VRPTW is done, as well as its variants; in the third section, the results obtained from simulating the model across the instances are presented; in the last section, a conclusion and future work are presented.

2 VRP and VRPTW Literature Review

In this section, the concepts of the Vehicle Routing Problem (VRP) and the Vehicle Routing Problem with Time Windows (VRPTW) are presented. VRP was presented in 1959 as a problematic problem in managing transports, introduced by Dantzig and Ramser [2], and it consisted of organizing a route of gasoline transport trucks between the terminal and the stations fed by itself [3, 4]. This concept emerged from the generalization of the Traveling Salesman Problem (TSP), which is a circuit characterized by the existence of several vertices (clients) through which a vehicle must pass. The VRP consists in finding a set of routes for k vehicles that start their route from the depot in order to visit all of the vertices exactly once, and the objective is to minimize the total cost of the trips made [5]. Under the restrictions, an optimal route of delivery should be found starting from one or more specific points in order to answer the clients, localized in different points [3]. The objectives can depend on the strand explored, as will be seen further in this document when analyzing the Vehicle Routing Problem with Time Windows. Over time and due to evolution, new aspects related to VRP have emerged, which are the subject of academic study and practical applications. Some examples of these extensions are the capacitated vehicle routing problem, CVRP, the vehicle routing problem with a time window, VRPTW, the split delivery vehicle routing problem, SDVRP, and the dynamic vehicle routing problem, DVRP. In the CVRP, the objective is not exceeding the maximum capacity of the vehicle while transporting a greater number of goods in a single trip [6]. In the VRPTW, the delivery is only available during a certain period of time, and the vehicle should supply the demands to the client only when the time window opens [5]. When a client can be visited by several vehicles, the problem is a VRP variant known as SDVRP [7, 8]. An updated review of the DVRP literature can be found in [9]. VRP problems can have different resolution methods, divided between exact algorithms and heuristic algorithms. The exact algorithms used can be of four types: branch and bound approach, cutting planes approach, network flow approach, and dynamic programming approach. With the increase in requirements from companies, it is more and more important that the schedules for deliveries and services performed are according to the defined time windows service. A delay in delivery can jeopardize the defined schedule for the vehicle as a consequence of penalties or inefficiencies due to the loss of its place in the queue and return when rescheduled. Hence, a study about the VRPTW is pertinent, as these are combinatorial optimization problems and know how certain instances can impact the problems within this topic. The main focus of VRPTW is to find an optimal set of delivery or collection routes from a depot to a certain amount of clients, taking into consideration all of the constraints and the defined time windows [10]. The VRPTW is NP-hard problem [11] and can be classified into two types: an approach to the problem using flexible time windows, where a delivery that is made out of the time window, sooner or later, implies a penalty, or there can be an approach using hard time windows, where delays are not allowed and if the delivery is ready before the time it should wait for the time window to start [12, 13]. Usually, routing problems with time windows (VRPTW) consider hard time windows. If the problem to be analyzed is related to flexible time windows, another denomination can be used, that being Vehicle Routing Problem with Soft Time Windows (VRPSTW) or Vehicle Routing Problem with Flexible Time Windows (VRPFlexTW) [14]. The works that mostly use hard time windows are the ones that are of maximum responsibility. As an example, we have: patrol security service, banks, post and food deliveries, school bus routes, distribution of newspapers, and garbage collection [12, 14]. In most distribution services, it is necessary to have more than one route or more than one vehicle. Therefore, based on the carried out research, we set out to analyze the impact of changing the time windows in a VRP. We decided to develop a VRPTW model in order to run various instances and present the results. Based on the developed model presented next, the solution to each instance utilizes hard time windows. This way, the deliveries cannot be made before or after the defined time window previously defined. The developed model includes sets of data related to the company and to the clients (sets and parameters) as well as decision variables in order to save the values that are important to the route decision-making.

Sets:
C clients $\{2, ..., c\}$;
N vertices in the delivery network $\{1, 2, 3, ..., c + 1\}$;
N_0 vertices from which the vehicle can leave $\{1, 2, 3, ..., c\}$;
N_+ vertices to which the vehicle can travel to $\{2, 3, ..., c + 1\}$;
V vehicles $\{1, 2, 3, ..., m\}$.

Parameters:
e_i lower bound of the time window for client's i delivery, $i \in C$;
l_i upper bound of the time window for client's i delivery, $i \in C$;
t_{ij} travel time between i and j, $i \in N_0$, $j \in N_+$;
q_i client's i demand, $i \in C$;
a_k vehicle's k capacity, $k \in V$;
c_{ij} travel cost per kilometer from client i to j, $i \in C$ and $j \in C$;
s_i time service per order on client i, $i \in C$.

The decision variables are:

$$x_{ij}^k \begin{cases} 1 \textit{ if the vehicle k travels from i to j;} \\ 0 \textit{ otherwise.} \end{cases}$$

b_i^k variable that identifies the moment when the service starts in the client i, $i \in C$, by the vehicle k, $k \in V$;
y_i^k fraction of client's i demand, $i \in C$, delivered by the vehicle k, $k \in V$.

Objective function:

$$\min \sum_{i \in N_0} \sum_{j \in N_+} \sum_{k \in V} c_{ij} \cdot 1.25 \cdot t_{ij} \cdot x_{ij}^k \quad (1)$$

subject to

$$\sum_{j \in N_+} x_{1j}^k \leq 1 \qquad \forall k \in V \quad (2)$$

$$\sum_{i \in N_0} x_{ip}^k - \sum_{j \in N_+} x_{pj}^k = 0 \qquad \forall p \in C, \forall k \in V \quad (3)$$

$$\sum_{k \in V} y_i^k = 1 \qquad \forall i \in C \quad (4)$$

$$\sum_{k \in V} \sum_{i \in N_0} x_{ij}^k \geq 1 \qquad \forall j \in N_+ \quad (5)$$

$$\sum_{i \in C} q_i \cdot y_i^k \leq a_k \qquad \forall k \in V \quad (6)$$

$$y_i^k \leq \sum_{j \in N+} x_{ij}^k \qquad \forall i \in C, \forall k \in V \quad (7)$$

$$b_i^k + s_i \cdot q_i \cdot y_i^k + t_{ij} - M(1 - x_{ij}^k) \leq b_j^k \qquad \forall i \in C, \forall j \in C, \forall k \in V \tag{8}$$

$$e_i \leq b_i^k \leq l_i \qquad \forall i \in C, \forall k \in V \tag{9}$$

The objective function (1) minimizes the total routing costs, where 1.25 is a factor that converts the travel time into travel kilometers, considering a kilometer average of 70 kilometers per hour travelled. The first restriction presented (2) guarantees that each vehicle, if necessary, will leave the depot and will get to a specific client. Equation (3) makes sure that each vehicle will leave the last client and get to the depot. Equation (4) it is secured that every single demand from a client is attended to. The Eq. (5) is guaranteed that each client will be visited at least once by at least one vehicle. In Eq. (6), vehicle capacity will not be exceeded. The Eq. (7) ensures that each client's demand will only be met if a particular vehicle passes by that client. The Eq. (8) defines a minimum time to start the service of client j on a given route and eliminate sub-routes.

The Eq. (9) ensures that all clients are served within the time windows. The Eq. (10) defines the decision variables domain.

3 Results

The model was implemented in IBM ILOG CPLEX Optimization Studio 21.1.0.0 software. In order to have a better perception of how the routing with time windows works, different instances were made. The number of clients was fixed along the tests (10 clients), as well as the number of vehicles used (7 vehicles) and the travel time between clients (presented in Table 1).

Table 1 Travel time in minutes between clients

		Clients										
		1	2	3	4	5	6	7	8	9	10	11
Clients	1	0	30	20	25	20	17	26	30	50	46	32
	2	30	0	45	25	40	62	13	46	31	29	52
	3	20	45	0	70	32	52	43	30	33	20	29
	4	25	25	70	0	25	30	40	62	34	70	49
	5	20	40	32	25	0	25	54	64	43	32	31
	6	17	62	52	30	25	0	23	48	45	32	45
	7	26	13	43	40	64	23	0	61	45	32	14
	8	30	46	30	62	64	48	61	0	26	48	34
	9	50	31	33	34	43	45	45	26	0	54	24
	10	46	29	20	70	32	32	32	48	54	0	45
	11	32	52	29	49	31	45	14	34	24	45	0

Table 2 Clients' demand in tons

	Client ID									
	2	3	4	5	6	7	8	9	10	11
Low Clients' Demand	3	2	1	1	2	3	2	3	3	1
Medium Clients' Demand	4	6	11	10	5	7	8	5	9	4
High Clients' Demand	12	14	20	18	11	16	10	21	17	15

$$x_{ij}^{k} \in \{0, 1\}, b_{i}^{k} \geq 0, y_{i}^{k} \geq 0 \quad (10)$$

It should be mentioned that client 1, throughout all instances, is considered as the depot.

In order to make a better analysis of the obtained results, it was decided to consider three different demands and three different vehicle capacities. The demands from the clients are presented in Table 2. Regarding vehicle capacity, low-capacity vehicles (three tons), medium-capacity vehicles (eleven tons), and high-capacity vehicles (twenty-four tons) were considered. Throughout the instances, as can be seen by analyzing Table 3, different demands and capacities were interspersed, obtaining the results duly presented.

The ΔTW stands for the time window variation. First, when analyzing the results obtained along each instance individually, it can be seen that the larger the time window used, the smaller the total distance travelled and the lower the total cost. In instances 1 and 3, both the distance and the total costs remain the same as the time windows are increased since the capacity of the vehicles used is relatively close to client demand. For this decrease to occur, client demand would have to be lower, the capacity of the vehicles used would have to be higher or the fleet of vehicles would have to be larger. Let us now analyze the results obtained between different instances. When there is a low demand, it can be seen that the total distance travelled and the total cost is lower for a higher capacity. Although, initially, for time windows of 30 and 90 minutes, the parameters analyzed are the same, in the case of a high or medium capacity, as the time windows increase, the values become more distant, leading to a better optimization for a fleet with a capacity of 24 tons. Analyzing the influence of different capacities for average demand, it is possible to denote that the distance and cost are lower for a fleet of vehicles with a higher capacity (24 tons), similar to what was verified for low demand. Thus, it can be concluded that with the increase in capacity, in general, and if the quantity of products to be transported justifies it, the total distance and the total cost decrease.

Table 3 Instances' results

Instance				Solution		
Number	Vehicle's capacity (Ton.)	Demand	ΔTW	Number of utilized vehicles	Distance (Km)	Cost (€)
1	3	Low	30	7 out of 7	672.5	**988.68**
	3	Low	90	7 out of 7	672.5	**988.68**
	3	Low	180	7 out of 7	672.5	**988.68**
	3	Low	240	7 out of 7	672.5	**988.68**
	3	Low	480	7 out of 7	672.5	**988.68**
2	11	Low	30	3 out of 7	405	595.35
	11	Low	90	3 out of 7	402.5	591.78
	11	Low	180	3 out of 7	391.25	575.19
	11	Low	240	2 out of 7	355	521.85
	11	Low	480	2 out of 7	340	**499.8**
3	11	Medium	30	7 out of 7	585	**2825.55**
	11	Medium	90	7 out of 7	585	**2825.55**
	11	Medium	180	7 out of 7	585	**2825.55**
	11	Medium	240	7 out of 7	585	**2825.55**
	11	Medium	480	7 out of 7	585	**2825.55**
4	24	Low	30	3 out of 7	405	595.35
	24	Low	90	3 out of 7	402.5	591.78
	24	Low	180	2 out of 7	376.25	553.14
	24	Low	240	2 out of 7	337.5	496.23
	24	Low	480	1 out of 7	315	**463.05**
5	24	Medium	30	4 out of 7	448.75	2167.29
	24	Medium	90	4 out of 7	437.5	2113.47
	24	Medium	180	4 out of 7	416.25	2010.66
	24	Medium	240	3 out of 7	397.5	1920.27
	24	Medium	480	3 out of 7	390	**1883.7**
6	24	High	30	7 out of 7	662.5	7142.52
	24	High	90	7 out of 7	650	7007
	24	High	180	7 out of 7	650	7007
	24	High	240	7 out of 7	650	7007
	24	High	480	7 out of 7	642.5	**6926.92**

4 Conclusion

The literature review was made in order to understand and frame the VRP and VRPTW problem, adding that nowadays the access to digital technology enables the use of these models to support decision-making and operations concerning transportation services on a supply chain. The development and evolution of these are essential for both the supplier and the client, as both benefit from obtaining a better solution. On the one hand, the client will be able to increase efficiency and decrease the operational costs related to the delivery process, using better models for routing problems. On the other hand, the client will have better service quality, this way, the satisfaction level, as well as retention in the company. Finally, when it comes to the results obtained through the developed analysis, it is possible to verify the supplier delivery cost impact with increasing and decreasing time windows. For a better analysis, the data relating to the delivery vehicles are not changed (it can be used the same number of vehicles and they have the same capacity) and client data (that being the number of clients and their demand are maintained). Through the obtained results, it is concluded that, on the one hand, the time windows should not be too small, as it can cause the optimal cost to be too high. In this way, a balance should be found between the interested parties. Concerning future work, it is intended to improve the developed model in order to take into consideration the level of client satisfaction, seeking to combine the minimization of the cost of charges with the maximization of service quality.

References

1. Ojeda Rios, B. H., Xavier, E. C., Miyazawa, F. K., Amorim, P., Curcio, E., & Santos, M. J. (2021). Recent dynamic vehicle routing problems: A survey. *Computers & Industrial Engineering, 160*, 107604.
2. RamachandranPillai, R., & Arock, M. (2021). Spiking neural firefly optimization scheme for the capacitated dynamic vehicle routing problem with time windows. *Neural Computing and Applications, 33*(1), 409–432.
3. Kumar, S. N., & Panneerselvam, R. (2012). A survey on the vehicle routing problem and its variants. *Intelligent Information Management, 4*(3), 66–74.
4. Okulewicz, M., & Mandziuk, J. (2019). A metaheuristic approach to solve dynamic vehicle routing problem in continuous search space. *Swarm and Evolutionary Computation, 48*, 44–61.
5. Zhang, H., Ge, H., Yang, J., & Tong, Y. (2022). Review of vehicle routing problems: Models, classification and solving algorithms. *Archives of Computational Methods in Engineering, 29*(1), 195–221.
6. Praveen, V., Keerthika, P., Sivapriya, G., Sarankumar, A., & Bhasker, B. (2022). Vehicle routing optimization problem: A study on capacitated vehicle routing problem. *Materials Today: Proceedings, 64*, 670–674.
7. He, P., & Hao, J.-K. (2022). General edge assembly crossover-driven memetic search for split delivery vehicle routing. *Transportation Science, 57*(2), 482–511.
8. Archetti, C., & Speranza, M. G. (2012). Vehicle routing problems with split deliveries. *International Transactions in Operational Research, 19*(1–2), 3–22.
9. Rios, B., Xavier, E., Miyazawa, F., Amorim, P., Curcio, E., & Santos, M. J. (2021). Recent dynamic vehicle routing problems: A survey. *Computers & Industrial Engineering, 160*, 107604.
10. Feillet, D., Dejax, P., & Gendreau, M. (2004). An exact algorithm for the elementary shortest path problem with resource constraints: Application to some vehicle routing problems. *Networks: An International Journal, 44*(3), 216–229.
11. Qing, Z., Limin, Q., Yingchun, L., & Shanjun, Z. (2006). An improved particle swarm optimization algorithm for vehicle routing problem with time windows. In *2006 IEEE Congress on Evolutionary Computation, CEC 2006* (pp. 1386–1390).

12. Miranda, D. M., & Conceição, S. V. (2016). The vehicle routing problem with hard time windows and stochastic travel and service time. *Expert Systems with Applications, 64*, 104–116.
13. Desaulniers, G., Madsen, O. B. G., & Ropke, S. (2014). Chapter 5: The vehicle routing problem with time windows. *Society for Industrial and Applied Mathematics, 2014*, 119–159.
14. Tas, D., Jabali, O., & Van Woensel, T. (2014). A vehicle routing problem with flexible time windows. *Computers & Operations Research, 52*(Part A), 39–54.

Reducing Error in Manufacturing in Industry 4.0: A Systematic Literature Review

Jacqueline Humphries, Alan Ryan, and Pepijn Van de Ven

Abstract Error has always presented a challenge for the quality team, but never more so than under Industry 4.0. In Industry 4.0, the operator is required to work on more complex products, be flexible across processes, and be interconnected to machines, dealing with streams of in-process data. This change in work has led to higher error and, therefore, greater challenges for the quality function. In order to understand how best to apply the capacity available under Industry 4.0 to ease this burden, it is important first to understand the current approach to quality and where the research gaps lie.

In this paper, an analysis of how error and quality has been managed in manufacturing, in the first 10 years of Industry 4.0, is completed, using a systematic literature review.

Seven thematic clusters are identified in this research from 64 selected papers. They are Quality Management Systems, Total Quality Management, Measurement and Analysis, Human Factors, Technology, Human-System Integration, and Training. A synthesis of the thematic clusters reveals sustaining innovations across all activities. The summary of current approaches will assist quality teams better to improve their approach to reduce error and will identify where there are future research opportunities.

Keywords Error · Quality · Manufacturing · Industry 4.0

J. Humphries (✉)
Technological University of the Shannon, Limerick, Ireland

Confirm Manufacturing Research Centre, Limerick, Ireland
e-mail: jacqueline.humphries@tus.ie

A. Ryan
University of Limerick, Limerick, Ireland

P. Van de Ven
University of Limerick, Limerick, Ireland

Confirm Manufacturing Research Centre, Limerick, Ireland

J. Reis et al. (eds.), *Driving Quality Management and Sustainability in VUCA Environments*, Springer Proceedings in Business and Economics,
https://doi.org/10.1007/978-3-031-52723-4_14

1 Introduction

Errors are recurring factors in manufacturing [1]. Yet if errors are undetected, they may be propagated downstream, diminish the quality of final products, increase the cost of production, and present ongoing challenges in relation to production validation and quality checking [2]. While Industry 4.0 has brought about disruptive changes and increased capacity in manufacturing [3, 4], it has not had the same impact on the reduction of error for the quality function. Indeed, the advances under Industry 4.0 have brought about further quality control issues [5]. The large range of products has a corresponding range of particular quality specifications, with streams of in-process data [6]. Furthermore, Industry 4.0 causes increased complexity for humans. Not only do they have more complex standard operating procedures (SOPs) to follow, but they are also required to collaborate in teams and with machines, be flexible, and work under dynamic conditions [5]. Flexibility results in shifting the human between different processes, and this has an impact on quality of work [7]. One reason is that highly flexible assemblies have an inherent uncertainty in the process which increases the risk of error [8]. This requires the human to locate and follow unfamiliar SOPs. The very nature of alternating attention between reading assembly instructions on the SOP and executing the assembly process is seen to generate new defect sources and increase the likelihood of error [2, 9]. Where the assembly instructions are particularly complex, human error becomes more of a problem [10, 11].

The reasons why errors occur, such as lack of expertise or fatigue, and classifications of error types are well defined in previous works [12, 13]. There is, however, a research gap as to how to manage error in the disruptive era of Industry 4.0. The objective of this paper is to understand how error is being managed in manufacturing, within the context of the first 10 years of Industry 4.0. The impact of this paper is that this knowledge will inform future research on where opportunities lie to further reduce error.

2 Method

Relative to other fields, there are a limited number of literature reviews in the field of engineering, with systematic literature reviews (SLR) rarely used [14]. The benefit of an SLR is that it provides an objective search process. Therefore, an SLR is conducted to answer the research question: what has been done in the first 10 years of Industry 4.0 to reduce error?

The approach undertaken to execute the SLR is influenced by the strategies in Wu, Goepp, and Siadat [14] and uses the Preferred Reporting Items for Systematic Reviews and Meta-analysis Protocols (PRIMSA) 2015 [15]. The process includes (i) selection of a database, (ii) selection of keywords and construction of query string, (iii) collection of articles, (iv) development of exclusion criteria and filtering of papers, (v) development of codes and identification of codes in all remaining papers, and (vi) critical analysis of full text within each coded category.

The Institute for Scientific Information Web of Science database, as recommended by Wu, Goepp, and Siadat [14], is chosen as it is commonly aligned with this topic and provides an interdisciplinary perspective. The database contains 23,000 journals and 110,000 proceedings, with 40 million source items and 700 million cited references. It contains a significant number of renowned publications in this field such as the Institute of Electrical and Electronic Engineers (IEEE), Springer, Taylor and Francis, and Elsevier [16]. With regard to the quality of the content, content for the database is subject to an evaluation and selection process based on impact, influence, peer review, timeliness, and location.

The keywords in the search string are abstracted following a review of the success of pilot literature searches, analyzing the research question, considering all synonyms and alternative spellings, and inclusion of keywords found in new relevant papers. As to context, focus is on manufacturing and production that use humans in the process. The industrial context is wide-ranging and includes in the search terms features of Industry 4.0 such as Smart Collaborative technologies. The focus is the first 10 years of Industry 4.0, and for that reason, only articles published in the period between 2011 and 2021 are considered. The following search string is used:

TITLE = ((“smart factory” OR “manufacturing” OR “*Factor*” OR “Industry*” OR “production” OR “cyber*physical” OR “collaborat*” OR “e*Manu facturing”)) AND.
TITLE = ((“quality” OR “error”)) AND.
TOPIC = ((“operator*” OR “human” OR “labo*r” OR “non*automated”)).

The search returns 175 papers. The data records for these are then exported in a text file format and analyzed in Microsoft Excel. Preliminary screening is conducted to evaluate the papers on the exclusion criteria by a review of the abstract, introduction, and conclusion. The exclusion criteria used are as follows:

1. The work is not related to a manufacturing system that uses the human in the process.
2. The search terms are mentioned as background or as context and do not form a substantial part of the text.
3. The study is not a technological scientific paper.
4. There is no access to full text.
5. The work is similar to, or a duplicate of, another article.
6. The work has been retracted.

One hundred and eleven papers are excluded. Of the remaining papers, a quality review has been conducted by review of the whole paper. In total, 64 papers are deemed relevant for the review. A hybrid inductive-deductive thematic analysis is conducted using a semantic approach.

Firstly, inductive analysis is completed, which allows for themes to emerge from the data. The papers are scan read to familiarize ourselves with the texts. Using the semantic approach, the explicit content of the papers are analyzed. Codes are identified within each paper. There is some revision of the codes when similarities with other papers are uncovered. Key themes are identified on the codes. The themes are given a descriptive name which reflects the identified codes.

Secondly, deductive analysis is completed on the results. The themes relevant to the research question are identified. The themes emerging are compared with theory and existing knowledge, which ensures that the themes are redefined using the industry nomenclature. For instance, one theme identified in the inductive phase is "continuous improvement." In industry, the term "Total Quality Management" is used, and so our themes are renamed. The themes found are Measurement and Analysis, Total Quality Management, Human Factors, Quality Management Systems, Technology, Human-System Integration, and Training. The deductive phase allowed for regrouping of the themes into approaches, into the traditional "industrial engineering" approach, and a more modern "interdisciplinary" approach. Following descriptive statistics, each of these themes is synthesized below.

3 Descriptive Results

This review identifies seven thematic clusters in the literature, illustrated in Fig. 1. The traditional industrial engineering approaches to managing quality account for 74% of all papers. Industrial engineering is made up of Quality Management Systems (6%), Total Quality Management (20%), Measurement and Analysis (27%), Human Factors (14%), and Training (8%). The remaining 26% of the total papers are interdisciplinary approaches, which includes Technology (15%) and Human-Technology Integration (11%). Note the percentages are rounded to the nearest whole number.

Journal publications are made up 38% of works. Book series are made up 45% of the works. The remaining 17% were in conference proceedings. Of the contributions by country, derived from conference location, the USA has the greatest number with

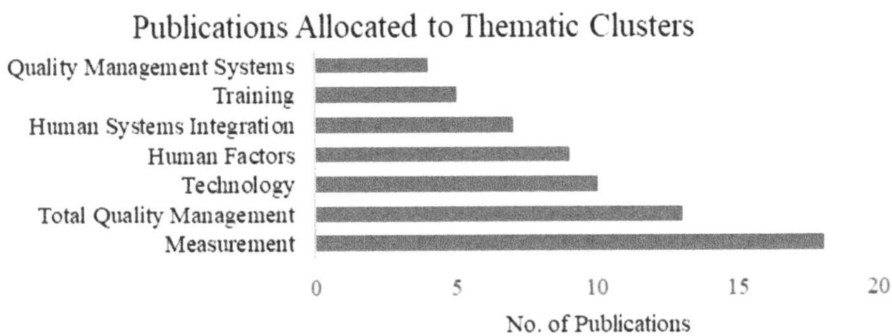

Fig. 1 Number of publications allocated to each thematic cluster

22% of the total, followed by the Peoples Republic of China with 11% and with India, the UK, and Indonesia with 7% each. However, when the Asian countries are grouped, the total from the region makes up 38% of the total.

There is an upward trend in research activity in the years since and including 2016, which accounts for 75% of all papers. What is found is that the papers in the latter 5 years are applying newer technologies such as machine learning and big data to the traditional quality management methods; thus, there is sustaining innovation across all of the industrial engineering approaches.

4 Measures to Address Quality in Manufacturing

Each of the thematic clusters is now described.

4.1 Quality Management Systems

The first thematic cluster relates to Quality Management Systems (QMS). The QMS is important in describing how consistently high quality can be achieved. The QMS detail the processes to be followed, the range of tools and systems to be used, the quality policies and procedures, the quality plan, and how the Quality Control (QC) function will assess, inspect, and audit products [17, 18]. The importance of the QMS is evidenced in how it is a crucial element of compliance as set down by regulatory bodies [19]. The challenge for the QMS is that Industry 4.0 brings about the capability, capacity, and demand for flexible manufacturing, yet the QMS is relatively static [20]. A gap exists to consider how the QMS can be more responsive where flexibility in production is expected.

4.2 Total Quality Management

Whereas QMS set out the processes and systems to be followed, Total Quality Management (TQM) considers how the QMS can be improved. Thus, the second thematic cluster relates to TQM, which is the most common practice in industry for reducing the opportunities for human error [21]. Lean mechanisms and Poka-Yoke, or mistake proofing, still feature strongly [12, 22–26]. The research shows that examining current processes to identify where improvements can be made can result in creative solutions to make the process more effective, efficient, or transparent [19, 27–29]. Fundamental to TQM activities is deferring to the human who knows the job well to identify where the continuous improvement opportunities lie [30]. TQM then feeds back up to the QMS the improved method of working which minimizes the risk for error.

Proposals for improvement cannot be considered in isolation. For example, in [31], a change in methods that improves the speed of production led to a knock-on effect of increasing human error. An alternative to TQM in the literature is to substitute the human by automation [32]. Quality can be improved by letting the human focus on what they do best and allow a machine to do the rest, thus reducing the opportunity for error. However, substitution must add value. There is a risk that the highly relevant skills and competencies of the skilled human, quality manager, and their quality circles are being substituted by cheaper labor and technologies [19]. The loss of these skills and competencies is detrimental to the quality function [33].

4.3 Measurement and Analysis

The third thematic cluster references measurement and analysis of error. Measurement and analysis of error inform efforts to reduce it in TQM, thus ensuring the preservation of quality throughout the whole process. Measurement and analysis was the most prolific cluster in the results, with 27% of all papers considering the statistical processes to sample, measure, and evaluate risks for error. The first consideration is when sampling occurs [34, 35]. QC often uses regular or irregular sampling inspection techniques. However, because of the time interval between samples being taken, a large number of defective products are often produced [7]. After determining where the measurement will take place, the next step is how the work is assessed. In some use-cases, timely quality inspection during the production process can now be performed using geometric measurement where dimensions of the product may be measured, such as volume or size [36]. Automated quality checks can also be done by comparison of the actual product with a computer-aided design model [37] and the use of high-precision measurement instruments such as coordinate measuring machines; however, these are not immune to human error themselves [38].

An updated taxonomy of faults that considers advances under Industry 4.0, and includes those originating from human error, is proposed in [13]. An updated taxonomy of human error causation is proposed in [39], and classification methods for human error are synthesized in [40]. Case studies applying the traditional methodologies for measuring error regularly appear in the literature [41–46]. However, challenges with measurement with respect to data collection and validation are reported [45]. Human error quantification techniques are based either on error databases or based on subjective expert judgment [41]. These are thought to be flawed as the assessment may be subjective [47], and the error databases are not complete due to unavailability of reliable data [45]. Thus, the common issue around measurement is the difficulty in conclusively assessing and correctly labeling the cause of the error. It is reported in [40] that up to 90% of failures have a direct or indirect relationship with the human at some part in the chain. However, two of the fathers of quality, Juran and Deming, posit that in fact 80% and 85%, respectively, of all quality problems are in fact management problems, i.e. symptoms of a malfunctioning system. Yet, as in [40], there is a tendency to over blame the human worker [39]. The likelihood is that the error is caused by something outside of their control, such as a procedural or system error or a source of error inherent in the design of the system such as speed [44, 48, 49]. Often the reason for the error is multivariate [50]. Statistically significant associations between human error and unsafe conditions which include machinery, tools, equipment, and organizational factors exist. The likelihood of a human error occurring is 4.032 times more likely when another contributing factor is present, than when it is not [51]. The research gap remains as to how intelligently to determine the cause of error so that TQM can be applied to address it.

4.4 Technology

The fourth thematic cluster considers technology and in particular digitalization of the factory floor, a key enabler for achieving quality in processes. High-quality data collected from sensors is processed using prediction tools to assist the human through displays or warnings [52]. Multistage QC based on sensor data improves single stage quality assurance checks [53], with quality issues being flagged quicker and earlier than by the human [54]. Commonly, deep learning and/or machine vision is being used for visual quality inspection and measurement [55, 56]. Machine vision provides an alternative to human quality inspection by means of an automated, nondestructive, fast, and cost-effective technique to conduct accurate, fast, and objective quality checks. Automated visual inspection, at the assembly workstation, shifts inspection from centralized to in-line. In [32] a computer vision quality system has an accuracy of 98.9%, 10% higher than the average reported for the human. Defective parts can be highlighted by means of a light beam on those parts as they move along a conveyor belt as in [57]. Alternatively, the error probability can be displayed to the human who decides if there is in fact an error or not [58]. Human activity recognition and/or eye tracking can evaluate whether the human correctly executed the production process, rather than perform a visual inspection of parts [59–61].

Machine learning is being used to manage the quality of process documentation. For example, computerized systems can automate the documentation of work steps, thus creating an audit trail [20]. This audit trail can be captured in documentation, image, tagging, or video format. When a human completes a task, in the background, machine learning can flag that as having been completed without error and within the quality tolerances. A simple gesture recognition system permits the human to identify detected defects on the object by pointing to it with their finger. This gesture attaches a tag or highlights the offending area of the product for a quality check [62]. Frame markers can be used to highlight errors which are then picked up later [63].

Technology appears to be being used to substitute the human on tasks or check the work of the human in a monitoring capacity. Kamble, Gunasekaran, and

Gawankar (2018) warn to expect resistance from the human for these interactions [64]. This is confirmed in the research of Kaasinen et al. (2020) who interviewed 44 operators and report that all participants are resistant toward systems that monitor their work through data collection systems [65]. A research gaps exists to find a way to use technology to augment the human to do their work better, rather than as being used to substitute the operator or in a performance-monitoring capacity.

4.5 Human-Systems Integration

The fifth thematic cluster considers Human-Systems Integration (HSI), an interdisciplinary approach to optimizing all human-related considerations in the design, development, production, and operation of systems. Although Industry 4.0 calls for new levels of socio-technical interaction between the human and technology [4, 66], most of the research returned in this thematic cluster is concerned with interface design. There are sustaining innovations in adaptive interfaces, codifying and personalizing information for particular users, designed in line with principles stemming from the Human-Machine Cooperation Systems approach [67, 68]. Interfaces range from haptic and visual signals such as alarms, warning lights, pick by light, and screen-based displays through to augmented and virtual reality environments [67, 69]. The more explanation the assistance system provides about its analysis of the process, rather than just displaying images of current and projected views, the more the human detects and corrects the error [67, 70, 71]. Trust in the system is important [72]. Research activity in HSI needs to broaden away from primarily interaction and interface design, toward a wider application of HSI to help achieve human-computer symbiotic integration.

4.6 Human Factors

The sixth thematic cluster considers Human Factors Engineering, a scientific discipline that considers the design and development of methods, systems, interfaces, user-experience, and tools from the user experience perspective, adapting the process to fit to its users [73]. Good human factors design is shown to have many benefits, including improving the user experience and reducing the number of human errors being made [10, 41, 74–76]. The corollary holds true. Poor human factors design causes quality losses in production [77]. Further, quality and production improvement initiatives often have the adverse effect of worsening working conditions from the human factor perspective. Complex technology systems or additional automation in the production processes can lead to increased stress, demanding cognitive workload, and fatigue for the human [78]. This then has a negative impact on quality of work being done. This impact is quantified in Farid and Neumann (2020) [75]. Over a 5-year simulation period with unresolved human factors, human error increased by 40%. With over 200 different human factor variables associated with quality [75], attempts are being made to identify the most influential factors on human error and product quality and where opportunities lie for quality improvements in TQM. Statistical models of human factors and their impact on quality are being developed [79, 80]. Of particular significance to human error are those human factors related to production system design, workstation design, training, and issues related to operation management [75]. A link between physical workload and poor quality is reported in 26% of studies [77]. Having said that, human fatigue is consistently seen as the most significant contributing factor to quality deficits in half of studies reviewed with visual fatigue being the most commonly reported problem [74, 75, 77, 81]. The research activity around human factors appears to be siloed from efforts in TQM and the other thematic clusters, where the human factor is ignored or treated as a risk. The research gap remains as to how to drive innovation in the other clusters from the human factor perspective.

4.7 Training

The seventh and final cluster concerns training. Training is only one element of the human capital capacity which is at the foundation of high-quality work [82]. Training prepares and improves the quality and competence of human resources and has a positive effect on the quality of work [49, 83]. Training in quality processes is found to reduce the risk of the human making procedural errors and allows them better to collaborate in the quality team and interpret requirements of the user [84]. Training is found to increase error detection awareness, when the training focuses on the types of error that are commonly made and how those errors can be identified. Error awareness training allows the human to develop prevention or mitigation mechanisms so they avoid committing the same error again [85]. However, the literature warns that training for quality has its limitations. While learning has a positive impact on productivity, it does not always have a positive impact on quality [83].

5 Conclusion

The theoretical contribution of this paper is that it is the first objective review that considers approaches to reducing error and improving quality during the first 10 years of Industry 4.0. Seven thematic clusters of research activity are identified. These clusters can be further categorized as belonging to a traditional industrial engineering approach and an interdisciplinary approach. The traditional approach accounts for 74% of all papers and includes Measurement and Analysis, Total Quality Management, Quality Management Systems, Human Factor Engineering, and Training initiatives. An interdisciplinary approach to error is present in 26%

of the literature, further classified into Technology and Human-System Integration. From a practical contribution, the importance of the traditional approaches is emphasized in the results. These tried and tested methods do work to some extent to manage error. They have been improved with sustaining innovations leveraging Industry 4.0 capacity. In particular, data-driven intelligence, through digitalization, integration, collection, and mining, is evident across the whole spectrum of activity. Most of the focus in the technology category, 21% of all works, is in the design and testing of artificial intelligence and machine learning models. They are often used in computer vision for quality checking or to automate the work, thus substituting the human. Thus, the digital transformation capacity is slowly making inroads into the quality function. However, it is concerning that quality management initiatives are using sustaining innovations to manage the disruptive innovation of Industry 4.0 on the manufacturing floor. The disruption to the manufacturing function is not being experienced to the same extent in the quality function. Quality 4.0 has a long way to go to realize its potential.

5.1 Future Research Perspectives

It is evident in the research that an integrated view of quality management is essential. Quality efforts need the rigor of the Industrial Engineering approaches—Quality Management Systems, Total Quality Management, Measurement and Analysis, Human Factors, and Training. Yet they need the innovation offered by Technology and new methods of Human-Technology Integration. A new model of Quality Management is required that incorporates the thematic clusters revealed herein. The traditional models do not reflect practice. Future research may constrain the models of best practices to organizations with high levels of Industry 4.0 readiness. A new model will inform future quality initiatives and provide guidance as to the skills and competencies required for those in the quality function.

Further, there is a significant research gap with respect to the role of the human across many of the thematic clusters. Although the human introduces error, human performance is seen as critical element to ensuring quality. The literature reveals a tendency to label errors as "human error," when the reality is that the error is likely to be as a result of multiple factors, such as inadequate systems or processes, poor human factors, or organizational culture. It is critical that the systems become human-centric. The importance of human centricity is evident in that it is seen as a key goal of Industry 5.0. Future research should consider how to apply that innovation in better ways, to create trusted socio-technical solutions that keep the human central in the quality process.

5.2 Limitations

This paper is based on a literature review of one database, Web of Science, which limits the results. As it is an interdisciplinary database, the results may have been skewed in that way. This should be validated by expert analysis or the search query replicated on alternative databases. This review did not consider contextual differences of the industries or the Industry 4.0 readiness level. It is to be expected that there are huge variations in practices across organizations.

The query string used in the search may have limited the results. There is an uncertainty around the terms used to represent the context of Industry 4.0. Further consideration of papers in the English language only may further limit the results. Validation of the results by use of more synonyms and other languages could be undertaken.

References

1. Su, Q., Liu, L., & Whitney, D. E. (2010). A systematic study of the prediction model for operator-induced assembly defects based on assembly complexity factors. *IEEE Transactions on Systems, Man, and Cybernetics – Part A: Systems and Humans, 40*, 107–120. https://doi.org/10.1109/TSMCA.2009.2033030
2. Torres, Y., Nadeau, S., & Landau, K. (2021). Classification and quantification of human error in manufacturing: A case study in complex manual assembly. *Applied Sciences, 11*, 749. https://doi.org/10.3390/app11020749
3. World Economic Forum. (2021). *Fourth industrial revolution*. In: World Econ. Forum website. https://www.weforum.org/focus/fourth-industrial-revolution. Accessed 22 June 2021.
4. Kagermann H, Wahlster W, Helbig J, et al. (2013). *Recommendations for implementing the strategic initiative INDUSTRIE 4.0: Securing the future of German manufacturing industry*.
5. Longo, F., Nicoletti, L., & Padovano, A. (2017). Smart operators in industry 4.0: A human-centered approach to enhance operators' capabilities and competencies within the new smart factory context. *Computers and Industrial Engineering, 113*, 144–159. https://doi.org/10.1016/j.cie.2017.09.016
6. Giannetti, C., & Ransing, R. S. (2016). Risk based uncertainty quantification to improve robustness of manufacturing operations. *Computers and Industrial Engineering, 101*, 70–80. https://doi.org/10.1016/j.cie.2016.08.002
7. Tseng, C. C. (2020). The practice path of high quality intelligent manufacturing in Taiwan. In *Proceedings of the 2020 4th international conference on management engineering, software engineering and service sciences* (pp. 277–281). ACM.
8. Kristiansen, E., Nielsen, E. K., Hansen, L., & Bourne, D. (2020). A novel strategy for automatic error classification and error recovery for robotic assembly in flexible production. *Journal of Intelligent and Robotic Systems, 100*, 863–877. https://doi.org/10.1007/s10846-020-01248-3
9. Krugh, M., & Mears, L. (2018). A complementary cyber-human systems framework for industry 4.0 cyber-physical systems. *Manufacturing Letters, 15*, 89–92. https://doi.org/10.1016/j.mfglet.2018.01.003
10. Comberti, L., Demichela, M., & Leva, M. C. (2018). In S. Haugen, A. Barros, C. van Gulijk, et al. (Eds.), *A multi-discipline method to assess the human performance in manufacturing industry for safety and quality optimization*. CRC Press.

11. Galetto, M., Verna, E., & Genta, G. (2020). Accurate estimation of prediction models for operator-induced defects in assembly manufacturing processes. *Quality Engineering, 32*, 595–613. https://doi.org/10.1080/08982112.2019.1700274
12. Sondermann, J. P., & Yoke, P. (2013). *Prinzip und Techniken für Null-Fehler_Prozesse*. Hanser.
13. Badr, I., & Vogel-Heuser, B. (2018). Taxonomy of faults influencing the quality of automation production systems. *IFAC-PapersOnLine, 51*, 95–102. https://doi.org/10.1016/j.ifacol.2018.06.243
14. Wu, X., Goepp, V., & Siadat, A. (2020). Concept and engineering development of cyber physical production systems: A systematic literature review. *International Journal of Advanced Manufacturing Technology, 111*, 243–261. https://doi.org/10.1007/s00170-020-06110-2
15. PRISMA. (2015). *Transparent reporting of systematic reviews and meta-analyses*. http://www.prisma-statement.org/. Accessed 12 May 2021.
16. Reuters, T. (2010). A versatile workflow solution ISI Web of knowledge. Integrated access to high quality, multidisciplinary research literature.
17. Albers, A., Gladysz, B., Pinner, T., et al. (2016). Procedure for defining the system of objectives in the initial phase of an industry 4.0 project focusing on intelligent quality control systems. *Procedia CIRP, 52*, 262–267. https://doi.org/10.1016/j.procir.2016.07.067
18. Daneshjo, N., Malega, P., & Drábik, P. (2021). Techniques for production quality control in the global company. *Advances in Science and Technology Research Journal, 15*, 174–183. https://doi.org/10.12913/22998624/131558
19. Romero, D., Gaiardelli, P., Powell, D., et al. (2019). *Total quality management and quality circles in the digital lean manufacturing world* (pp. 3–11).
20. Bihi, T., Luwes, N., & Kusakana, K. (2018). Innovative quality management system for flexible manufacturing systems. In *2018 Open innovations conference (OI)* (pp. 40–46). IEEE.
21. Hassan, A. S., & Jaaron, A. A. M. (2021). Total quality management for enhancing organizational performance: The mediating role of green manufacturing practices. *Journal of Cleaner Production, 308*, 127366. https://doi.org/10.1016/j.jclepro.2021.127366
22. Salleh, N. A. M., Kasolang, S., & Jaffar, A. (2012). Simulation of integrated Total quality management (TQM) with lean manufacturing (LM) practices in forming process using Delmia quest. *Procedia Engineering, 41*, 1702–1707. https://doi.org/10.1016/j.proeng.2012.07.371
23. Lyu, J., Chen, P.-S., & Huang, W.-T. (2021). Combining an automatic material handling system with lean production to improve outgoing quality assurance in a semiconductor foundry. *Production Planning and Control, 32*, 829–844. https://doi.org/10.1080/09537287.2020.1769217
24. Cabrera, J., Corpus, O., Maradiegue, F., & Merino, J. C. A. (2020). Improving quality by implementing lean manufacturing, SPC, and HACCP in the food industry: A case study. *South African Journal of Industrial Engineering, 31*. https://doi.org/10.7166/31-4-2363
25. Sulistiyowati, W., Adamy, M. R., & Jakaria, R. B. (2019). Product quality control based on lean manufacturing and root cause analysis methods. *Journal of Physics Conference Series, 1402*, 022038. https://doi.org/10.1088/1742-6596/1402/2/022038
26. Kurata, Y. B, Navales, M. R. & Cedron, D. B., et al. (2018). *Quality assurance production based problem: A process improvement in the rolling mill line for steel manufacturing company in The Philippines*, pp. 74–82.
27. Modgil, S., & Sharma, S. (2016). Total productive maintenance, total quality management and operational performance. *Journal of Quality in Maintenance Engineering, 22*, 353–377. https://doi.org/10.1108/JQME-10-2015-0048
28. Antsev, V. Y., Vitchuk, N. A., & Miroshnikov, V. V. (2017). Improvement in production process for pipelines manufacturing based on quality management method. *Procedia Engineering, 206*, 950–957. https://doi.org/10.1016/j.proeng.2017.10.577
29. Yang, H. (2012). Control technique research of key quality characteristic during the mechanical and electrical production manufacturing process. *Advances in Materials Research, 442*, 215–219. https://doi.org/10.4028/www.scientific.net/AMR.442.215
30. Blaga, P. (2020). The importance of human resources in the continuous improvement of the production quality. *Procedia Manufacturing, 46*, 287–293. https://doi.org/10.1016/j.promfg.2020.03.042

31. Emami-Mehrgani, B., Neumann, W. P., Nadeau, S., & Bazrafshan, M. (2016). Considering human error in optimizing production and corrective and preventive maintenance policies for manufacturing systems. *Applied Mathematical Modelling, 40*, 2056–2074. https://doi.org/10.1016/j.apm.2015.08.013
32. Ng, C. K., Wu, C. H., Ip, W. H., et al. (2011). A real time quality monitoring system for the lighting industry: A practical and rapid approach using computer vision and image processing (CVIP) tools. *International Journal of Engineering Business Management, 3*, 22. https://doi.org/10.5772/45670
33. Pfeiffer, S. (2016). Robots, industry 4.0 and humans, or why assembly work is more than routine work. *Societies, 6*, 16. https://doi.org/10.3390/soc6020016
34. Madadi, N., & Wong, K. Y. (2013). A deterministic aggregate production planning model considering quality of products. *IOP Conference Series: Materials Science and Engineering, 46*, 012015. https://doi.org/10.1088/1757-899X/46/1/012015
35. Wang, S.-M., Lee, C.-Y., Gunawan, H., & Yeh, C.-C. (2021). On-line error-matching measurement and compensation method for a precision machining production line. *International Journal of Precision Engineering and Manufacturing*. https://doi.org/10.1007/s40684-021-00336-5
36. Oezden, R., Popa, M. S., Kortekamp, M., & Aclan, M. (2021). Correlation of geometrical specifications for flexible quality control in the manufacturing of plastic products. *Teh Vjesn – Tech Gaz, 28*, 1456+.
37. Panda, B. N., Bahubalendruni, R. M., Biswal, B. B., & Leite, M. (2017). A CAD-based approach for measuring volumetric error in layered manufacturing. *Proceedings of the Institution of Mechanical Engineers, Part C: Journal of Mechanical Engineering Science, 231*, 2398–2406. https://doi.org/10.1177/0954406216634746
38. Cajal, C., Santolaria, J., Samper, D., & Velazquez, J. (2016). Efficient volumetric error compensation technique for additive manufacturing machines. *Rapid Prototyping Journal, 22*, 2–19. https://doi.org/10.1108/RPJ-05-2014-0061
39. Martinez, R. M. R., de la Riva, R. J., de la Escapita, R., et al. (2017). *Development of a taxonomy of human error causation of accidents involving injuries to hands in the harnesses industry* (pp. 317–324). Advances in Social & Occupational Ergonomics.
40. Shi, Z., Xie, L., & Li, X. (2015). Research of human error evaluation technique of production system based on PFMEA. In *2015 IEEE 10th conference on industrial electronics and applications (ICIEA)* (pp. 1815–1820). IEEE.
41. Sheikhalishahi, M., Azadeh, A., Pintelon, L., et al. (2016). Maintenance scheduling optimization in a multiple production line considering human error. *Human Factors and Ergonomics in Manufacturing & Service Industries, 26*, 655–666. https://doi.org/10.1002/hfm.20405
42. Sembiring, N., Tambunan, M. M., & Febriani, M. (2019). Human error analysis on production process of door products with SHERPA and HEART method. *IOP Conference Series: Materials Science and Engineering, 505*, 012025. https://doi.org/10.1088/1757-899X/505/1/012025
43. Dhillon, B. S. (2014). Human error in maintenance: An investigative study for the factories of the future. *IOP Conference Series: Materials Science and Engineering, 65*, 012031. https://doi.org/10.1088/1757-899X/65/1/012031
44. Di Pasquale, V., Franciosi, C., Lambiase, A., & Miranda, S. (2016). Methodology for the analysis and quantification of human error probability in manufacturing systems. In *2016 IEEE student conference on Research and Development (SCOReD)* (pp. 1–5). IEEE.
45. Böllhoff, J., Metternich, J., Frick, N., & Kruczek, M. (2016). Evaluation of the human error probability in cellular manufacturing. *Procedia CIRP, 55*, 218–223. https://doi.org/10.1016/j.procir.2016.07.080
46. Selvik, J. (2016). On use of the international standard ISO 14224 on reliability data collection in the oil and gas industry: How to consider failure causes from a human error perspective. In *Risk, reliability and safety: Innovating theory and practice* (Vol. 33487-2742, pp. 1003–1009). CRC Press, Taylor & Francis Group.

47. Zhang, M., Zhang, D., Yao, H., & Zhang, K. (2020). A probabilistic model of human error assessment for autonomous cargo ships focusing on human–autonomy collaboration. *Safety Science, 130*, 104838. https://doi.org/10.1016/j.ssci.2020.104838
48. Ahmed, A., Guozhu, J., Salam, A., & Nawaz, M. K. (2019). Reliability and quality control approach for collaborative assembly process. In *2019 16th international Bhurban conference on applied sciences and technology (IBCAST)* (pp. 210–217). IEEE.
49. Anu, V., Walia, G., Bradshaw, G., & Alqudah, M. (2019). Developing and evaluating learning materials to introduce human error concepts: Results from industry and academia. In *2019 IEEE Frontiers in education conference (FIE)* (pp. 1–9). IEEE.
50. Huang, Y., Dai, W., Chi, Y., & Zhao, Y. (2016). Quality entropy model of the numerical control manufacturing process. In *2016 11th international conference on reliability, maintainability and safety (ICRMS)* (pp. 1–8). IEEE.
51. Reyes, R. M., de la Riva, J., Maldonado, A., et al. (2015). Association between human error and occupational accidents' contributing factors for hand injuries in the automotive manufacturing industry. *Procedia Manufacturing, 3*, 6498–6504. https://doi.org/10.1016/j.promfg.2015.07.936
52. Lee, S. H., & Haider, A. (2012). Applying six sigma methodology to improve quality of information: Case of a manufacturing organisation. In *Technology Management for Emerging Technologies (PICMET)* (pp. 3368–3378). Vancouver.
53. Bakker, O. J., Ratchev, S. M., & Popov, A. A. (2015). Towards a condition-monitoring framework for quality Assurance in Intelligent Multistage Manufacturing EnvironmentÕ. *IFAC-PapersOnLine, 48*, 2089–2094. https://doi.org/10.1016/j.ifacol.2015.06.397
54. Peres, R. S., Barata, J., Leitao, P., & Garcia, G. (2019). Multistage quality control using machine learning in the automotive industry. *IEEE Access, 7*, 79908–79916. https://doi.org/10.1109/ACCESS.2019.2923405
55. Ravi, V. R., Baig, A. M. A., Udhaya Karthick, G., & Thiruchittrambalam, S. (2014). A real time quality monitoring system for engineering industry: A practical and rapid approach using embedded vision system with LabVIEW. *Applied Mechanics and Materials, 592–594*, 2129–2133. https://doi.org/10.4028/www.scientific.net/AMM.592-594.2129
56. El Hachem, C., Perrot, G., Painvin, L., & Couturier, R. (2021). Automation of quality control in the automotive industry using deep learning algorithms. In *2021 international conference on computer, control and robotics (ICCCR)* (pp. 123–127). IEEE.
57. Pierer, A., Wiener, T., Gjakova, L., & Koziorek, J. (2021). Zero-error-production through inline-quality control of presshardened automotive parts by multi-camera systems. *IOP Conference Series: Materials Science and Engineering, 1157*, 012074. https://doi.org/10.1088/1757-899X/1157/1/012074
58. Villalba-Diez, J., Schmidt, D., Gevers, R., et al. (2019). Deep learning for industrial computer vision quality control in the printing industry 4.0. *Sensors, 19*, 3987. https://doi.org/10.3390/s19183987
59. Wang, Z., Qin, R., Yan, J., & Guo, C. (2019). Vision sensor based action recognition for improving efficiency and quality under the environment of industry 4.0. *Procedia CIRP, 80*, 711–716. https://doi.org/10.1016/j.procir.2019.01.106
60. Niemann, J., Fussenecker, C., & Schlösser, M. (2019). Eye Tracking for Quality Control in Automotive Manufacturing. In: EuroSPI 2019: Systems. *Software and Services Process Improvement*, 289–298.
61. Li, G., Shi, J., Luo, H., & Tang, M. (2013). A computational model of vision attention for inspection of surface quality in production line. *Machine Vision and Applications, 24*, 835–844. https://doi.org/10.1007/s00138-012-0429-1
62. Wagner, P. G., Lengenfelder, C., Holzbach, G., et al. (2021). Secure and privacy-respecting documentation for interactive manufacturing and quality assurance. *Applied Sciences, 11*, 7339. https://doi.org/10.3390/app11167339
63. Csokmai, L., Moldovan, O., Tarca, I. C., & Tarca, R. (2013). Software framework for advanced error troubleshooting in flexible manufacturing system. *Applied Mechanics and Materials, 397–400*, 21–24. https://doi.org/10.4028/www.scientific.net/AMM.397-400.21

64. Kamble, S. S., Gunasekaran, A., & Gawankar, S. A. (2018). Sustainable industry 4.0 framework: A systematic literature review identifying the current trends and future perspectives. *Process Safety and Environment Protection, 117*, 408–425. https://doi.org/10.1016/j.psep.2018.05.009
65. Kaasinen, E., Schmalfuß, F., Özturk, C., et al. (2020). Empowering and engaging industrial workers with operator 4.0 solutions. *Computers and Industrial Engineering, 139*, 105678. https://doi.org/10.1016/j.cie.2019.01.052
66. Stadnicka, D., Litwin, P., & Antonelli, D. (2019). Human factor in industry of the future: Knowledge acquisition and motivation. *FME Transactions, 47*, 823–830. https://doi.org/10.5937/fmet1904823S
67. Pacaux-Lemoine, M. P., Poulain, T., & Debernard, S. (2013). Human-Machine Cooperation and manufacturing system. In *IECON 2013 - 39th Annual Conference of the IEEE Industrial Electronics Society* (pp. 8288–8293). IEEE.
68. Trentesaux D, & Millot P (2016). *A human-centred design to break the myth of the "magic human" in intelligent manufacturing systems*, pp. 103–113.
69. Borisov, N., Weyers, B., & Kluge, A. (2018). Designing a human machine Interface for quality Assurance in car Manufacturing: An attempt to address the "functionality versus user experience contradiction" in professional production environments. *Advanced Human-Computer Interact, 2018*, 1–18. https://doi.org/10.1155/2018/9502692
70. Pacaux-Lemoine, M.-P., & Trentesaux, D. (2019). Ethical risks of human-machine Symbiosis in industry 4.0: Insights from the human-machine cooperation approach. *IFAC-PapersOnLine, 52*, 19–24. https://doi.org/10.1016/j.ifacol.2019.12.077
71. Petersen, N., & Stricker, D. (2015). Cognitive augmented reality. *Computers and Graphics, 53*, 82–91. https://doi.org/10.1016/j.cag.2015.08.009
72. Kallinen, K. (2017). The effects of transparency and task type on trust, stress, quality of work, and co-worker preference during human-autonomous system collaborative work. In *Proceedings of the companion of the 2017 ACM/IEEE international conference on human-robot interaction* (pp. 153–154). ACM.
73. Neumann, W. P., & Dul, J. (2010). Human factors: Spanning the gap between OM and HRM. *International Journal of Operations & Production Management, 30*, 923–950. https://doi.org/10.1108/01443571011075056
74. Liu, L., & Guo, C. (2016). Research on the improvement of human error in the process of production operation—Taking the fatigue operation in the production as the object of study. In *2016 international conference on logistics, Informatics and Service Sciences (LISS)* (pp. 1–6). IEEE.
75. Farid, M., & Neumann, W. P. (2020). Modelling the effects of employee injury risks on injury, productivity and production quality using system dynamics. *International Journal of Production Research, 58*, 6115–6129. https://doi.org/10.1080/00207543.2019.1667040
76. Yang, C.-C., & Yang, K.-J. (2013). An integrated model of the Toyota production system with Total quality management and people factors. *Human Factors and Ergonomics in Manufacturing & Service Industries, 23*, 450–461. https://doi.org/10.1002/hfm.20335
77. Neumann, P., Kolus, A., & Wells, R. (2016). Human factors in production system design and quality performance—A systematic review. *IFAC-PapersOnLine, 49*, 1721–1724. https://doi.org/10.1016/j.ifacol.2016.07.830
78. Giagloglou, E., Brankovic, S., & Macuzic, I. (2015). Improving quality of work life through electrophysiology: An idea accepted by industry. *Int J Qual Res, 9*, 643–656.
79. Asgari, E., Homri, L., Siadat, A., et al. (2017). Proposing an assignment mathematical model in assembly line manufacturing system with considering human factors' role in product quality. In *2017 IEEE international conference on industrial engineering and engineering management (IEEM)* (pp. 1551–1555). IEEE.
80. Bhalaji, R. K. A., Bathrinath, S., Samantra, C., & Saravanasankar, S. (2020). *Analysis of human error of EHS in healthcare industry using TISM*. pp. 55–66.

81. Yung, M., Kolus, A., Wells, R., & Neumann, W. P. (2020). Examining the fatigue-quality relationship in manufacturing. *Applied Ergonomics, 82*, 102919. https://doi.org/10.1016/j.apergo.2019.102919
82. Lee, J. Y., Swink, M., & Pandejpong, T. (2011). The roles of worker expertise, information sharing quality, and psychological safety in manufacturing process innovation: An intellectual capital perspective. *Production and Operations Management, 20*, 556–570. https://doi.org/10.1111/j.1937-5956.2010.01172.x
83. Ko, C.-P., & Ko, C.-C. (2012). The relationship of training quality in E-learning and organizational commitment to organizational performance in the Taiwan service industry. *IERI Procedia, 2*, 821–827. https://doi.org/10.1016/j.ieri.2012.06.177
84. Mincolelli, G., Imbesi, S., & Zallio, M. (2020). Collaborative quality function deployment. *A Methodology for Enabling Co-design Research Practice*, 91–99.
85. Manjunath, K., Anu, V., Walia, G., & Bradshaw, G. (2018). Training industry practitioners to investigate the human error causes of requirements faults. In *2018 IEEE international symposium on software reliability engineering workshops (ISSREW)* (pp. 53–58). IEEE.

A New Model for Quality 4.0

Jacqueline Humphries, Alan Ryan, and Pepijn Van de Ven

Abstract Traditionally, efforts to ensure quality were grounded in a rigorous industrial engineering approach. However, there has been a dramatic shift on how manufacturing is carried out, due to the disruption inherent in Industry 4.0. The traditional methods of managing quality are not flexible enough to cope with the pace of change in modern manufacturing and the Factories of the Future. Discourse around modern quality management primarily focuses on the new digital tools available under Industry 4.0. Digital tools, in of themselves, will not address the complexity of the quality management problem. A new model for Quality 4.0 is required.

In this research, a literature review was completed to investigate how quality is being managed in the era of Industry 4.0. A new model for Quality 4.0 is proposed based on the findings. The model has at its foundation three pillars of Process, Human, and Technology. The model integrates these pillars within the traditional rigorous industrial engineering approach and systems for quality management, which include Quality Management Systems, Total Quality Management, and Measurement and Analysis.

The impact of this research is a generalizable model for real-world quality systems fit for the future. The model highlights the digitalization skills and quality structures required for a flexible data-driven modern Quality 4.0 system.

Keywords Error · Quality · Manufacturing · Industry 4.0 · Quality 4.0

J. Humphries (✉)
Technological University of the Shannon, Limerick, Ireland

Confirm Manufacturing Research Centre, Limerick, Ireland
e-mail: jacqueline.humphries@tus.ie

A. Ryan
University of Limerick, Limerick, Ireland

P. Van de Ven
University of Limerick, Limerick, Ireland

Confirm Manufacturing Research Centre, Limerick, Ireland

J. Reis et al. (eds.), *Driving Quality Management and Sustainability in VUCA Environments*, Springer Proceedings in Business and Economics,
https://doi.org/10.1007/978-3-031-52723-4_15

1 Introduction

The traditional methods of achieving quality and the management of error date back to the beginning of production. Major change was brought about in the 1960s, when the Toyota Production System and the Poka Yoke approach, or mistake-proofing, made huge impact on improving processes to reduce error [1, 2]. This evolved into a more holistic approach that valued the contribution of humans in resolving error. New theories, models, and methods were developed in the 1970s and 1980s, which Kehoe [3] refers to as the halcyon days of the quality revolution. The Deming System of Profound Knowledge [4] still has significant influence to this day. He presented his 14 Principles of Management, which increases productivity, quality, and peoples' joy in work. He proposed that quality transformation is everyone's job. This is achieved by creating a culture in which pride in one's job is fostered. Taguchi, Elsayed, and Hsiang [5] outlined their methods of dealing with quality problems based on calculation, product, and process design. Rasmussen, Duncan, and Leplat [6] advocated a multidisciplinary approach to quality through integration of training, with improvements to systems functioning and design. They noted that technology, while reducing the opportunity for error, makes errors harder to identify and control. Also, it was considered that one of the main inhibitors to quality was that there was a general lack of insight across disciplines [2]. However, there have been significant technical advances in Digital Quality Management [7, 8] in the first 10 years of Industry 4.0 which have the potential to address these problems.

The impact of these technological advances, such as artificial intelligence (AI), digital twins, automation, and three-dimensional (3D) printing, on reducing the opportunity for error is common in the research [2 , p. 39]. There is, however, a research gap with respect to drawing these fragmented research efforts together into a holistic view. The objective of this research is to understand how best to achieve quality in modern day manufacturing and to construct a modern quality model based on best practices.

2 Method

A narrative literature review is conducted to explore the current approaches to managing and addressing quality in modern manufacturing.

The Institute for Scientific Information Web of Science database is chosen, as recommended by Kamble [9]. Of interest is an analysis of quality research activity in the first 10 years of Industry 4.0, and for that reason, inclusion criteria included the timeline 2011 to 2021. Journal papers, conference articles, and book chapters are included. The inclusion criteria also used English as the publication language. The search query used the terms ("quality" *or* "error"). The search query situated the research within the context of modern manufacturing which is identified using the terms "smart factory" *or* "manufacturing" *or* "*Factories*" *or* "Industry*" *or* "production" *or* "cyber*physical" *or* "collaborat*" *or* "e*Manufacturing". Terms within the topic constrained the papers to manufacturing where humans are involved in the process using the terms "operator*" *or* "human" *or* "labo*r" or "non*automated".

Data is exported to an Excel spreadsheet for data analysis. A quality review of the papers is conducted. The first round of elimination included reading the title, abstract, and conclusions and classifying the papers. The second round of elimination involved reading the full text of all articles and retaining articles that focus on the research objective. Research not related to manufacturing systems that use humans in the process are excluded. The findings of the literature form the basis of the discussion and the proposed new model herein.

3 Literature Review

Quality 4.0 is considered to be the modern day approach to managing quality [10]. The term aligns with the revolution in production using the nomenclature of Industry 4.0. The opportunities of Industry 4.0 innovations to support quality are practically limitless. This leads to an extension of the traditional approach to quality management. This extension is powered by big data, big data tools, system integration, and other technologies available under Industry 4.0. The extension expands the scope and activities of the Quality Management function [8, 10].

There have been some recent works on defining Quality 4.0 and works on how to transition to Quality 4.0 [7, 8, 10–17]. There are a small number of Quality 4.0 models, which are critiqued hereafter. A new model for Quality 4.0 based on the literature is presented.

Chiarini (2020) [18] proposes that there are four elements in a Quality 4.0 model. The first element relates to creating value through the automated collection of quality data, predictive analytics, and artificial intelligence. The second element relates to investing in people to create the correct skillset and a culture for quality. The third element is customer value co-creation, which he describes in terms of the automatic collection of data across the value stream to feed back into quality processes. The final element relates to system integration. System integration automatically collects data from cyber-physical systems and sensors and production systems with respect to measurement and nonconformance. It automates the documentation control processes and management of data through enterprise systems. Chiarini's [18] model captures the human and technological pillars of quality and advances under Industry 4.0. However, it does not include the rigor of the traditional industrial engineering (IE) approaches to quality, such as Quality Management Systems (QMS), Measurement, and Total Quality Management (TQM), which are emphasized in the literature.

The Maganga and Taifa (2023) [19] Conceptual Framework for Quality 4.0 is also based on four branches – Industry 4.0 Technologies, TQM Principles, Big Data Management, and Real Time Quality Management. Like Chiarini, they exclude existing practices and systems. Maganga and Taifa posit that their model considers many of the shortcomings of existing models such as lack of clarity and comprehensiveness. In the TQM Principles branch, there is a strong emphasis on leadership and people involvement. However, the other three branches have overlapping elements, thus introducing their own shortcomings around clarity. For example, artificial intelligence and machine learning appear in both the Industry 4.0 Technologies and Big Data Management branches. Sensor data appears in Quality Management branch and as Internet of Things (IoT) in Big Data Management branch. While this model lists many of the technologies required for the collection, processing, and predictive analytics of telemetry data, it is indicative of the dearth of technological, and particularly digitalization, skills in the Quality domain.

The LNS Research model identifies 11 axes of Quality 4.0 [20 , p. 8]. It considers the impact of digitalization across the three pillars of Quality 4.0 – technology, processes, and people. This model consists of 11 axes: data, analytics, connectivity, collaboration, application development, scalability, Quality Management Systems, compliance activities, culture of quality, quality leadership, and competency. It is a comprehensive model that captures a wide spectrum of considerations. The LNS model is criticized however in that it does not provide guidance as to how to transition into Quality 4.0 [21]. Further, the LNS model assumes that a digitalization strategy is already in place at the organization. Herein lies a challenge, in that there is a wide spectrum of maturity level with respect to Industry 4.0. Only 16% of organizations have implemented Quality 4.0 initiatives, and 63% have not even started planning [8]. There are two elements missing from the LNS model. Firstly, the Socio-Technical aspects, Human Technology Integration, and Interaction Design are missing [22]. The design of the interface of systems has a significant impact on "intention to use." Intention to use is fundamental for the human to be engaged with, and augmented by the system, so that errors can be reduced. People need to be able to trust systems [23]. Secondly, the Human Factors perspective is missing. The literature emphasizes the importance of good human factor design and continuous improvement to create an environment where the human is likely to make less error [24]. The LNS model does include a "risk" category under the management systems axis, but the literature demonstrates that there needs to be a move away from considering humans as risk and move toward empowerment of the human [25].

Lim (2020) [16] highlights the links between traditional Operation Technology and Information Technology. He describes quality management approaches using real-world case studies. His work also aligns with the proposal by Naveh and Lei [2] that digital transformation can enhance the traditional IE approach to quality. Operation Technology consists of Quality Management Systems, quality engineering, process engineering, and the traditional perspectives. These are connected by artificial intelligence and through human intervention between the physical and cyber worlds.

The literature reveals that despite organizations recognizing the transformative power of Quality 4.0, those that have implemented Quality 4.0 initiatives find a wide range of challenges. Challenges that impede adoption of Quality 4.0 include technology, data-related issues, culture, strategy, and people [11]. Secondly, the literature reveals that Quality 4.0 discourse tends to ignore the IE practices that are currently employed in every manufacturing scenario. Much of the Quality 4.0 discourse focuses on transformative technologies. A new model is required that marries the traditional and proven IE practices to quality management, with newer socio-technical advances.

4 A New Model for Quality 4.0

Based on a review of the literature, an alternative model is proposed. Our model captures the practices and recommendations for managing error in manufacturing (see Fig. 1). Our Quality 4.0 model has new reaches into strategy, scalability, and compliance [26, 27]. Our model aligns with the discussion around Quality 4.0 which has at its core the pillars of process, human, and technology [26, 28]. The three pillars will now be considered in turn.

The process pillar refers to the acceptance that processes will be more flexible under Industry 4.0, and new processes of conducting the quality control function are required. Essential is that process support is provided for each step in the production so as to augment the human with the context-specific information they need to reduce error. Also within this pillar are the assumptions that the parts and tools made available are of the quality expected and that the Critical to Quality specification for that process is available. A Design for Quality approach and the continuous improvement of processes address much of the inherent quality pain points.

The human pillar captures the unique and valuable contribution of people and the importance of a quality culture in the prevention of error [25]. Human resources and their associated intellectual capital are fundamental to reducing error. Training, motivation, fostering a sense of belonging, and a culture of quality are key enablers [1, 27, 29, 30]. When individuals are empowered, they will take full responsibility for their own work and will assume the role of quality assessor, being proactive in ensuring the quality of their own work. They will also embrace organizational goals such as achieving higher quality [31]. This empowerment can be achieved by giving the human accountability, responsibility, and choice (ibid). This respect for the human intellect and engagement with humans fosters a culture of getting quality right first time [1, 32]. The interdisciplinary approach of Human Systems Integration should be used to ensure that the systems are human-centric [33, 34]. Human factors should be considered in any initiatives to reduce error [35]. The most significant human factors such as human fatigue, ergonomics of workstation, and production systems must be key considerations, along with a recognition that the human factors extend beyond risk and that the human brings unique contributions in terms of decision-making, problem-solving, creativity, and intuition that must be capitalized and made central to the process. This model captures Deming's 14 Points on Quality Management, which is the fallacy that controls, fears, and merits systems reduce human error [36]. What is required is to create a culture for constant purpose toward quality, in which all workers are involved in the transformation [2, 37–39]. Leadership is key to achieving this.

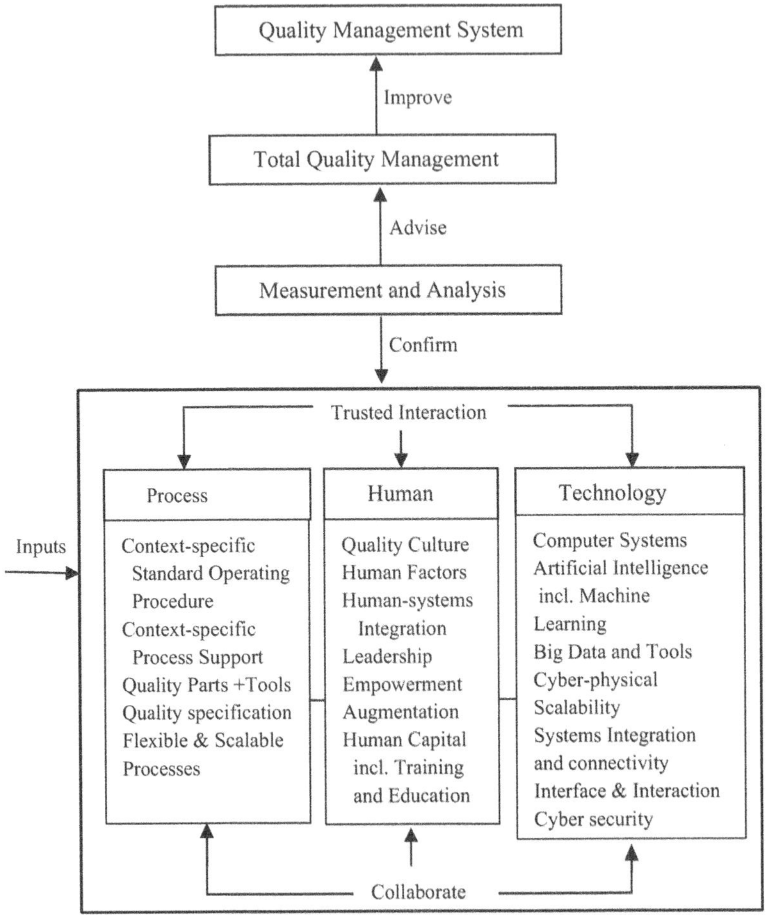

Fig. 1 New proposed model for reducing human error in Industry 4.0

The final pillar, namely, the technology pillar, captures the capacity available under Industry 4.0 that enables access to new knowledge in a predictive manner. The key technologies include artificial intelligence in particular machine learning, deep learning and computer vision, big data, and data science tools. Scalability is enabled through cloud computing, virtualization, and containerization of code. Connectivity refers to the trusted integration of systems and sharing of data enabled through communication infrastructure and secured through blockchain and cybersecurity tools [26, 27]. The aim is for full connectivity between the manufacturers' systems such as Manufacturing Execution Systems (MES), Total Productive Maintenance (TPM), Product Lifecycle Management (PLM) systems, and Enterprise Resource Planning (ERP) systems. Cyber-physical systems closely linked with IoT build on the availability of high volume and high velocity multivariate data, generated by sensors throughout the manufacturing floor. This data can be used to detect and predict quality [40, 41]. Lastly, human interface design, through personalization of systems adaptive to the human and the codification of knowledge, should be based on the human-machine cooperation approach [22]. The interface design may take the form of alternative realities, such as augmented or virtual, or may be simple haptic or visual feedback. The key is that the interface design suits the purpose, augments the human with timely context-sensitive information, and does not interfere in the process [42].

In the proposed model, there are two ternary relationships shown between the pillars of process, the human and technology. The first relates to collaboration. Efforts should be made to create situations where humans work together collaboratively within multidisciplinary teams using technology to perform processes which positively impact performance [29, 43–45]. The act of collaboration enables the co-creation of high-quality products. The second ternary relationship, between those same three pillars, is a trusted interaction. Accessibility to systems and data that support the human to make fewer errors is only as effective as the system's usability, loyalty, and intention to use [46]. Ideally the interaction should mimic human-human collaboration where the multiple agents, in this case the human, the technology, and the process, work together and thereby achieve better quality outcomes. This can be achieved by providing the environment for, and the sharing of, appropriate information in a seamless trusted manner [23].

Sitting at the top of the model is the overarching QMS. QMS informs the whole production process to be followed, the range of tools and systems to be used, and how the product or service will be assessed, inspected, and audited [47, 48]. QMSs are documented in quality policies and procedures, work instructions, records, and relevant objectives that are then followed by the human [49]. QMSs are built on quality-by-design (QbD) principles. QbD ensures that the essential quality characteristics of the product are considered from the outset and feed into the design stage of all manufacturing processes and touch-points with the product [50]. How the QMS is applied to a specific process is set out in the Quality Plan [51]. The Quality Plan sets out how the quality objectives will be achieved; who is responsible; what processes, controls, audits, or actions must be conducted; and how the plan can be modified [51]. To ensure that the products meet the specific requirements, the audits and controls are completed by the quality control (QC) function [48, 52]. The QMS should aim to use the capabilities in the technology pillar to be intelligent and be flexible so as to be able to adapt quickly to new situations [53].

Essential to reducing error is the notion of continuous improvement, which is achieved through TQM. Best practice is to map the current production process, analyze, and improve [39]. Data is taken from measurement. The basic tools of quality are used for the mapping activity [54]. These tools include root cause analysis, cause-and-effect diagrams, histograms, time-and-motion studies, check sheets, spaghetti diagrams, and control sheets [52, 55–63]. Once the mapping is complete, analysis begins. To achieve optimal impact, the processes where the most human error occurs are identified and then targeted for improvement. Tools, such as Pareto charts and value stream mapping, can also be used to help identify which processes to prioritize [58, 62, 64, 65]. Thus, problematic steps in the process, once identified, can be redesigned and improvements fed back to the QMS. These then are communicated through the system.

Sitting between the pillars and the systems is the Measurement and Analysis activity. Measurement and analysis of error provides information for the TQM processes as to where improvements can be made [66]. Accessibility to a wide variety of data, including real-time telemetry data from sensors, coupled with advances in data mining, artificial intelligence including machine learning, and statistical methods creates a proactive predictive function.

The inputs to our model are the customer requirements for the products, which will have requirements across any of the quality attributes: performance, characteristics, reliability, conformity, longevity, serviceability, esthetics, and perceived quality [48]. The next inputs are the compliance requirement to good manufacturing practice (GMP), as set down by the Food and Drug Administration (FDA) or other regulatory bodies [67]. The next set of inputs are the international or industry-specific standards of certifications such as the International Organization for Standardization (ISO) 9000 or ISO 9001 and standards related to particular production methods [68, 69]. Compliance with regulatory body and international and industry standard certification systems ensures quality through the manufacturing process [70]. The next inputs are the principles of Industry 4.0, which are human-centric and sustainable. The principles of Industry 4.0 include a requirement for vertical networking of smart production systems enabling rapid response to error and horizontal integration across new value chains. Data should be available at all stages of the product's life cycle [71, 72]. The final inputs are the lean principles based on the Toyota Production System, which considers defining value to the customer, eliminating waste, and continuous improvement.

5 Conclusion

A new model for Quality 4.0 is proposed, based on the learnings from a literature review. This model will assist in reducing error and improving quality in real-world domains. The model is developed based upon the practices and research activity in the smart manufacturing domain such as in medical device manufacturing. The model is generalizable to situations where humans are involved in delivering the service or product. In medical device manufacturing, the human worker plays a key role in both the assembly and quality of products, and that is encapsulated here with the human as one of the key pillars for the system. Although the model has been developed based on best practices within industry, it may also be applied in other real-world contexts, such as education. While the inputs may change – so Industry 4.0 principles may be replaced with Pedagogy principles – the basic model has relevance.

For the model to have an impact, it must also influence behavior. Our new Quality 4.0 model captures the disruption in the quality management role, with a move from a traditional industrial engineering approach to an interdisciplinary approach. There is a need for the Quality Team to be leaders, not bystanders in the digitalization of quality. Thus, the need for new competencies such as digitalization skills for the Quality Team is apparent. These competencies are those listed in our model under the Technology pillar, which include artificial intelligence and machine learning, big data analytics and exchange of data, system integration, and connectivity of systems. Under the Human pillar, new competencies are required such as an emphasis on the quality culture, human empowerment, human system integration, and human augmentation by technology. Approaching Quality from the traditional perspective of process, continuous improvement, measurement, human factors engineering, and training is no longer fit for purpose.

5.1 Future Research Perspectives

The human is a pillar of the Industry 4.0 agenda. There is a recurring message in the literature that the human is a critical success factor to quality production [73]. Indeed, the premise in Industry 4.0 is that the humans are interconnected to and work together in a complementary way with systems reaching toward human-machine symbiosis [9, 72, 74]. The terms Human 4.0 and Operator 4.0 are used to capture this [68, 75]. This stems from the lean principle of "respect for people" which recognizes the unique contribution of the human to production. However, the research suggests that the notion of Operator 4.0 is being considered as hyperbole. Research around the value of Operator 4.0 tends to be primarily conceptual and theoretical, with little by way of empirical evidence. Much of the research uncovered in this review is primarily focused on the measurement and prediction of human error, quality checking of the human's work, substituting the human by automation, considering human factors as risks, and the application of a Poka Yoke approach to TQM. This goes against the principles of removing quotas, merit systems, and numerical goals, as proposed by Deming [36]. In order to leverage the capacity proposed under new Quality 4.0 models, a cultural shift is required toward a full appreciation of the value of Operator 4.0 to achieving quality. Empirical evidence is crucial to achieve this.

The model proposed herein is based on the themes that emerged from the literature and which were compared with theory and existing knowledge. Further research should validate this model with expert opinion. In addition, the effectiveness and the efficiency of the model to solve real-world quality problems should be evaluated.

5.2 Limitations

This narrative review is conducted on a search of one database, and in English. This will have limited the results. Further, there is a lack of clarity around the terms and synonyms used to represent Quality 4.0. A generic search was conducted so as to retrieve research that focused on quality on a granular level. In doing so, it is possible that some relevant research has not been included. The results could be validated by review with expert opinion, or a repetition of the review through the use of more synonyms, other languages, and other databases.

References

1. C.-C. Yang and K.-J. Yang, "An integrated model of the Toyota production system with Total quality management and people factors," Human Factors and Ergonomics in Manufacturing & Service Industries, vol. 23, no. 5, pp. 450–461, Sep. 2013, doi: https://doi.org/10.1002/hfm.20335.
2. Naveh, E., & Lei, Z. (2019). Coping with errors in organizations: Challenges, opportunities, and Frontiers for operations management research. *Foundations and Trends® in Technology, Information and Operations Management, 12*(4), 349–433. https://doi.org/10.1561/0200000055
3. Kehoe, D. F. (1995). *The fundamentals of quality management*. Springer.
4. Deming, W. E. (1982). *Out of the crisis*. MIT Center for Advanced Engineering Study.
5. Taguchi, G., Elsayed, E. A., & Hsiang, T. C. (1989). *Quality engineering in production systems*. McGraw-Hill Companies.
6. Rasmussen, J., Duncan, K., & Leplat, J. (1987). *New technology and human error*. Wiley.
7. Zonnenshain, A., & Kenett, R. S. (2020). Quality 4.0 – The challenging future of quality engineering. *Quality Engineering, 32*(4), 614–626. https://doi.org/10.1080/08982112.2019.1706744
8. Broday, E. E. (2022). The evolution of quality: From inspection to quality 4.0. *International Journal of Quality and Service Sciences, 14*(3), 368–382. https://doi.org/10.1108/IJQSS-09-2021-0121
9. Kamble, S. S., Gunasekaran, A., & Gawankar, S. A. (2018). Sustainable industry 4.0 framework: A systematic literature review identifying the current trends and future perspectives. *Process Safety and Environment Protection, 117*, 408–425. https://doi.org/10.1016/j.psep.2018.05.009
10. Sony, M., Antony, J., & Douglas, J. A. (2020). Essential ingredients for the implementation of quality 4.0. *TQM Journal, 32*(4), 779–793. https://doi.org/10.1108/TQM-12-2019-0275
11. Küpper, D., Knizek, C., Ryeson, D., & Noecker, J. (2019). *Quality 4.0 takes more than technology*. [Online]. Available: https://images.magnetmail.net/images/ clients/ASQ_/attach/BCG-Quality-40-Takes-More-Than-Technology-July-2019(1).pdf
12. Sader, S., Husti, I., & Daroczi, M. (2021). A review of quality 4.0: Definitions, features, technologies, applications, and challenges. *Total Quality Management and Business Excellence, 33*, 1–19. https://doi.org/10.1080/14783363.2021.1944082
13. Sureshchandar, G. S. (2021). Quality 4.0 – A measurement model using the confirmatory factor analysis (CFA) approach. *International Journal of Quality & Reliability Management, 40*, 280. https://doi.org/10.1108/IJQRM-06-2021-0172
14. Dias, A. M., Carvalho, A. M., & Sampaio, P. (2022). Quality 4.0: Literature review analysis, definition and impacts of the digital transformation process on quality. *International Journal of Quality & Reliability Management, 39*(6), 1312–1335. https://doi.org/10.1108/IJQRM-07-2021-0247
15. Maganga, D. P., & Taifa, I. W. R. (2022). Quality 4.0 transition framework for Tanzanian manufacturing industries. *TQM Journal, 35*, 1417. https://doi.org/10.1108/TQM-01-2022-0036
16. Lim, J. S. (2019). *Quality Management in Engineering*. Taylor & Francis., A CRC title, part of the Taylor & Francis imprint, a member of the Taylor & Francis Group, the academic division of T&F Informa, plc, 2020.: CRC Press. https://doi.org/10.1201/9780429281600

17. Fonseca, L. (2022). The EFQM 2020 model. A theoretical and critical review. *Total Quality Management and Business Excellence, 33*(9–10), 1011–1038. https://doi.org/10.1080/14783363.2021.1915121
18. Chiarini, A. (2020). Industry 4.0, quality management and TQM world. A systematic literature review and a proposed agenda for further research. *TQM Journal, 32*(4), 603–616. https://doi.org/10.1108/TQM-04-2020-0082
19. Maganga, D. P., & Taifa, I. W. R. (2023). Quality 4.0 conceptualisation: An emerging quality management concept for manufacturing industries. *TQM Journal, 35*(2), 389–413. https://doi.org/10.1108/TQM-11-2021-0328
20. Jacob, D. (2017). *Quality 4.0 impact and strategy handbook: Getting digitally connected to transform quality management.*
21. Sisodia, R., & Villegas Forero, D. (2019). Quality 4.0–How to Handle Quality in the Industry 4.0 Revolution, Report No. E2019:128," Göteborg, Sweden, 2019. [Online]. Available: https://odr.chalmers.se/bitstream/20.500.12380/300650/1/E2019_128.pdf
22. Pacaux-Lemoine, M. P., Poulain, T., & Debernard, S. (2013). Human-machine cooperation and manufacturing system. In *IECON 2013 – 39th annual conference of the IEEE industrial electronics society* (pp. 8288–8293). IEEE. https://doi.org/10.1109/IECON.2013.6700521
23. van Waveren, S., Carter, E. J., & Leite, I. (2019). Take one for the team. In *Proceedings of the 19th ACM international conference on intelligent virtual agents* (pp. 151–158). ACM. https://doi.org/10.1145/3308532.3329475
24. Neumann, P., Kolus, A., & Wells, R. (2016). Human factors in production system design and quality performance – A systematic review. *IFAC-PapersOnLine, 49*(12), 1721–1724. https://doi.org/10.1016/j.ifacol.2016.07.830
25. Pfeiffer, S. (2016). Robots, industry 4.0 and humans, or why assembly work is more than routine work. *Societies, 6*(2), 16. https://doi.org/10.3390/soc6020016
26. American Society for Quality (ASQ), Quality 4.0," 2020. https://asq.org/quality-resources/quality-4-0. Accessed 2 Apr 2022.
27. Thekkoote, R. (2022). Enabler toward successful implementation of quality 4.0 in digital transformation era: A comprehensive review and future research agenda. *International Journal of Quality & Reliability Management, 39*, 1368. https://doi.org/10.1108/IJQRM-07-2021-0206
28. Armani, C. G., de Oliveira, K. F., Munhoz, I. P., & Akkari, A. C. S. (2021). Proposal and application of a framework to measure the degree of maturity in quality 4.0: A multiple case study. In *Advances in mathematics for industry 4.0* (pp. 131–163). Elsevier. https://doi.org/10.1016/B978-0-12-818906-1.00006-1
29. Lee, J. Y., Swink, M., & Pandejpong, T. (2011). The roles of worker expertise, information sharing quality, and psychological safety in manufacturing process innovation: An intellectual capital perspective. *Production and Operations Management, 20*(4), 556–570. https://doi.org/10.1111/j.1937-5956.2010.01172.x
30. Gaffova, Z., & Satanova, A. (2017). Human resource management and quality management in woodworking industry companies in Slovakia. In *10th annual international scientific conference on more wood, better management, increasing effectiveness: setting points and perspectives*. Prague, Czech Republic.
31. Onyemeh, N. C., & Lee, C. W. (2014). Improving quality of operations via industry-specific empowerment antecedents: A study of the oil and gas industry. In *2014 IEEE international conference on industrial engineering and engineering management* (pp. 143–147). IEEE. https://doi.org/10.1109/IEEM.2014.7058617
32. Ooi, K.-B., Lee, V.-H., Chong, A. Y.-L., & Lin, B. (2013). Does TQM improve employees' quality of work life? Empirical evidence from Malaysia's manufacturing firms. *Production Planning and Control, 24*(1), 72–89. https://doi.org/10.1080/09537287.2011.599344
33. M.-P. Pacaux-Lemoine, T. Poulain, and S. Debernard, "Human-Machine Cooperation and manufacturing system," in IECON 2013 – 39th Annual Conference of the IEEE Industrial Electronics Society, IEEE, 2013, pp. 8288–8293. doi: https://doi.org/10.1109/IECON.2013.6700521.

34. Trentesaux, D., & Millot, P. (2016). A Human-Centred Design to Break the Myth of the 'Magic Human'. *Intelligent Manufacturing Systems*, 103–113. https://doi.org/10.1007/978-3-319-30337-6_10
35. Giagloglou, E., Brankovic, S., & Macuzic, I. (2015). Improving quality of work life through electrophysiology: An idea accepted by industry. *International Journal for Quality Research, 9*(4), 643–656.
36. Deming, W. E. (2018). *The new economics for industry, government, education* (3rd ed.). MIT Press.
37. Emspak, F. (1992). Total quality management and skills based automation. *IFAC Proceedings, 25*(27), 69–73. https://doi.org/10.1016/S1474-6670(17)49560-7
38. Tata, J., Prasad, S., & Thorn, R. (1999). The influence of organizational structure on the effectiveness of TQM programs. *Journal of Managerial Issues, 11*(4), 440–453.
39. Modgil, S., & Sharma, S. (2016). Total productive maintenance, total quality management and operational performance. *Journal of Quality in Maintenance Engineering, 22*(4), 353–377. https://doi.org/10.1108/JQME-10-2015-0048
40. Peres, R. S., Barata, J., Leitao, P., & Garcia, G. (2019). Multistage quality control using machine learning in the automotive industry. *IEEE Access, 7*, 79908–79916. https://doi.org/10.1109/ACCESS.2019.2923405
41. Hemming, S., de Zwart, F., Elings, A., Petropoulou, A., & Righini, I. (2020). Cherry tomato production in intelligent greenhouses – Sensors and AI for control of climate, irrigation, crop yield, and quality. *Sensors, 20*(22), 6430. https://doi.org/10.3390/s20226430
42. Bicho, E., Erlhagen, W., Louro, L., Costa e Silva, E., Silva, R., & Hipolito, N. (2011). A dynamic field approach to goal inference, error detection and anticipatory action selection in human-robot collaboration. *New Frontiers in Human-Robot Interaction*, 135–164.
43. Çaglayan, B., & Bener, A. B. (2016). Effect of developer collaboration activity on software quality in two large scale projects. *Journal of Systems and Software, 118*, 288–296. https://doi.org/10.1016/j.jss.2016.03.055
44. Mincolelli, G., Imbesi, S., & Zallio, M. (2020). Collaborative quality function deployment. A methodology for enabling co-design research practice, 91–99. https://doi.org/10.1007/978-3-030-20444-0_9
45. Hassan, A. S., & Jaaron, A. A. M. (2021). Total quality management for enhancing organizational performance: The mediating role of green manufacturing practices. *Journal of Cleaner Production, 308*, 127366. https://doi.org/10.1016/j.jclepro.2021.127366
46. Orehovački, T., & Babić, S. (2016). Mobile quality of social web applications designed for collaborative writing, 368–379. https://doi.org/10.1007/978-3-319-39483-1_34
47. Albers, A., Gladysz, B., Pinner, T., Butenko, V., & Stürmlinger, T. (2016). Procedure for defining the system of objectives in the initial phase of an industry 4.0 project focusing on intelligent quality control systems. *Procedia CIRP, 52*, 262–267. https://doi.org/10.1016/j.procir.2016.07.067
48. Daneshjo, N., Malega, P., & Drábik, P. (2021). Techniques for production quality control in the global company. *Advances in Science and Technology Research Journal, 15*(1), 174–183. https://doi.org/10.12913/22998624/131558
49. Razmochaeva, N. V., Semenov, V. P., & Bezrukov, A. A. (2019). Role of process automation in quality management of enterprises in perfumery and cosmetic industry. In *IEEE Conference of Russian Young Researchers in Electrical and Electronic Engineering (EIConRus)* (pp. 1449–1452). https://doi.org/10.1109/EIConRus.2019.8657085
50. Papantoniou, I., Chai, Y. C., Luyten, F. P., & Schrooten, J. (2013). Process quality engineering for bioreactor-driven manufacturing of tissue-engineered constructs for bone regeneration. *Tissue Engineering. Part C, Methods, 19*(8), 596–609. https://doi.org/10.1089/ten.tec.2012.0526
51. Misztal, A., & Bachorz, S. (2014). Quality planning of parts machine production based on housing of cylinder head milling machines. *Applied Mechanics and Materials, 657*, 986–990. https://doi.org/10.4028/www.scientific.net/AMM.657.986

52. Jovic, F., Jovic, A., & Krmpotic, D. (2013). Quality control engineering in automated ceramic tile production using a signal information content approach. *Advanced Engineering Informatics, 27*(1), 93–107. https://doi.org/10.1016/J.AEI.2012.11.007
53. Bihi, T., Luwes, N., & Kusakana, K. (2018). Innovative quality management system for flexible manufacturing systems. In *In 2018 open innovations conference (OI)* (pp. 40–46). IEEE. https://doi.org/10.1109/OI.2018.8535610
54. Tague, N. R. (2004). *The quality toolbox* (2nd ed.). American Society for Quality.
55. Singh, M., Khan, I. A., Grover, S., & Gupta, S. C. (2011). Assessing quality of manufacturing organizations - a graph theoretic approach. In *2011 IEEE international conference on industrial engineering and engineering management* (pp. 1740–1744). IEEE. https://doi.org/10.1109/IEEM.2011.6118214
56. Salleh, N. A. M., Kasolang, S., & Jaffar, A. (2012). Simulation of integrated Total quality management (TQM) with lean manufacturing (LM) practices in forming process using Delmia quest. *Procedia Engineering, 41*, 1702–1707. https://doi.org/10.1016/j.proeng.2012.07.371
57. Yang, H. (2012). Control technique research of key quality characteristic during the mechanical and electrical production manufacturing process. *Advances in Materials Research, 442*, 215–219. https://doi.org/10.4028/www.scientific.net/AMR.442.215
58. Antsev, V. Y., Vitchuk, N. A., & Miroshnikov, V. V. (2017). Improvement in production process for pipelines manufacturing based on quality management method. *Procedia Engineering, 206*, 950–957. https://doi.org/10.1016/j.proeng.2017.10.577
59. Kurata, Y. B., Navales, M. R., Cedron, D. B., Marcelino, M. J., & Pening, T. N. (2018). *Quality Assurance Production Based Problem: A Process Improvement in the Rolling Mill Line for Steel Manufacturing Company in the Philippines* (pp. 74–82). https://doi.org/10.1007/978-3-319-60474-9_7
60. Blaga, P. (2020). The importance of human resources in the continuous improvement of the production quality. *Procedia Manufacturing, 46*, 287–293. https://doi.org/10.1016/j.promfg.2020.03.042
61. Blalack, W. J. (2020). Additive manufacturing applications for quality improvement and cost reduction. In *2020 31st annual SEMI advanced semiconductor manufacturing conference (ASMC)* (pp. 1–4). IEEE. https://doi.org/10.1109/ASMC49169.2020.9185342
62. Cabrera, J., Corpus, O., Maradiegue, F., & Merino, J. C. A. (2020). Improving quality by implementing lean manufacturing, SPC, and HACCP in the food industry: A case study. *South African Journal of Industrial Engineering, 31*(4). https://doi.org/10.7166/31-4-2363
63. Sulistiyowati, W., Adamy, M. R., & Jakaria, R. B. (2019). Product quality control based on lean manufacturing and root cause analysis methods. *Journal of Physics Conference Series, 1402*(2), 022038. https://doi.org/10.1088/1742-6596/1402/2/022038
64. Dhillon, B. S. (2014). Human error in maintenance: An investigative study for the factories of the future. *IOP Conference Series: Materials Science and Engineering, 65*, 012031. https://doi.org/10.1088/1757-899X/65/1/012031
65. Lyu, J., Chen, P.-S., & Huang, W.-T. (2021). Combining an automatic material handling system with lean production to improve outgoing quality assurance in a semiconductor foundry. *Production Planning and Control, 32*(10), 829–844. https://doi.org/10.1080/09537287.2020.1769217
66. Razmochaeva, N. V., Semenov, V. P., & Bezrukov, A. A. (2019). Role of process automation in quality Management of Enterprises in perfumery and cosmetic industry. In *2019 IEEE conference of Russian young researchers in electrical and electronic engineering (EIConRus)* (pp. 1449–1452). IEEE. https://doi.org/10.1109/EIConRus.2019.8657085
67. Kilic, P. (2021). *Quality Management Systems (QMSs) of human-based tissue and cell product manufacturing facilities* (pp. 263–279). https://doi.org/10.1007/7651_2020_341
68. Romero, D., Gaiardelli, P., Powell, D., Wuest, T., & Thürer, M. (2019). Total quality management and quality circles in the digital lean manufacturing world, 3–11. https://doi.org/10.1007/978-3-030-30000-5_1

69. Nutburn, M. (2019). *TQM vs QMS: The difference between Total quality management and quality management systems*. British Assessent Bureau. https://www.british-assessment.co.uk/insights/tqm-vs-qms-the-difference-between-total-quality-management-and-quality-management-systems/. Accessed 4 Feb 2022
70. Giacomarra, M., Galati, A., Crescimanno, M., & Tinervia, S. (2016). The integration of quality and safety concerns in the wine industry: The role of third-party voluntary certifications. *Journal of Cleaner Production, 112*, 267–274. https://doi.org/10.1016/j.jclepro.2015.09.026
71. Schlaepfer, R. C., & Koc, M. (2015). *Industry 4.0: Challenges and solutions for the digital transformation and use of exponential technologies*. [Online]. Available: https://s3.amazonaws.com/academia.edu.documents/42978372/Deloittes_study_on_industry_4.0.pdf
72. Bahrin, M. A. K., Othman, M. F., Azli, N. N., & Talib, M. F. (2016). Industry 4.0: A review on industrial automation and robotic. *Jurnal Teknologi, 78*(6–13), 137–143.
73. Kagermann, H., & Wahlster, W. (2022). Ten years of Industrie 4.0. *Science, 4*(3), 26. https://doi.org/10.3390/sci4030026
74. Frohm, J., Lindström, V., Stahre, J., & Winroth, M. (2008). Levels of automation in manufacturing. *International Journal of Human Factors and Ergonomics, 30*(3).
75. S. Mattsson, Å. Fast-Berglund, D. Li, and P. Thorvald, "Forming a cognitive automation strategy for operator 4.0 in complex assembly," Computers and Industrial Engineering, vol. 139, no. August 2018 2020, 2020, doi: https://doi.org/10.1016/j.cie.2018.08.011, 105360.

A Comparative Study of Service Quality in Portuguese and Angolan Higher Education Institutions

Ana Rolo, Margarida Saraiva, Gracieth Leandro, Teresa Nogueiro, and Rui Alves

Abstract Quality in education is an important topic as it is reflected in the educational levels of a society. Many studies consider students as stakeholders; however, families, employers, and society in general can also be a set of stakeholders in this educational process. This article addresses this issue from an international perspective that results from a collaborative work carried out by Portuguese and Angolan researchers. He intends to study service quality in three higher education institutions (HEIs), two in Portugal and one in Angola. The study was based on the HEISQUAL model, a modern approach to measuring the quality of service in higher education institutions. In past studies, we used SERVQUAL model. It was possible to concluded that, for the institutions under study located in Angola, the evaluation made by the students is bottom. The quantitative methodology was used, supported by the new instrument called HEISQUAL, a questionnaire that aims to measure the quality of service provided by an HEI. The current survey follows a convenient sampling technique.

Based on these results, good practices were identified to be implemented to improve the quality performance indicators in Angola. The half-yearly monitoring will allow following the evolution of this implementation. In addition to the scientific component, this work is based on a greater concern, namely, the contribution to Sustainable Development Goal (SDG) 4 – Quality Education. It is expected that the application of good practices will result in the improvement of the quality of education perceived by the students.

Keywords Education · HEI · Quality

A. Rolo (✉) · R. Alves
Instituto Politécnico de Setúbal, Setúbal, Portugal
e-mail: ana.rolo@esce.ips.pt

M. Saraiva · T. Nogueiro
Universidade de Évora, Évora, Portugal

G. Leandro
Universidade Agostinho Neto, Luanda, Angola

J. Reis et al. (eds.), *Driving Quality Management and Sustainability in VUCA Environments*, Springer Proceedings in Business and Economics,
https://doi.org/10.1007/978-3-031-52723-4_16

1 Introduction

Competition among higher education institutions (HEIs) is increasingly high. According to Wilkins [14] and Rolo et al. [11], higher education has become a very competitive market, and there is great pressure on HEIs to improve performance and quality.

A school is an organization, and as such, it must act like a company in managing its resources and attracting its customers. Rolo et al. [11] refer that this effort is increasingly important due to current pressures related to the decrease in birth rates and the increase in the formative offer (number of courses, programs, and degrees). HEIs had to react to this new reality, trying to attract students. For this reason and more than ever, quality in HEIs has received enormous attention from many stakeholders, such as the government, university directors, students and their parents, researchers, and academics, as they are responsible for turning students into valuable resources for businesses.

Therefore, it is very important to measure and monitor the quality of services provided.

Measuring quality in higher education is a complex and multidimensional process that involves a variety of dimensions and therefore requires a comprehensive and holistic approach that takes the different dimensions into account [4]. For this purpose, it is essential to identify quality indicators and use an adequate measurement instrument.

Harvey and Green [6] identified four dimensions of quality in education: input quality (referring to the characteristics of students, faculty, resources, and facilities that are needed for effective delivery of education), process quality (referring to the teaching and learning processes and the quality of student-teacher interactions that will influence student engagement and satisfaction), output quality (refers to the knowledge and skills students acquire as a result of their education; graduation rates and time to graduation), and outcome quality (the broader impact of higher education on individuals and society, as employability and civic engagement, employer satisfaction, and hiring rates; contribution to local and global communities).

Shah and Nair [12] also identified four dimensions but different from the previous ones, namely, social responsibility, sustainability, diversity and inclusion, and entrepreneurship.

The authors argued that these dimensions are increasingly important in higher education, as institutions seek to adapt and respond to changing needs in society.

Simultaneously, several instruments were being tested to measure service quality in HEIs: SERVQUAL, by Parasuraman, Zeithaml, and Berry [8–10] and Fuchs and Fangpong [5]; SERVPERF, by Cronin and Taylor [3]; and HEdPERF, by Abdullah [2]. However, these instruments had some limitations despite having been modified and adapted.

HEISQUAL is presented as an instrument that measures quality of service exclusively in HEIs, from the perspective of students. It also eliminates the operational (process) and technical (end results) deficiencies identified in previously tested scales [1]. These reasons justify its choice to test in this comparative study.

This study aims to present a comparative study on service quality in higher education institutions. It was conducted in Portugal and Angola to make a cross-comparison between two different higher education institutions, one belonging to the polytechnic higher education network in Portugal and a university in Angola. In this first exploratory study, the institutions involved will not be identified, because the authorizations did not arrive in time for the submission of this article.

This article presents the following structure: introduction, presentation of the HEISQUAL model, methodology, presentation and discussion of results, conclusions, and limitations and proposals for future research.

2 HEISQUAL Model

Despite the lack of consensus regarding the definition of service quality dimensions, there are some consensus that quality should be determined by stakeholders, such as the students, their parents and family, academic and administration staff, and society [13].

The main shortcomings of previous tools, such as SERVQUAL and SERVPERF, are the fact that they are generic and not specific to the area of education. To overcome the shortcomings of that tools, a new model that resulted from the modification and improvement of existing models was developed, aiming to increase precision and applicability in the specific contexts of higher education. SERQUAL and SERVPERF models were designed to be applied to services and not specifically to education, a specific service that is very different from the others.

Abbas [1] proposes a new instrument called HEISQUAL to measure the quality of services in HEIs. This model adopts a specific language for the sector under analysis. It is composed of 7 main dimensions, 16 subdimensions, and several items (Table 1). The dimensions identified by Abbas are as follows:

- **Teachers' profile:** this dimension was further divided into four subdimensions, Namely, teachers' subject knowledge (four items), communication skills (four items), teaching style (four items), and behavior with students (five items).
- **Curriculum**: this factor with a single dimension, which contained four items, includes "The curriculum taught at my university is comprehensive and easy to understand" and "The curriculum taught at my university improves students' intellectual abilities (decision-making, problem-solving)."
- **Infrastructure and facilities:** this theme is divided into three dimensions, namely, learning facilities (four items), supportive facilities (three items), and cleanliness and maintenance (three items).
- **Management and support staff**: this theme is further divided into two dimensions, Namely, behavior with students (three items) and administrative work (six items).

Table 1 HEISQUAL framework—survey data summary

Dimensions	Sub-Dimensions	ITEMS	Portuguese HEI	Angolan HEI	Dif. GAP
Teachers' Profile	Subject Knowledge	My instructors have comprehensive knowledge of their field	4,8	4,1	0,7
		The knowledge possessed by my instructors is up-to-date	4,7	3,9	0,8
		My instructors have adequate knowledge to answer students' questions	4,4	3,5	0,9
		My instructors give real-life examples to enhance students' understanding	4,75	4,6	0,15
	Communication Skills	The communication skills of my instructors are excellent	4,83	4,25	0,58
		My instructors have effective command over the course language (English, German, Spanish)	4,5	3,05	1,45
		I can easily understand the concepts explained by my instructors	4,76	4,6	0,16
		My instructors communicate in a polite and respectful manner	4,1	4,25	-0,15
	Teaching Style	My instructors explain the complex concepts in an easy and understandable manner	4,2	3,65	0,55
		My instructors create an interactive environment in the class and engage students' interest	3,9	3,7	0,2
		My instructors ensure active class participation of students	4,6	3,9	0,7
		My instructors use different tools and techniques to enhance students' learning (games, class activities)	4,3	4,05	0,25
	Behaviour with Students	My instructors demonstrate non-biased and non-discriminating behaviour	4,5	3,3	1,2
		Instructors in my university instil confidence in students	4,45	3,0	1,45
		My instructors show interest in understanding and solving students' problems	4,6	3,4	1,2
		My instructors follow fair grading criteria	4,7	4,1	0,6
		My instructors regularly give feedback to students on their academic performance	4,2	3,8	0,4
Curriculum	Curriculum Quality	The curriculum taught at my university is comprehensive and easy to understand	3,6	3,55	0,05
		The curriculum taught at my university helps students to think in creative and proactive way	4,15	3,85	0,3
		The curriculum taught at my university improves students' intellectual abilities (decision making, problem-solving	4,4	3,93	0,47
		My university designs curricula by considering future job perspectives for students	3,8	3,35	0,45
Infrastructure and Facilities	Learning Facilities	My university has adequate library resources (books, magazines, newspapers, study space, furniture, access to online databases)	4,85	3,2	1,65
		My university ensures availability of supportive tools and equipment for learning e.g. internet/Wi-Fi, whiteboards, projector, air conditioner	4,75	2,9	1,85
		The number of students per class is maintained at low to medium levels	4,25	3,3	0,95
		Lectures in my university are held as per the timetable and course content	4,15	3,4	0,75
	Supportive Facilities	My university provides satisfactory transportation, cafeteria, and bookshop services to students and staff members	4,65	4,15	0,5
		My university provides adequate housing facilities to students and staff members	3,2	2,3	0,9
		Prices and quality of goods and services in my university are reasonable	4,5	4,2	0,3
	Cleanliness and Maintenance	My university has sufficient and well-maintained classrooms	4,15	4,0	0,15
		The campus is kept neat and clean	4,85	4,15	0,7
		My university campus and buildings are visually appealing and eye-catching	4,9	4,25	0,65
Management and Support Staff	Behaviour with Students	Management and support staff of my university deal with students in an appropriate manner	4,1	3,6	0,5
		Management and support staff deal with all students fairly and equally	4,15	3,25	0,9
		Management of my university gives value to students' feedback	4,3	4,25	0,05
	Administrative Work	Admin and support staff of my university have effective communication skills	3,15	2,7	0,45
		Admin and support staff of my university have comprehensive knowledge of their job	3,0	2,85	0,15
		Admin and support staff of my university keep accurate and up-to-date records	2,85	2,65	0,2
		Administrative processes at my university are clear and well structured	2,5	2,75	-0,25
		Admin and support staff (willing to) solve students' problem on time	3,15	3,1	0,05
		The operating hours of my university are convenient for all	2,5	2,7	-0,2
Employment Quality	Links with Employers	My university has strong relations with the industry for its students' employment	4,55	4,1	0,45
		My university helps its graduates in finding jobs	3,15	2,3	0,85
		My university regularly organizes different job and industry interaction events	4,65	3,5	1,15
	Employability Training	My university has active job placement/work experience service for students	4,7	3,65	1,05
		Different employability seminars are offered by my university to graduating students	4,65	3,3	1,35
		The graduates of my university are highly employable and have high demand in the industry	4,2	3,35	0,85
		My university has a good reputation in the industry based on their graduates' job performance	4,55	3,6	0,95
Safety and Security	Security Measures	My university ensures high standards of safety and security on campus	4,35	3,4	0,95
		Security staff in my university are well trained and professional	4,2	3,5	0,7
		Security staff is provided with suitable and modern equipment	4,1	3,45	0,65
		My university has installed an adequate number of security cameras and metal detectors	3,5	2,9	0,6
	Safety Equipment	My university has installed an adequate number of fire extinguishers	3,8	2,5	1,3
		My university has an adequate number of first aid equipment	2,6	2,3	0,3
		People within my university campus feel secure and confident	4,5	3,7	0,8

(continued)

Table 1 (continued)

Students' Skills Development	Extra-Curricular Activities	My university ensures adequate recreational and sports facilities	4,1	3,3	0,8
		My university provides financial and non-financial support for extra-curricular activities	3,9	3,4	0,5
		My university organizes different extra-curricular activities on a regular basis	4,3	3,25	1,05
	Personal Development	The atmosphere in my university instils leadership and active team player attributes in students	4,0	3,7	0,3
		The atmosphere at my university develops students' soft skills (communication, critical and logical thinking, problem-solving)	4,5	3,95	0,55
		My university environment develops self-confidence and emotional stability in students	4,15	3,8	0,35
		The atmosphere in my university promotes acceptance and equality in students	4,0	3,55	0,45
		My university environment develops the intellectual abilities of students (creativity, decision making, high-quality work)	4,5	3,9	0,6
		My university environment promotes conscientiousness (taking responsibility, time management, organizing things) in students	4,5	3,9	0,6

- **Employment quality:** this theme is further divided into two dimensions, namely, links with employers (three items) and employability training (four items).
- **Safety and security**: this theme is also divided into two dimensions, namely, security measures (four items) and safety equipment (three items).
- **Students' skills development**: this theme is divided into two dimensions, namely, extracurricular activities (three items) and personal development (six items).

Once the scale has been fine-tuned and validated, it meets the necessary requirements for assessing the quality of the service, thus constituting an instrument that we will test.

3 Methodology

The present research follows a mixed method technique to compare the quality of service (SQ) between two HEIs from two different countries (Portugal and Angola), using the HEISQUAL model, consisting of seven dimensions that consider both operational and technical elements (and are subdivided into subdimensions) composed of items that respondents (students from both institutions) evaluated on a five-point Likert scale, where 1, strongly disagree; 2, disagree; 3, neutral; 4, agree; and 5, strongly agree.

The research was conducted in January 2023. Respondents were students, from 18 till 47, of two HEIs, one from Portugal and other from Angola. The resulting sample comprised 365 valid questionnaires from Portuguese HEI (response rate about 15%) and 187 valid questionnaires from Angolan HEI (response rate about 27%). The research instrument was a structured questionnaire based on a five-point Likert scale, developed by Abbas [1] to measure the service quality in HEI. Means of the items were calculated (Table 1) to have a comparison between the two samples and the mean value of each dimension (Table 2 and Fig. 1).

This study aims to investigate, firstly, which are the "strongest" and "weakest" dimensions and to explain some values identified in Table 1 with a circle, namely,

Table 2 HEISQUAL—mean by dimension

HEI service quality dimensions	Portuguese HEI mean	Angolan HEI mean
Teachers' profile	4.2	3.6
Curriculum	4.0	3.7
Infrastructure and facilities	4.4	3.6
Management and support staff	3.3	3.1
Employment quality	4.4	3.4
Safety and security	3.9	3.1
Students' skills development	4.3	3.8

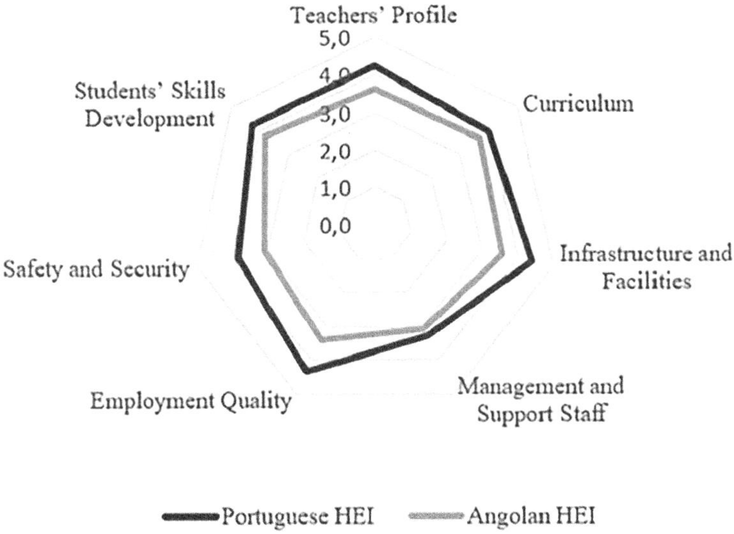

Fig. 1 Comparative HEISQUAL dimensions mean

all means equal to or less than 2.5 (considered "negatives") and all items with GAP above 1.0 (the gap above one point, between the different HEIs, is considered very expressive, and we need to find explanations for such a difference) and the negative GAPs (which represent a greater "satisfaction" of the Angolan students, something that reflects an unusual situation). Focus group sessions were organized in order to collect information/explanations about these results. The participants were 32 from the Portuguese HEI and 16 from Angolan HEI; the ages were between 19 and 33 years. The participants were briefly informed about the objectives of the research in general and were given the option of a focus group session or an individual interview; 12 Portuguese students volunteered for focus group sessions and 10 opted for an individual interview. Nine Angolan students choose focus group sessions, and seven opted for an individual interview.

4 Data Analysis and Results

In addition to filling out the questionnaires that the summarized means were registered in Table 1, the following questions were asked (following [1] example):

1. What are the factors that constitute good service quality in higher education institutions?
2. What are the areas in which your institution is performing well?
3. What are the areas where your institution should improve?

The information obtained from the interviews and focus group, summarized and crossed with the information obtained from the questionnaire, is presented below:

- **Teachers' profile:** almost all participants in the interviews and focus groups highlighted the fundamental role of teachers, although, in terms of averages, the dimensions that stood out the most were "infrastructure and facilities" and "employment quality," with an average 4.4 for Portuguese students and "students' skills development" (3.8) and "curriculum" (3.7) for Angolan students (Table 2). In this dimension, gaps greater than 1 (1.45) were highlighted in items related to language skills and confidence, as Angolan students identified these weaknesses in their teachers.
- **Curriculum:** most of the participants focused on the importance of curriculum concerning understandability, comprehensiveness, and role in developing the skills required.
- **Infrastructure and facilities:** this is one of the dimensions with the highest average for Portuguese students, which demonstrates the importance given to it; however, Angolan students classify some of the items less well, namely, those related to library resources and equipment (air-conditioning, projectors, interactive whiteboards). This classification gave rise to gaps greater than one point and reflects the Angolan reality of the lack of educational infrastructure. Still in terms of accommodation for students and staff, both groups of students classify the accommodation offer as weak, especially the group of Angolan students that gave a negative mark (2, 3).
- **Management and support staff**: similar to teachers, students daily have to interact with the management and support staff of their institution. In this regard, the behavior of management and support staff and the efficiency and effectiveness in their processes are highly important. This topic is the one that is worst evaluated by both groups (3.3 average of Portuguese students and 3.1 of Angolans). Administrative processes and service opening hours are the most critical items.
- **Employment quality:** because of the highly competitive job market and the dynamics of the industry, a large number of participants highlighted the important role of their educational institution in developing employability skills, providing employment training, and the links of their educational institutions with potential employers.

Angolan students identified some shortcomings at this level, evaluating between 3.3 and 3.65, which contrasts with the good evaluation attributed to this service by Portuguese students, which results in gaps above one point.

- **Safety and security:** some students mentioned their concerns about the safety and security services offered by their HEI, particularly concerning facilities in the case of an emergency. Some students mentioned their concerns about the safety and security equipment offered by their HEI, particularly concerning facilities in the case of an emergency (first aid equipment and fire extinguishers)—Angolan students classified these items negatively.
- **Students' skills development:** this theme is focused on the social and personal development aspects of students, as many students believe that, along with academic development, HEIs should also concentrate on students' personality development.

Portuguese students classify this dimension well, which has the second best average.

For Angolan students, this is the dimension with the highest average, given the importance they recognize.

By calculating the averages per dimension, we have the perspective of which dimensions are stronger and which are weaker.

Observing Table 2, we can conclude, and answer question 2, that the most important dimensions, for Portuguese students, are infrastructure and facilities and job quality (4.4) followed very closely by students' skills development (4.3) which is the dimension most valued by Angolan students, for whom the curriculum is the second major dimension.

The areas where both institutions should improve are the administrative processes and the support facilities, especially residences. Particularly, the Angolan HEIs are facing other challenges related with the linguistic competences of professors, some difficulties with administrative staff, and the link with the employers to help students to find a job.

5 Conclusion

Since education is considered by many to be the best tool to promote sustainable development, it is necessary to focus on "quality education" (SDG4) and on access to education for all, reducing inequalities (SDG10).

In this research, the factors that constitute a good quality of service in higher education institutions were identified: the dimensions in which your institution are performing well and the areas in which your institution should improve. We are thus in possession of important information for decision-making. We can suggest measures that may be implemented to improve these situations through benchmarking.

For Angola HEI, it is suggested that there be a clear focus on training teachers and researchers to be fluent in English. The administrative processes must be reviewed and optimized. It is suggested, too, the creation of an internship and employability office that ensures communication between the university and employers, promoting job offers among students. We also propose monitoring the implementation of these procedures, as well as their evolution.

6 Limitations and Future Research Proposals

The limitations of this research work relate to the cultural differences between the two countries [7] and limitations related to data collection (convenience sampling). Data were collected in Portugal and Angola, so the inclusion of other countries can provide opportunities for cross-comparison between different cultures in higher education.

The present study should be interpreted as a starting point for the validation of this instrument for measuring the quality of HEIs (HEISQUAL), from the perspective of the students. Future work intends to increase the sample size to enable more robust analyses. It is also recommended that the same study be replicated in countries with different levels of development to confirm the identified dimensions.

Thus, as future research projects, we propose to extend the analysis to more HEIs, in these same two countries, in an initial phase, with the later possibility of extending it geographically to more countries, namely, Brazil, Cape Verde, and Guinea-Bissau, countries where many of the current master's students come from.

The annual follow-up will be proposed too, to allow the analysis of the impacts of the suggested and implemented good practices, and it would be possible to identify factors and good practices that could be adopted.

Future research can be carried out considering other stakeholders, for example, employers, government, society, etc., as HEIs must also meet their needs.

References

1. Abbas, J. (2020). HEISQUAL: A modern approach to measure service quality in higher education institutions. *Studies in Educational Evaluation, 67*(100933), 1–11.
2. Abdullah, F. (2006). The development of HEdPERF: A new measuring instrument of service quality for the higher education sector. *International Journal of Consumer Studies, 30*(6), 569–581.
3. Cronin, J. J., & Taylor, S. A. (1992). Measuring service quality: A reexamination and extension. *Journal of Marketing, 56*(3), 55–68.
4. Dwaikat, N. Y. (2021). A comprehensive model for assessing the quality in higher education institutions. *The TQM Journal, 33*(4), 841–855.
5. Fuchs, K., & Fangpong, K. (2021). Using the SERVQUAL framework to examine the service quality in higher education in Thailand. *Education Quarterly Reviews, 4*(2), 363–370.
6. Harvey, L., & Green, D. (1993). Defining quality. *Assessment and Evaluation in Higher Education, 18*, 9 34.

7. Langa, P. (2013). *Higher education in Portuguese speaking African countries*. African Minds.
8. Parasuraman, A., Zeithaml, V., & Berry, L. (1985). A conceptual model of service quality and its implications for future research. *Journal of Marketing, 49*, 41–50.
9. Parasuraman, A., Zeithaml, V., & Berry, L. (1988). SERVQUAL: A multiple-item scale for measuring customer perceptions of service quality. *Journal of Retailing, 64*, 12–40.
10. Parasuraman, A., Berry, L., & Zeithaml, V. (1991). Refinement and reassessment of the SERVQUAL scale. *Journal of Retailing, 67*(4), 57–67.
11. Rolo, A., Alves, R., Saraiva, M., & Leandro, G. (2023). The SERVQUAL instrument to measure service quality in higher education – A case study. *SHS Web of Conferences, 160*, 01011.
12. Shah, M., & Nair, C. S. (2016). *A global perspective on private higher education* (1st ed.). Chandos Publishing.
13. Trivellas, P., & Dargenidou, D. (2009). Organisational culture, job satisfaction and higher education service quality. *The TQM Journal, 21*(4), 382–399.
14. Wilkins, S. (2020). The positioning and competitive strategies of higher education institutions in The United Arab Emirates. *International Journal of Educational Management, 34*(1), 139–153.

Exploring the Potential Contribution of Blockchain Technology to Sustainability Reporting

Veysel Akman

Abstract This chapter aims to create a space where contemporary arguments on sustainability, corporations' social responsibility, and blockchain technology can be all discussed together. It offers a new corporate communication tool with a blockchain solution. Corporations are under social pressure for the environmental and social damages they cause. However, their sustainability initiatives and reports are criticized due to the vagueness of their content. This criticism led organizations to practice under the name "Corporate Social Responsibility" (CSR) and standardize performance transparency. But CSR reports may not also provide accurate results because traditional reporting techniques do not allow clear communication for various reasons: lack of standardization, neutrality, and trustworthiness. Considering executives can distort corporate information to include merely positive ones and CSR reporting is only voluntary, it is natural to raise doubts about the reports. After all, CSR activities attempt to achieve more ethically sound practices. Within this framework, taking advantage of developing technology is vital. Blockchain technology promises to meet sustainability requirements and ensure report reliability. The outcomes of pilots show using blockchain technology in the sustainability area offers potential. Its key features—decentralized trust, traceability, and immutability—make blockchain a feasible option for CSR reporting too. In this chapter, we identify the problems of CSR reports and show the suitability of using blockchain in the reporting area. We followed the article called *"A Ten-Step Decision Path to Determine When to Use Blockchain Technologies"* and answered their questions about using blockchain technology for sustainability reporting. This question set helps to identify needs, participants' interactions, behaviors, and community rules. Our analysis shows that using blockchain for CSR reporting is a viable option. Blockchain-based sustainability reporting mechanisms may force companies to be more transparent in their initiatives. Such a corporate communication tool prevents distortion of public information and increases company visibility.

V. Akman (✉)
University of Warsaw, Warsaw, Poland

J. Reis et al. (eds.), *Driving Quality Management and Sustainability in VUCA Environments*, Springer Proceedings in Business and Economics,
https://doi.org/10.1007/978-3-031-52723-4_17

Keywords Blockchain · Corporate Social Responsibility · Sustainability reporting

1 Introduction

The exact definition of sustainability has been a matter of debate for a long time. The only certainty about the notion is that it fluctuates based on time scale and place. Bell and Morse [2] summarize this dependence as *"spatial and time scales are key components of achieving sustainability; ..., they are problematic in the sense that careful selection of scale or reference point can be used to prove almost anything."* In such an age, the word sustainability has become one of the wildcard concepts that policymakers, economists, and scientific papers regularly refer to [29].

Politically, sustainable development is one of the main goals of almost all governments. Governments are willing to evaluate development's performance not only in terms of economics but also in terms of improvements in environmental and social conditions [3]. On the other side, an increasing argument is how economic sectors or corporations contribute to these objectives [52]. Businesses, which are the pioneers of economic and social development, have been criticized for the environmental and social damages they cause.

This criticism has led institutions to adopt some practices under the name of "Corporate Social Responsibility" (CSR). CSR is a voluntary management technique that aims to make corporate operations more ethical. With systematic and standardized CSR, corporate transparency, reliability, strong participation of stakeholders, and cooperation between sectors can be achieved [34]. Stakeholders have led organizations to strive for a comprehensive assessment of environmental and social conditions as well as economic impacts. However, an issue that sustainability performance is difficult to assess [3].

Today, most multinational companies prepare periodic reports in which they explain their social responsibility activities. As a natural extension of financial reporting, social responsibility reports, also called sustainability reports, expand the scope of business reporting by explaining the environmental, economic, and social performance of the business [12].

However, these sustainability reports, intentionally or unintentionally, may not always provide useful results. Existing CSR reporting methods do not allow successful communication for various reasons, such as lack of thoroughness, neutrality, relevance, and trustworthiness. Reporting sustainability information is seen as a symbolic technique to positively manage stakeholder views [20].

CSR reports from corporations show that the reports come in a variety of formats, in terms of both content and quality. The reliability of CSR reports raises doubts due to uncontrolled and irregular reporting. Considering that managers distort corporate information to include only positive ones and that CSR reporting is based on a voluntary basis, it is natural to raise doubts about the reports [43].

Despite the above-mentioned issues, it should not be overlooked that CSR is an attempt to achieve a more equitable and environmentally friendly goal. While CSR reports may appear to many to be empty self-aggrandizement, experts argue that excellent benefits may be reached by refining the model [51]. A new technology that promises to meet sustainability requirements and is believed to be essential for industries to ensure their reliability in changing market conditions is blockchain technology [23].

Although the origins of blockchain technology may be traced back to the early 1980s, the notion was first conceptualized in 2009. The blockchain concept was combined with various technologies in 2008 to produce modern cryptocurrencies: electronic transactions secured by cryptographic processes rather than a central repository or authority. With the publication of the Bitcoin network in 2009, the first generation of modern cryptocurrencies, this technology became well known. The transfer of digital information that represents electronic cash occurs in a distributed system like Bitcoin and comparable systems [60].

Rejeb and Rejeb [45] define blockchain technology as a distributed database of transactions executed and shared among networked participants. Blockchain technology is reliable as it provides a system of records that makes it very difficult or almost impossible to change information. Each new transaction record must be added to many ledgers before each transaction can be verified and recorded in the system. This requires changing the data in each ledger, i.e., each node, to change anything in the transaction. In this sense, the system makes it almost impossible to hack the blockchain [24].

Although Nakomoto [37] focused his attention purely on the monetary market, the repercussions of the concept showed in the 2010s that the potential of blockchain is not limited to the financial realm. Many people have recognized the success of blockchain technology in financial reporting, and many well-known firms continue to explore the application areas of the technology. It is also suggested that blockchain technology, which has already been shown to be useful in various industries, has the potential to provide environmental and social advantages in line with the UN's Sustainable Development Goals [56].

Taking into account the flaws of current sustainability reporting methodologies, the use of blockchain technology in the field of sustainability offers the potential for solving problems. As mentioned earlier, the key features of trust, traceability, and immutability make blockchain a viable solution for reporting in this industry. However, it would be optimistic to expect companies to transparently share information about their operations publicly. A trustable gateway between external world data and blockchain is an enduring issue called the Oracle problem [6]. In other words, when the source of the information is untrustworthy, storing it on a blockchain may not seem like a solution. The added value of blockchain technology, on the other hand, stems from its source traceability and immutability. Once information is published, whether accurate or false, it cannot be changed. This irrevocable publication puts the corporation under pressure to submit accurate data. A corporation that is aware that the information given will be available in perpetuity will be incentivized to provide the most accurate information possible.

The remainder of this chapter is organized as follows: In Sect. 2, we introduce the concepts of sustainability and corporate social responsibility. In Sect. 3, we provide a technological background on blockchain technology and its relationship with sustainability. Following that, we explored the relationship between sustainability and blockchain technology in Sect. 4. In Sect. 5, we follow a decision path to analyze the suitability of blockchain technology in the area of reporting. Lastly, we concluded the research in Sect. 6.

2 Sustainability and Corporate Social Responsibility

The historical roots of sustainability, and especially the understanding of the concept of limited resources, go back to ancient times. Deforestation, soil salinization, and productivity losses experienced in Ancient Egypt, Mesopotamia, Greek, and Roman civilizations are today discussed under the title "sustainability problems" [16]. Ecologically conscious consumption has been an important practice throughout history in order to ensure the continuity of the human species.

However, international recognition of sustainability was achieved at the 1972 UN Conference on the Human Environment, the so-called Stockholm Conference. Although the term did not appear until the 1980s, 114 government delegates at the Stock Conference agreed that development and the environment should be considered mutually, as they are linked at a critical point. The Stockholm Declaration, which includes 26 principles, has placed environmental issues at the forefront of international concerns [57]. The establishment of such an internationally recognized organization has been a turning point in the field of sustainable development and paved the way for the institutionalization of environmental awareness.

Following this awareness, by showing a responsibility beyond the economic performance of companies, pressures have increased on all stakeholders to show sustainability performance [44]. The corporate sustainability discussion is rooted in companies' approaches to sustainability. This term is used with some concepts, such as sustainable development and CSR, in some articles. Corporate sustainability acknowledges the growth and profitability of the company but also states that the company's social objectives related to sustainable development, such as environmental protection, social justice, and equity, and economic development, must be fulfilled [59].

According to Garcia et al. [20], "*CSR reporting demonstrates firms' willingness to discharge accountability and respond to stakeholders' information demands.*" Companies are expected to explain their economic, social, and environmental objectives and plans with transparency as a result of the social consciousness that has arisen.

Additionally, in today's society, customer preferences do not only depend on how environmentally friendly a product is. Green marketing also has the potential to be problematic and should not be overlooked. People tend to be skeptical of green claims at first. So much so that if a green claim is found to be false or inconsistent with other services, it can seriously damage the company's brand reputation. As consumer society becomes more conscious, it tends to make more informed decisions [4].

Sustainability reports, whether intentionally or unintentionally, do not always provide useful results. Current CSR reporting processes do not allow successful communication for various reasons, such as lack of completeness, objectivity, relevance, and reliability. Firms provide comprehensive narrative information. As a result, sharing sustainability information is seen as a symbolic technique for positively managing stakeholder views [20].

When the CSR report examples are examined, it is clear that the reports are in various formats in terms of both content and quality. Non-standardized and voluntary reporting techniques have forced businesses to take advantage of CSR reports. The reliability of CSR reports raises doubts due to uncontrolled and irregular reporting. Moreover, managers' incentives to distort corporate information in a positive way can hinder the credibility of CSR reports [32, 43]. As a result, firms face difficulties in increasing the credibility of their CSR reports and thus their capital market impact [36]. It has become increasingly difficult to fill in performance indicators that are impossible to track, especially future disclosures.

Despite the above-mentioned issues, it should not be overlooked that CSR is an effort to achieve a more equitable and environmentally friendly goal. Although CSR reports, which are a frequently discussed topic in the literature, may appear to many as empty self-glorification, experts argue that excellent benefits can be achieved by refining the model [52]. The main problem is thought to be that the concept has not yet been fully developed.

Faber and Hadders [18] discuss available reporting formats in their article. Accordingly, it is seen that existing reporting methods have a significant vulnerability and do not have reference points [18]. Grigoris (2016) classified the challenges in CSR reports in his article and divided them into nine categories. The highlights are lack of transparency, proposal of general criteria, negligence of the main business stakeholder, and failure to introduce criteria related to CSR results [22]. The most important challenge affecting the categorization process is the lack of transparency and traceability.

Due to the characteristics of the CSR reports outlined above, all stakeholders questioned their reliability. Reports for a specific audience have deviated from their targets in recent years due to the pages full of reports that cannot be tracked. Enlightening all stakeholders, including the local community, which is the main purpose of the reports, is not feasible due to the highly complex datasets. Unfortunately, due to hard-to-understand information, it is not possible to monitor whether companies fulfill their commitments.

To summarize the above discussion, some of the problems with CSR reports are as follows:

- Lack of completeness, objectivity, and reliability
- Non-standardized and voluntarily based
- Lack of credibility and transparency

In today's sustainability understanding, the activities of enterprises related to environmental and social concerns attract everyone's attention. As a result, reports should be written at a level that everyone can understand and follow. An easy-to-access version of CSR can play a role in reducing these concerns and increasing the credibility of companies. In order to do this, it is vital to take advantage of developing technology. A new technology that promises to meet sustainability requirements and is believed to be essential for industries to ensure their reliability in changing market conditions is blockchain technology [23].

3 Blockchain Technology

The 2007–2008 financial crisis had a significant impact on all countries. The so-called Great Recession had such an impact that the public's opinion of institutions such as banks has been brought into doubt. Bankruptcy announcements by well-known institutions also caused ripple effects. This chain of events raised concerns about the stability and transparency of such organizations [26].

The fact that the value of money in circulation is not based on a finite resource was the primary cause of the financial crisis. The global financial crisis exposed the reality that money may be created an infinite number of times without attaching any value to it. Yeldan [61] highlights the crisis that *"the 2008 global crisis was accompanied by a financialization process characterized by artificial values, in which money was not tethered to any gold or precious metal standard in human history"* [61]. On the same subject, Schinckus [48] points out the information asymmetry between the financial industry and clients as the main cause of the crisis.

For many practitioners, the emergence of Bitcoin in the same year that the financial crisis happened is not a coincidence [54]. But it should be emphasized that digital currencies existed before Bitcoin. In 1982, cryptographer David Chaum proposed the concept of electronic money. Taking the initial steps toward the token concept, Chaum created a "blinding formula" to encrypt information transmitted between persons. It has been accomplished in this manner without the necessity to reveal the personal information of persons who transfer between individuals [7]. Chaum conceptualized his ideas and established his own company, Digicash, in 1989. The ideas and concepts were so influential that the base formulas and methods were milestones in the evolution of current digital currencies, although DigiCash went bankrupt in 1998 [8, 54].

During the same decade, cryptographer Nick Szabo's BitGold was the first to envision a peer-to-peer network, mining, a ledger or registry, and cryptography, all of which are fundamental characteristics of most blockchain applications today. The most notable characteristic of BitGold was its decentralized network. Szabo attempted to demonstrate that a value transaction is achievable without the use of an intermediary. Later, the BitGold project, like many other prior Bitcoin attempts, failed, but his work provided the groundwork for Bitcoin [54].

In blockchain technology, information is recorded in blocks. Blockchain, which is a distributed ledger, stores any transaction list in a peer-to-peer network [42]. Blockchain, an immutable database, is a distributed ledger that securely transfers data. The term blockchain is a combination of two words. The word “block” represents batches, while the word “chain” represents cryptographically linked blocks. After the transactions are confirmed, the block is created, timestamped, and added to the previous block [38]. The data in the blockchain are stored in blocks in an encrypted form. Each block contains encrypted summary information about the previous block. In this way, all data are linked to each other in the form of a chain. The fact that the blocks are connected to each other in this way creates a more reliable structure against hacker attacks. Unlike classical database systems, blockchain stores data by distributing it on the Internet instead of collecting it in a single center.

A blockchain is a database without central control. Transparency is ensured by distributing the data to all devices in the network in an encrypted manner. Information registered to the blockchain cannot be changed or deleted [14]. The inability to change and delete information may raise the question of what to do when there are incorrect information records. In this case, the data are re-added to the system, and both the incorrect information and the updated information are stopped in the record. That is, the old wrong and the new correct record are recorded in the system.

One of the key advantages of blockchain is that it enables consensus in an unreliable environment. With the intelligent system it provides, individuals or institutions can reach an agreement on a proven blockchain without having to trust each other. In traditional systems, a third-party central authority is needed to approve transactions, which creates several bottlenecks in the industry. Blockchain technology, on the other hand, eliminates such third parties, cuts expenses, and saves time. Fundamental features of blockchain technology, such as immutability, reliability, security, collaboration, and decentralized structure, are among the most required features in the industry [53].

The fundamental features of blockchain are summarized as below in [63]:

- **Decentralization:** The validation of a transaction between two parties that do not trust each other must be provided in some way. Traditionally, validation is performed by a centrally trusted third party, such as a bank, notary, etc. In other words, centralized authority ensures trust between parties. This traditional validation incurs additional costs and may lead to complications. A third-party validator is not required in a blockchain transaction. The validation procedure is specified and executed automatically by the nodes, which is referred to as the consensus mechanism. The records may be seen and validated by any node in the blockchain network. Control power is distributed over the network, which can save costs while also speeding up procedures.
- **Data Consistency:** Once the transaction is completed and approved by the network, it is not possible to alter the data in the chain. All nodes have access

to the full network's transaction history, and any effort to modify the validated data is easily detectable.
- **Anonymity:** Each node in the network has two unique credentials, similar to an ID or a bank account number. One is similar to an account number that can be shared with everyone, while the other is similar to a password that is exclusive to the individual. Accounts can be replicated as needed, and data security is provided by a technique that is fully independent of one another and cannot be predicted. Because of a characteristic called the Avalanche effect, there is no relationship or similarity between the data submitted to the blockchain and the block number generated. This approach protects users' anonymity and prohibits organizations from misusing personal data.
- **Traceability:** As the title indicates, tracing the history of blocks in the blockchain, which is a type of interconnected data block, is extremely simple. Data history tracking is relatively simple due to the time when blocks connected to each other in a specific order and encryption mechanism are recorded and signed.

4 Blockchain for Sustainability: A New Relationship

The applicability of blockchain to sustainability has started to be discussed in the last few years, in both academia and business. Many early studies are focused on the topic of direct blockchain ecosystems' sustainability. The reason for this dominance in the literature is mainly due to disproved critiques of bitcoin-mining-related high energy consumption [11, 15, 30, 31, 33, 55, 58]. Later successful De-Fi applications bring popularity to the underlying technology and open the path for further development.

In order to understand the connection between blockchain and sustainability for environmental and social pillars, Jayawardhana and Colombage [24] highlighted the importance of a unique definition of the terms and summarized the relationship as follows:

> *Good governance leading to responsible actions is required to prevent any adverse effect and/or promote positive effects of blockchain technology or cryptocurrencies on any of the three types of sustainability.* [24]

From this perspective, it can be inferred that a positive combination of blockchain and sustainability can be revealed through the proper governance of the affected system. Leading institutions in the sustainability area called governments and companies for action on blockchain technology [17].

To categorize the benefits and channel the relevant efforts, SDGs are suitable tools. Schinckus [48] conducted research to match the advantages of blockchain applications and the objectives they serve. He summarized that the decentralized, incorruptible, and transparent nature of blockchain technology may decrease fraud (goals 8, 10) and improve food and overall supply chain security (goals 2, 3, 12, 14, and 15) [48].

In a similar vein, Truby [55] discusses the blockchain's traceability feature and potential contribution to sustainable development. He highlights that blockchain applications can solve the problem of addressing negative externalities. On the same topic, Le Sève et al. [27] argued on the externality problem in their article "Delivering blockchain's potential for environmental sustainability." The authors concluded that "blockchain is intuitively suited to decentralized or collective management of common pool resources, given that the technology itself functions through distributed consensus rather than central control" [27].

The Civic Ledger Company is a well-known example of how blockchain can support externality problems. Water ledger provides a decentralized marketplace for water management systems [28]. The project aims to streamline the water trade and inform farmers, industrial users, etc. Besides professionals, the Water Ledger marketplace is also publicly accessible for environmental information.

Moreover, IBM Corporation launched several projects with blockchain technology to explore the potential of IoT sensors. In the project, IoT remote sensors were used to monitor groundwater resources in real time. All the data are in the blockchain database [10]. One of the most important outcomes of the project is that the information asymmetry in the ecological data is clarified, and a space for stronger market mechanisms is allocated.

Carbon footprint tracking is another area where blockchain can contribute with its inherited features [39, 55]. The World Economic Forum introduced a digital carbon currency in order to enhance the efficiency of carbon credit mechanisms [50]. With this blockchain initiative, information asymmetry during the carbon credit approval problem is targeted. From a similar perspective, Franke et al. [19] conducted a feasibility study for using blockchain technology in the carbon market mechanism. In the research, it is concluded that current blockchain applications promise improvement in carbon credit [19]. Moreover, Patel et al. [40] have proposed a token-based economic model for carbon trading mechanisms at the corporate level. With the help of blockchain technology, a transparent and decentralized carbon market at the global level is aimed at. In the study, it is concluded that implementing a blockchain-based solution to the carbon trading mechanism has a tremendous impact on the participation of parties, usefulness, and cost [40]. As explained, one of the most important contributions of such a mechanism is to be able to deduce the effects of pollution based on the location, and this helps to internalize the externalities.

5 The Decision Path to Use Blockchain in CSR Reporting

As mentioned earlier, the key features of blockchain make it a viable solution for sustainability reporting. At the same time, it is known that there are several advantages and disadvantages to blockchain applications [21]. Blockchain should not be used for every issue that we want to tackle. The article called *"A Ten-Step Decision Path to Determine When to Use Blockchain Technologies"* helps practitioners decide whether to use blockchain technology or not in their application area [41]. In this way, one can decide whether a blockchain implementation is beneficial or whether traditional applications will be satisfactory. In what follows, we will answer all the key questions from the article in the context of CSR reporting:

1. **Is There a Need for a Shared Common Database?**
   Yes, a shared common database is needed for CSR reporting since several parties should be involved and interact with each other. The potential interactions may include a preparation phase or checking investment opportunities for stakeholders with the reports.
   On the other hand, the reason for sharing a common database can come from the ecological consciousness of the customers. The research company Clutch conducted a survey with 420 consumers in the United States to explore the influence of CSR on the perception of brands and their consumption habits. The company found out that *"seventy-one percent (71 percent) of people think it's important for businesses to take a stance on social movements. ... More than half of people (59 percent) are likely to stop shopping at a company that supports an issue they disagree with"* [13]. The company concluded that it is crucial for businesses to demonstrate their social and environmental performance. This survey demonstrates that a company's production practices are influential on customer decisions. Therefore, the data for all operations must be in a public database.
2. **Are Multiple Parties Involved?**
   Yes, multiple parties should be involved in the CSR reporting process. As explained above, investors, consumers, and all possible business partners of the company need to have at least a read access to the company's performance data. Another important party that should be taken into consideration is international organizations, which set the rules of reporting for companies. Such institutions are responsible for ensuring the compliance of corporations. Another proof of CSR reporting requires multiple parties; some institutions provide the standards for CSR communications, such as GRI.
3. **Do the Involved Parties Have Conflicting Interests and/or Are they Trusted?**
   The need for new technology in the corporate reporting area has been raised due to public skepticism against current CSR tools. The cause of such skepticism is the outcome of firms separating their actual activities and masking them with marketing tools. The several scandals that have brought huge public awareness have raised questions about the purpose of CSR tools. See [5, 9, 25, 35].
   The above-mentioned scandals and many other corporate failures addressing communication with consumers resulted in the perception of CSR merely for financial purposes rather than being eco-friendly practices [49]. It is easy to say that trust between companies and consumers is broken, and traditional CSR tools are not satisfactory for the public anymore.

4. **Can or Do the Participants Want to Avoid a Trusted Third Party?**
Companies are working collaboratively with third-party auditors in order to overcome mistrust in CSR tools. However, this method is also criticized by many practitioners in the sense that *for assessing the accuracy of sustainability information is rarely explained in the assurance report, and many focus only on "quantitative indicators that can be verified using recalculation"* [1].
On the other side, Sarajoti et al. [47] highlight the importance of trusted third parties in the process. But their functions may be transformed, and their drawbacks may be overcome with blockchain integration into reporting [47].
5. **Do the Rules Governing System Access Differ Between Participants?**
In this step, the accessibility of the database is examined. When it comes to shared access, traditional databases can cover this requirement adequately. However, when it comes to giving different roles to involved parties such as validators, users, etc., the capabilities of blockchain are still a better alternative. In the idealized solution offered by the institutions above, third-party companies have validating access, the companies can have upload data access, and for sure, anyone who is interested in the data needs read access.
6. **Do the Rules for Transacting Remain Largely Unchanged?**
Once a rule is set for the blockchain application, it is not possible to change it again; even the admin user cannot alter the data technically. Therefore, pre-agreed standardization is crucial for blockchain applications. Due to the non-standardized nature of sustainability reporting, pre-agreed standardization is challenging. Several organizations have attempted to bring some standardization to the area, such as the Global Reporting Initiative, ISO, EcoVadis, etc., but CSR is a voluntary practice, and those guidelines are not followed collaboratively. Since there is no common practice for sustainability reporting, this topic makes immutability both a challenge and an advantage. Due to trust issues, the immutable database is definitely the preferred alternative; however, standardization is needed for sustainability reporting. The standardization of sustainability reporting is another area that requires a lot of attention; however, it is outside the scope of this research study.
7. **Is There a Need for an Objective, Immutable Log?**
Pederson et al. identify the benefits of tamper-proof data and its transaction history for auditing purposes [41]. The feature of immutable logs and data makes blockchain a suitable solution, especially when it is combined with other technologies such as RFID, IoT remote sensing, etc. The potential of this combination highlighted by Cruz and Cruz, especially for end-to-end visibility over supply chains [46].
The article "The Use of Blockchains to Enhance Sustainability Reporting and Assurance*" explores the capabilities of blockchain technology in the area of reporting sustainability. Blockchain solutions have huge potential for supply chain visibility thanks to their traceability features. Such solutions help to mitigate the counterfeiting and trafficking of illegal goods and drugs. The improvement areas with the help of blockchain are identified as real-time

reporting, human effort, and error during the reporting process, and all of these advancements are reducing costs significantly [1].

8. **Is Public Access Required?**
   Yes, public access to published information is definitely required. In an idealized solution, a company will be able to upload their social and environmental performance to the blockchain and any stakeholder, which means the public should have access to the mentioned information. This will pressure companies to be accurate and, at the same time, force mechanisms to follow their disclosures. Companies will be more accountable for their published data. Thanks to blockchain-based CSR reporting instruments, companies will be in front of the public, which will eventually overcome the main problems of CSR reports.

On the next steps of the decision path, the questions are solely focused on the technical details of the different blockchain types.

All of the above-given cases are applicable to CSR reporting activities, and a permissionless public blockchain represents an ideal solution for CSR reporting. Blockchain technology has huge potential to advance the reporting process.

6 Conclusion

This chapter discusses the potential of blockchain technology in the area of sustainability reporting. Conceptual research has been conducted in order to clarify the perceptions of sustainability and blockchain, respectively. It is summarized that blockchain technology's key characteristics are decentralized governance, data persistency, and auditability [62]. Various applications that use blockchain in the area of sustainability have been analyzed in the literature.

The responsibilities of companies in terms of sustainability are highlighted. In the same manner, the CSR concept is extended, and the vulnerabilities of current CSR reporting practices have been identified. We have concluded that accurate, consistent, and transparent operations data of companies must be shared with the public transparently.

One of the main challenges with sustainability reporting is the lack of a standardized format and data to be included in the reports. The content of the reports requires separate, thorough research, which is beyond the scope of this chapter. Nevertheless, blockchain facilitates a repository for any environmental data and aids in tracking the source owner to determine whether the data provided are accurate or false.

As explained before, traditional CSR reporting techniques lack credibility. This trust concern is caused by the above-explained scandals, the complexity of reports, disclosure-filling failures, etc. After the literature review, we have verified that innovation in the area of communicating CSR is needed. The inherited features of blockchain technology make it viable for the CSR reporting area. To test the suitability of blockchain technology for CSR reporting, a ten-step decision path has been used. In the article called *A Ten-Step Decision Path to Determine When to Use Blockchain Technologies*, authors claim about the paper that *"the series of simple yes-no questions will help practitioners decide when to use a blockchain and which type of blockchain technology to deploy"* [41]. This questionnaire is answered in a systematic way, and it is deduced that blockchain applications can be used for CSR reporting.

References

1. Bakarich, K. M., Castonguay, J. J., & O'Brien, P. E. (2020). The use of blockchains to enhance sustainability reporting and assurance. *Accounting Perspectives, 19*(4), 389–412.
2. Bell, S., & Morse, S. (2012). *Sustainability Indicators: Measuring the immeasurable?*. Routledge.
3. Böhringer, C., & Jochem, P. E. P. (2007). Measuring the immeasurable—a survey of sustainability indices. *Ecological Economics, 63*(1), 1–8.
4. Bukhari, S. S. (2020). Green marketing and its impact on consumer behavior. *European Journal of Business and Management, 3*(4), 375–383.
5. Burnett, M., & Welford, R. (2007). Case study: Coca-Cola and water in India: Episode 2. *Corporate Social Responsibility and Environmental Management, 14*(5), 298–304.
6. Caldarelli, G. (2020). Understanding the blockchain oracle problem: A call for action. *Information, 11*(11), 509.
7. Chaum, D., Rivest, R. L., & Sherman, A. T. (1982). In *Proceedings of Crypto 82*.
8. Chaum, D., Rivest, R. L., & Sherman, A. T. (1998). Crypto'82. In *Advances in Cryptology 1981–1997* (pp. 13–19). Springer.
9. Cherry, M. A., & Sneirson, J. F. (2010). Beyond profit: Rethinking corporate social responsibility and greenwashing after the bp oil disaster. *Tulane Law Review, 85*, 983.
10. Chohan, U. W. (2019). Blockchain and environmental sustainability: Case of IBM's blockchain water management. In *Notes on the 21st Century (CBRI)*.
11. Chowdhury, M. J. M., Colman, A., Kabir, M. A., Han, J., & Sarda, P. (2018). Blockchain versus database: A critical analysis. In *2018 17th IEEE International Conference on Trust, Security and Privacy in Computing and Communications/12th IEEE International Conference on Big Data Science and Engineering (TrustCom/BigDataSE)* (pp. 1348–1353). IEEE.
12. Clikeman, P. M. (2004). Return of the socially conscious corporation: How can you reap the rewards of good corporate citizenry? Be one–and broadcast it! *Strategic Finance* 22–27.
13. Cox, T. A. (2019). *How corporate social responsibility influences buying decisions*. Clutch.
14. Crosby, M., Pattanayak, P., Verma, S., Kalyanaraman, V., et al. (20161). Blockchain technology: Beyond bitcoin. *Applied Innovation, 2*(6–10), 71.
15. De Vries, A. (2019). Renewable energy will not solve bitcoin's sustainability problem. *Joule, 3*(4), 893–898.
16. Du Pisani, J. A. (2006). Sustainable development–historical roots of the concept. *Environmental Sciences, 3*(2), 83–96.
17. Dumitriu, P. (2020). Blockchain applications in the united nations system: Towards a state of readiness. United Nations, Report of the Joint Inspection Unit.
18. Faber, N. R., & Hadders, H. (2015). Rethinking the social contract: Measuring and reporting sustainability in context. In *Global Cleaner Production & Sustainable Consumption Conference*. Elsevier Science.
19. Franke, L., Schletz, M., & Salomo, S. (2020). Designing a blockchain model for the Paris agreement's carbon market mechanism. *Sustainability, 12*(3), 1068.

20. Garcia-Torea, N., Fernandez-Feijoo, B., & De La Cuesta, M. (2020). CSR reporting communication: Defective reporting models or misapplication? *Corporate Social Responsibility and Environmental Management, 27*(2), 952–968.
21. Golosova, J., & Romanovs, A. (2018). The advantages and disadvantages of the blockchain technology. In *2018 IEEE 6th Workshop on Advances in Information, Electronic and Electrical Engineering (AIEEE)* (pp. 1–6). IEEE.
22. Grigoris, G. (2016). The challenges of corporate social responsibility assessment methodologies.
23. Jamwal, A., Agrawal, R., Sharma, M., Kumar, V., & Kumar, S. (2021). Developing a sustainability framework for Industry 4.0. *Procedia CIRP, 98*, 430–435.
24. Jayawardhana, A., & Colombage, S. (2020). Does blockchain technology drive sustainability? An exploratory review. pp. 17–42. https://doi.org/10.1108/S2043-052320200000015002
25. Jung, J. C., & Sharon, E. (2019). The Volkswagen emissions scandal and its aftermath. *Global Business and Organizational Excellence, 38*(4), 6–15
26. Kazgan, G. (2012). 2008 küresel krizi: Nedenleri, etik ilkeleri ve iktisat eğitimi. Technical report, Discussion Paper.
27. Le Sève, M. D., Mason, N., Nassiry, D., et al. (2018). Delivering blockchain's potential for environmental sustainability. In *ODI Briefing Note*.
28. Ledger, C. (2021). Water ledger.
29. Lélé, S. M. (1991). Sustainable development: A critical review. *World Development, 19*(6), 607–621.
30. Loviscach, J. (2012). The environmental cost of bitcoin.
31. McCook, H. (2014). An order-of-magnitude estimate of the relative sustainability of the bitcoin network. *A critical assessment of the Bitcoin mining industry, gold production industry, the legacy banking system, and the production of physical currency*, 2, 25.
32. Mishra, S., & Modi, S. B. (2013). Positive and negative corporate social responsibility, financial leverage, and idiosyncratic risk. *Journal of Business Ethics, 117*(2), 431–448.
33. Mora, C., Rollins, R. L., Taladay, K., Kantar, M. B., Chock, M. K., Shimada, M., & Franklin, E. C. (2018). Bitcoin emissions alone could push global warming above 2 C. *Nature Climate Change, 8*(11), 931–933.
34. Moravcikova, K., Stefanikova, L., & Rypakova, M. (2015). CSR reporting as an important tool of CSR communication. *Procedia Economics and Finance, 26*, 332–338.
35. Mukonza, C., Hinson, R. E., Adeola, O., Adisa, I., Mogaji, E., & Kirgiz, A. C. (2021). *Green marketing in emerging markets: Strategic and operational perspectives*. Springer Nature.
36. Muslu, V., Mutlu, S., Radhakrishnan, S., & Tsang, A. (2019). Corporate social responsibility report narratives and analyst forecast accuracy. *Journal of Business Ethics, 154*(4), 1119–1142.
37. Nakamoto, S., & Bitcoin, A. (2008). A peer-to-peer electronic cash system. *Bitcoin, 4*, 2. https://bitcoin.org/bitcoin.pdf
38. Natalia Maslova, C. M. A., & PM P, C. T. P. (2018). Blockchain: Disruption and opportunity. *Strategic Finance, 100*(1), 24–29.
39. Parmentola, A., Petrillo, A., Tutore, I., & De Felice, F. (2022). Is blockchain able to enhance environmental sustainability? A systematic review and research agenda from the perspective of sustainable development goals (SDGs). *Business Strategy and the Environment, 31*(1), 194–217.
40. Patel, D., Britto, B., Sharma, S., Gaikwad, K., Dusing, Y., & Gupta, M. (2020). Carbon credits on blockchain. In *2020 International Conference on Innovative Trends in Information Technology (ICITIIT)* (pp. 1–5). IEEE.
41. Pedersen, A. B., Risius, M., Beck, R., et al. (2019). A ten-step decision path to determine when to use blockchain technologies. *MIS Quarterly Executive, 18*(2), 99–115.
42. Raj, P. (2021). Demystifying the blockchain technology. In *Advances in computers* (Vol. 121, pp. 1–42). Elsevier.
43. Ramanna, K. (2013). A framework for research on corporate accountability reporting. *Accounting Horizons, 27*(2), 409–432.

44. Ramos, T. B., Cecílio, T., Douglas, C. H., & Caeiro, S. (2013). Corporate sustainability reporting and the relations with evaluation and management frameworks: The Portuguese case. *Journal of Cleaner Production, 52*, 317–328.
45. Rejeb, A., & Rejeb, K. (2020). Blockchain and supply chain sustainability. *Logforum, 16*(3):363–372. https://doi.org/10.17270/J.LOG.2020.467
46. Rosado da Cruz, A., & Cruz, E. (2020). Blockchain-based traceability platforms as a tool for sustainability. In *Proceedings of the 22nd International Conference on Enterprise Information Systems. SCITEPRESS-Science and Technology Publications.*
47. Sarajoti, P., Chatjuthamard, P., Papangkorn, S., & Phiromswad, P. (2022). *CSR reporting and blockchain technology*. IntechOpen.
48. Schinckus, C. (2020). The good, the bad and the ugly: An overview of the sustainability of blockchain technology. *Energy Research & Social Science, 69*, 101614.
49. Schmeltz, L. (2014). Identical or just compatible? the utility of corporate identity values in communicating corporate social responsibility. *International Journal of Business Communication, 51*(3), 234–258.
50. Schwab, K., & Zahidi, S. (2020). The global competitiveness report: How countries are performing on the road to recovery. In *World economic forum.*
51. Sheehy, B. (2015). Defining CSR: Problems and solutions. *Journal of Business Ethics, 131*(3), 625–648.
52. Sheehy, B, & Farneti, F. (2021). Corporate social responsibility, sustainability, sustainable development and corporate sustainability: What is the difference, and does it matter? *Sustainability, 13*(11), 5965.
53. Shrimali, B., & Patel, H. B. (2022). Blockchain state-of-the-art: Architecture, use cases, consensus, challenges and opportunities. *Journal of King Saud University-Computer and Information Sciences, 34*(9), 6793–6807.
54. Taskinsoy, J. (2019). Blockchain: A misunderstood digital revolution. things you need to know about blockchain. In *Things you need to know about blockchain.*
55. Truby, J. (2018). Decarbonizing bitcoin: Law and policy choices for reducing the energy consumption of blockchain technologies and digital currencies. *Energy Research & Social Science, 44*, 399–410.
56. Truby, J. (2018). *Using bitcoin technology to combat climate change*. Nature Middle East.
57. *United Nations Conference on the Human Environment*, Stockholm 1972, 1972.
58. Vranken, H. (2017). Sustainability of bitcoin and blockchains. *Current Opinion in Environmental Sustainability, 28*, 1–9. Sustainability governance.
59. Wilson, M. (2003). Corporate sustainability: What is it and where does it come from. *Ivey Business Journal, 67*(6), 1–5.
60. Yaga, D., Mell, P., Roby, N., & Scarfone, K. (2019). Blockchain technology overview. arXiv preprint arXiv:1906.11078.
61. Yeldan, E. (2009). Kapitalizmin yeniden finansallaşması ve 2007/2008 krizi: Türkiye krizin neresinde. *Çalışma ve Toplum, 1*(20).
62. Zheng, Z., Xie, S., Dai, H.-N., Chen, X., & Wang, H. (2018). Blockchain challenges and opportunities: A survey. *International Journal of Web and Grid Services, 14*(4), 352–375.
63. Zheng, Z., Xie, S., Dai, H., Chen, X., & Wang, H. (2017). An overview of blockchain technology: Architecture, consensus, and future trends. In *2017 IEEE International Congress on Big Data (BigData Congress)* (pp. 557–564).

Erasmus+ 2021–2027 in Higher Education: Contributions to the 2030 Agenda and to the Sustainable Development Goals

Teresa Nogueiro, Margarida Saraiva, and Ana Rolo

Abstract The 2030 Agenda for Sustainable Development, approved by all UN member states in 2015, includes the 17 Sustainable Development Goals (SDGs), which are seen by Boeren (Int Rev Educ 65:277–294, 2019) as a program of actions not only for people but also for the planet and prosperity. These goals are essentially intended to help nations work together to build a cleaner world and a fairer global society while safeguarding the environment and are aligned with those advocated by the Erasmus+ program. Both the program and the SDGs have goals to be achieved and associated indicators that should be used as tools to achieve these goals that corroborate those outlined in the 2030 Agenda. This work aims to find out to which SDGs contribute most the Erasmus+ program. To this end, we used a qualitative analysis of the regulation that led to the creation of the Erasmus+ program for the period 2021–2027 and the 17 SDGs. It was concluded that the most relevant SDGs are 4, 5, 8, 9, 10, 13, 16, and 17 and that the Erasmus+ program seems to have a strong potential to drive and mobilize sustainable development in the world through the participation of people and the projects implemented and developed. It was also concluded that SDG4 contributes directly or indirectly to the other selected SDGs, but no SDG contributes to SDG4.

Keywords Erasmus+ 2021–2027 · Higher education · 2030 agenda · Sustainable Development Goals

---

T. Nogueiro (✉) · M. Saraiva
University of Évora, Évora, Portugal

A. Rolo
CICE—Instituto Politécnico de Setúbal, Setúbal, Portugal

J. Reis et al. (eds.), *Driving Quality Management and Sustainability in VUCA Environments*, Springer Proceedings in Business and Economics,
https://doi.org/10.1007/978-3-031-52723-4_18

1 Introduction

In 2015, through the approval of the 17 Sustainable Development Goals by UNESCO that define the 2030 Agenda that addresses sustainable development in various dimensions, effective institutions are promoted in addition to peace and justice.

The SDGs are thus a rallying point for all countries—rich, poor, and middle-income—to work together to increase prosperity while safeguarding the environment. They recognize that eradicating poverty requires methods that promote economic growth and meet a variety of social needs, such as education, health, social protection, and employment opportunities, all while combating climate change and protecting the environment.

The specific objectives of the Erasmus+ program are defined in a common way for the various fields. They are to be met through learning mobility (Key Action 1), cooperation between organizations and institutions (Key Action 2), and support for policy development and cooperation (Key Action 3) that have a mainly transnational or international character (EUR-Lex Access to European Union Law. Document 32021R0817, 2021). The program regulation currently in force aims to ensure full respect for the rights to gender equality and nondiscrimination based on sex, racial or ethnic origin, religion or belief, disability, age, or sexual orientation and to promote the application of Articles 21 (nondiscrimination) and 23 (equality between women and men) of the Charter of Fundamental Rights of the European Union (Charter of Fundamental Rights of the European Union, 2012). Academic freedom should be protected by nations receiving funds under the program, in accordance with Article 13 (Freedom of the Arts and Sciences) of the Charter (Charter of Fundamental Rights of the European Union, 2012).

This work aims to find out which SDGs contribute most to the Erasmus+ program. To this end, we used a qualitative analysis of the regulation that led to the creation of the Erasmus+ program for the period 2021–2027 and the 17 SDGs.

This paper presents the following structure: introduction, brief presentation of the themes, methodology, presentation and discussion of the results, conclusions, limitations, and future research proposals.

Next section briefly presents the themes of Erasmus+ program and the Sustainable Development Goals.

2 Brief Presentation of the Themes

The Erasmus+ program builds on over 25 years of experience in European education, training, and youth program, encompassing intra-European and international cooperation. Erasmus+ is the result of the merger of several previous European programs undertaken between 2007 and 2013 by the European Commission. It aims to go beyond these programs by promoting synergies and cross-enrichment in different fields of education, training, and youth, removing artificial barriers between actions and project formats, fostering new ideas, attracting new actors from the labor market and civil society, and stimulating new forms of cooperation.

The 2021–2027 program aims to be a direct instrument to support sustainable growth, quality jobs, and social cohesion, to boost innovation and strengthen European identity and active citizenship through lifelong learning and educational, professional, and personal development of individuals (Regulation (EU) 2021/817 [18] of the European Parliament and of the Council of 20 May 2021).

Bañon Gomis et al. [1] emphasize that sustainability is not just a "fad or trend" valued by circumstantial conditions. Its importance is linked to the ethics that guide human conduct, reflecting the values of courage, prudence, and hope. The term "sustainability" may even be popular today, but the concept itself has ancient and universal roots.

Feil and Schreiber [10] state that sustainability consists of a goal or parameter defined through scientific criteria, which measures and monitors the results generated by the use of Sustainable Development strategies. Sustainability is a way of being and living that requires aligning human practices with the limited potential of each biome and the needs of today's and tomorrow's generations [3].

The SDGs, which represent and define the global priorities for the 2030 Agenda signed by nearly 190 countries, seek to focus global efforts around a set of common goals and targets. The 17 SDGs are defined in areas that impact the quality of life of all citizens of the world, including future generations. According to the United Nations, the 2030 Agenda and the 17 Sustainable Development Goals are the common vision for humanity, a contract between world leaders and people, and a list of actions to be taken on behalf of the people and the planet [27].

In recent decades, a broad spectrum of decision-makers has become interested in the topic of Sustainable Development. People and society can only prosper if all countries and regions experience continuous, inclusive, and sustainable economic progress [11].

Figure 1 presents the 17 SDGs according to social, environmental, economic, and institutional dimensions.

The social dimension includes six SDGs, namely, those that aim to end poverty in all its forms everywhere (SDG1), end hunger, achieve food security and improved nutrition and promote sustainable agriculture (SDG2), ensure healthy lives and promote well-being for all at all ages (SDG3), ensure inclusive and equitable quality education and promote lifelong learning opportunities for all (SDG4), achieve gender equality and empower all women and girls (SDG5), and reduce inequality within and among countries (SDG10). The next step of the pyramid also presents six SDGs for the environmental dimension. These six SDGs are those that intend to ensure availability and sustainable management of water and sanitation for all (SDG6); access to affordable, reliable, sustainable, and modern energy for all (SDG7) and sustainable consumption and production patterns (SDG12); take urgent action to combat climate change and its impacts (SDG13); conserve and sustainably use the oceans, seas, and marine resources for sustainable development (SDG14); and protect, restore, and promote sustainable use of terrestrial ecosystems, sustainably

Fig. 1 SDG dimension pyramid

manage forests, combat desertification, and halt and reverse land degradation and halt biodiversity loss (SDG15). There are 3 SDGs that integrate the economic dimension of sustainable development. They are SDG 8 related to inclusive and sustainable economic growth, full and productive employment and decent work for all, SDG 9 which aims to build resilient infrastructure, promote inclusive and sustainable industrialization and foster innovati on and, finally, SDG 11 associated with inclusive, safe, resilient and sustainable cities and human settlements. At the top of the pyramid is the institutional dimension with 2 SDGs: SDG16—promote peaceful and inclusive societies for sustainable development, provide access to justice for all, and build effective, accountable, and inclusive institutions at all levels—and SDG17, strengthen the means of implementation and revitalize the global partnership for sustainable development [27].

According to the United Nations, the 17 SDGs presents 169 targets in total. Under the 17 SDGs and on this date, 3802 events were promoted, 1344 publications were made, and 7376 actions were taken [27].

However, and keeping in mind the Erasmus+ program regulation, it was possible to verify that the program does not contribute to all 17 SDGs. Thus, those to which the program does contribute were identified and are presented in the Presentation and Discussion of the Results section.

3 Methodology

The methodology focused on the qualitative analysis of the SDGs in the context of higher education, as well as the analysis of Regulation (EU) 2021/817 of the European Parliament and Council of May 20, 2021, establishing Erasmus+: the Union program for education and training, youth, and sport.

To perform this analysis, we first defined the keywords that would make the link between the content of Regulation (EU) 817/2021 and the SDGs. Then, based on the description of the goals and indicators of the 17 SDGs, we selected those that were considered relevant and with a clear link to higher education and the Erasmus+ program. We integrated our research findings with a literature review for the discussion. The discussion of the results obtained on the contribution of SDG4 to the other SDGs was based on random bibliography searched in open source. To perform the search, we used as a base word Education and then associated it with words like innovation, decent work, inclusion, and equality, among others. All qualitative analysis was carried out manually by the authors, which means that no qualitative data analysis software was used.

4 Presentation and Discussion of Results

This section presents and discusses the results obtained from this work. Thus, the following table presents a summary of the results.

By defining keywords associated with the SDGs for higher education, it was possible to identify these word(s) in the text of Regulation (EU) 2021/817 [18]. When extracting the expressions associated with these words, it was necessary to somehow correlate them with the selected targets and indicators. Analyzing Table 1, several results are highlighted.

Thus, in social dimension in which SDG4, 5, and 10 are integrated, it was possible to associate the following keywords: Inclusive, Quality, Lifelong learning, Inclusive + lifelong learning, and Equal and Inequalities (inclusion, equality, and nondiscrimination). Regarding the economic dimension, the keywords identified were Employment, Decent work (quality jobs), Economic growth (only expressions associated to sustainable growth), and Innovation (and research). SDG8 and 9 are linked to this dimension. To SDG13, integrated into the environmental dimension, the keyword Climate was associated. Finally, the keywords Peace and Justice (fundamental rights) and Partnerships are associated with SDG16 and 17, which are included in the institutional dimension.

In the context of higher education and the Erasmus+ program, not all indicators are relevant and truly fit these realities. In the scope of this study, it was possible to identify several indicators associated with higher education but not all of them have a direct or indirect relationship with items from the Regulation (EU) 2021/817 of the program. The same happened with the identified SDGs.

It was thus possible, in the framework of SDG4, to associate indicators 4.3.1 (participation rate of youth and adults in formal and nonformal education and training in the previous 12 months, by sex) items 1 and 13; to 4.4.1 (proportion of youth and adults with information and communications technology (ICT) skills, by type of skill) item 11; to 4.5.1 (parity indices (female/male, rural/urban, bottom/top wealth quintile, and others such as disability status, indigenous peoples, and conflict-affected, as data become available) for all education indicators) items 4, 7,

Table 1 Relationship between Sustainable Development Goal targets 4, 5, 8, 9, 10, 13, 16, and 17 and respective indicators and items from Regulation (EU) 2021/817

SUSTAINABLE DEVELOPMENT GOALS 4 (Social Dimension)
Keywords: Inclusive, Quality, Lifelong learning, Inclusive + lifelong learning
Target number 4.3—Equal access to affordable technical, vocational and higher education
4.3.1 is the participation rate of youth and adults in formal and non-formal education and training in the previous 12 months, by sex.
(1) (. . .) Investing in learning mobility for all, regardless of background and means, and in cooperation and innovative policy development in the fields of education and training, youth and sport is key to building inclusive, cohesive and resilient societies and sustaining the competitiveness of the Union . . .
(13) (. . .) The Program should support an inclusive and broad EU Youth Dialogue, the priorities . . .
Target number 4.4—Increase the number of people with relevant skills for financial success
4.4.1 is the proportion of youth and adults with information and communications technology (ICT) skills, by type of skill.
(11) (. . .) A coherent lifelong learning approach is central to managing the different transitions . . .
Target number 4.5—Eliminate all discrimination in education
4.5.1 is parity indices (female/male, rural/urban, bottom/top wealth quintile and others such as disability status, indigenous peoples and conflict-affected, as data become available) for all education indicators.
(4) (. . .) Principle 1 of the European Pillar of Social Rights provides that everyone has the right to quality and inclusive education, training and lifelong learning in order to maintain and acquire skills . . .
(7) (. . .) confirmed the key findings of the report on the mid-term evaluation of the 2014–2020 Program and emphasised the need to make the future Program more inclusive, to continue to focus priorities . . .
(8) (. . .) This would allow more people to move to another country to learn or work and would allow the program to focus on inclusiveness and on reaching more people with fewer opportunities . . .
(18) (. . .) In fulfilling its objectives, the Program should be more inclusive by improving participation among people with fewer opportunities . . .
(29) (. . .) Implementing bodies should promote measures to ensure that the DiscoverEU initiative is inclusive and geographically balanced with regard to both the travel passes allocated . . .
(30) (. . .) At the same time, in order to ensure broad and inclusive access to the Program, it is important that multilingualism be a key principle in the implementation of the Program . . .
Target number 4.7—Education for sustainable development and global citizenship
4.7.1 is the extent to which (i) global citizenship education and (ii) education for sustainable development, including gender equality and human rights, are mainstreamed at all levels in (a) national education policies; (b) curricula; (c) teacher education; and (d) student assessment.
(33) (. . .) In its communication of 18 July 2017 entitled 'Strengthening Innovation in Europe's Regions: Strategies for resilient, inclusive and sustainable growth', the Commission . . .
(12) (. . .) The Program is a key component of building a European Education Area. Following on from its communication of 14 November 2017 (. . .) in achieving the objectives of quality and inclusive education, training and lifelong learning, & in preparing the Union to face the digital and green transitions . . .
Target number 4.B—Expand higher education scholarships for developing countries

(continued)

Table 1 (contiuned)

4.B.1 is the volume of official development assistance flows for scholarships. (3) (...) Principle 1 thereof, which relates to education, training and lifelong learning. That communication stressed the need to step up mobility and exchanges, including through a substantially ... (6) (...) while calling for a number of improvements, such as making the Erasmus+ Program more inclusive, simpler and more manageable for beneficiaries ... Target number 4.C—Increase the supply of qualified teachers in developing countries 4.C.1 is the proportion of teachers in (a) pre-primary; (b) primary; (c) lower secondary; and (d) upper secondary education who have received at least the minimum organized teacher training (e.g. pedagogical training) pre-service or in-service required for teaching at the relevant level in a given country. (32) (...) The Program should support those European Universities in developing joint long-term strategies for high-quality education, research and innovation and for service to society (...)
SUSTAINABLE DEVELOPMENT GOALS 5 (Social Dimension) Keyword: Equal Target number 5.1—End discrimination against women and girls: End all forms of discrimination against all women and girls everywhere. 5.1.1 is whether or not legal frameworks are in place to promote, enforce and monitor equality and non-discrimination on the basis of sex. (4) (...) The European Pillar of Social Rights also makes clear the importance of good quality early childhood education and care and of ensuring equal opportunities for all ... (26) (...) The participation of recent graduates in learning mobility should be based on objective criteria, and equal treatment should be ensured ... (64) (...) In particular, this Regulation seeks to ensure full respect for the right to equality between women and men and the right to non-discrimination based on sex, racial or ethnic origin, religion or belief, disability, ... Target number 5.C—Adopt and strengthen policies and enforceable legislation for gender equality: Adopt and strengthen sound policies and enforceable legislation for the promotion of gender equality and the empowerment of all women and girls at all levels. 5.C.1 is the Proportion of countries with systems to track and make public allocations for gender equality and women's empowerment. (43) (...) The principles of transparency, equal treatment and non–discrimination set out in the Financial Regulation should be respected in the implementation of the Program ...
SUSTAINABLE DEVELOPMENT GOALS 8 (Economic Dimension) Keywords: Employment, Decent work (quality jobs), Economic growth (only expressions associated to sustainable growth), Target number 8.3—Promote policies to support job creation and growing enterprises 8.3.1 is the proportion of informal employment in non-agriculture employment, by sex. (59) (...) It is necessary to ensure the complementarity and consistency of the Program actions, (...) relating to education and training, culture and the media, youth and solidarity, employment and social inclusion ... (Chapter XI, Article 32) (...) The Program shall be implemented so as to ensure its overall consistency and complementarity with other relevant Union policies, Programs and funds, in particular those relating to education and training, culture and the media, youth and solidarity, employment and social inclusion ... (8) (...) This would allow more people to move to another country to learn or work and would allow the Program to focus on inclusiveness and on reaching more people with fewer opportunities ...

(continued)

Table 1 (contiuned)

(13) (. . .) a new EU Youth Strategy', the Program should support high-quality youth work, tools and systems for the training of youth workers, the validation of non-formal and informal learning and quality approaches to empower youth organisations . . . (12) (. . .)The Program should be equipped to contribute to the successor of the strategic framework for European cooperation in education and training and the updated European Skills Agenda for sustainable competitiveness, social fairness and resilience established in the communication of the Commission of 1 July 2020 with a shared commitment to the strategic importance of skills, key competences and knowledge for sustaining jobs and supporting growth, competitiveness, innovation and social cohesion (33) (. . .) The Bruges Communiqué of 7 December 2010 on enhanced European Cooperation in Vocational Education and Training for the period 2011–2020 called for support of vocational excellence for smart and sustainable growth. In its communication of 18 July 2017 entitled 'Strengthening Innovation in Europe's Regions: Strategies for resilient, inclusive and sustainable growth', the Commission called on Member States to link vocational education and training to innovation systems, as part of smart specialisation strategies at regional level. The Program should provide the means to respond to those calls and support the development of transnational platforms of centres of vocational excellence embedded in local and regional strategies for sustainable growth, innovation and competitiveness . . . (39) (. . .) In line with the European Green Deal as a blueprint for sustainable growth, the actions under this regulation should respect the 'do no harm' principle without changing the fundamental character of the Program . . . (Key action 2) (. . .) 2.2.2. partnerships of vocational education and training providers (platforms of centres of vocational excellence) embedded in local and regional strategies for sustainable growth, innovation and competitiveness that jointly work on high-quality transnational vocational Programs focused on meeting current and emerging sectoral skills needs . . . Target number 8.5—Full employment and decent work with equal pay 8.5.1 is the average hourly earnings of female and male employees, by occupation, age and persons with disabilities. (Article 3) (. . .) The general objective of the Program is to support, through lifelong learning, the educational, professional and personal development of people in the fields of education and training, youth and sport, in Europe and beyond, thereby contributing to sustainable growth, quality jobs and social cohesion, to driving innovation and to strengthening European identity and active citizenship . . . (Article 3) (. . .) The general objective of the Program is to support, through lifelong learning, the educational, professional and personal development of people in the fields of education and training, youth and sport, in Europe and beyond, thereby contributing to sustainable growth, quality jobs . . .
SUSTAINABLE DEVELOPMENT GOALS 9 (Economic Dimension) Keywords: Innovation (and research) Target number 9.5—Enhance research and upgrade industrial technologies 9.5.1 is research and development expenditure as a proportion of GDP. (12) (. . .) The Program should be equipped to contribute to the successor of the strategic framework for European cooperation in education and training and the updated European Skills Agenda for sustainable competitiveness, social fairness and resilience established in the communication of the Commission of 1 July 2020 with a shared commitment to the strategic importance of skills, key competences and knowledge for sustaining jobs and supporting growth, competitiveness, innovation and social cohesion, in line with the Council Recommendation of 22 May 2018 . . .

(continued)

Table 1 (contiuned)

(15) (. . .) The Program should be able to support any field of study and should, in particular, contribute to strengthening the Union's innovation capacity by supporting activities that help people develop the knowledge, skills, competences and attitudes they need in forward-looking study fields or disciplines such as science, technology, engineering, arts and mathematics (STEAM), climate change, environmental protection, sustainable development, clean energy, artificial intelligence, robotics, data analysis, design and architecture, and digital and media literacy. Innovation can be fostered through all learning mobility and cooperation actions, whether directly or indirectly managed . . . (16) (. . .) Synergies with the Horizon Europe Program established by Regulation (EU) 2021/695 of the European Parliament and of the Council (10)('Horizon Europe') should ensure that combined resources from the Program and Horizon Europe are used to support activities dedicated to strengthening and modernising European higher education institutions. Horizon Europe will, where appropriate, complement the Program's support for the European Universities initiative as part of the development of new joint and integrated long-term and sustainable strategies on education, research and innovation. Synergies with Horizon Europe will help to foster the integration of education and research, in particular in higher education institutions . . . (31) (. . .) thereby recognising the fundamental role of the institutions and organisations in equipping individuals with the knowledge, skills and competences needed in a changing world and helping the institutions and organisations to adequately fulfil their potential for innovation, creativity and entrepreneurship . . . (32) (. . .) the European Council called for cooperation between research, innovation and education to be encouraged, including through the European Universities initiative. The Program should support those European Universities in developing joint long-term strategies for high-quality education, research and innovation and for service to society . . . (33) (. . .) The Program should provide the means to respond to those calls and support the development of transnational platforms of centres of vocational excellence embedded in local and regional strategies for sustainable growth, innovation and competitiveness . . . (59) (. . .) It is necessary to ensure the complementarity and consistency of the Program actions, including those without a transnational or international character, with activities undertaken by the Member States and with other Union activities, in particular those relating to education and training, culture and the media, youth and solidarity, employment and social inclusion, research and innovation, industry and enterprise, digital policy . . . (Article 3) (. . .) contributing to sustainable growth, quality jobs and social cohesion, to driving innovation and to strengthening European identity and active citizenship. (. . .) the learning mobility of individuals and groups, and cooperation, quality, inclusion and equity, excellence, creativity and innovation at the level of organisations (. . .); non-formal and informal learning mobility and active participation among young people, and cooperation, quality, inclusion, creativity and innovation at the level of organisations (. . .); the learning mobility of sport staff, and cooperation, quality, inclusion, creativity and innovation at the level of sport organisations . . . (Key action 2) (. . .) partnerships for innovation to strengthen Europe's innovation capacity . . . (Article 32) (. . .) in particular those relating to education and training, culture and the media, youth and solidarity, employment and social inclusion, research and innovation, industry and enterprise . . . (Annex I) (. . .) 2.2.2. partnerships of vocational education and training providers (platforms of centres of vocational excellence) embedded in local and regional strategies for sustainable growth, innovation and competitiveness that jointly work on high-quality . . . (Key action 2) (. . .) 2.3.1. alliances: strategic cooperation between key players in the field of education and training and in business and research that foster innovation and modernisation of education and training systems;

(continued)

Table 1 (continued)

(Key action 2) 2.3.2. projects that foster innovation, creativity, e-participation, and social entrepreneurship in the fields of education and training and youth ... 9.5.2 is the number of researchers (in full-time equivalent) per million inhabitants.
SUSTAINABLE DEVELOPMENT GOALS 10 (Social Dimension) Keyword: Inequalities (inclusion, equality and non-discrimination) Target number 10.3—Ensure equal opportunities and end discrimination 10.3.1 is the proportion of population reporting having personally felt discriminated against or harassed in the previous 12 months on the basis of a ground of discrimination prohibited under international human rights law. (14) (...) There is a need to focus, in particular, on grassroots sport, taking into account the important role that sports play in promoting physical activity and a healthy lifestyle, interpersonal relations, social inclusion and equality (...) (18) (...) Therefore, within a Union-wide framework of such measures aimed at increasing participation among people with fewer opportunities, inclusion action plans should be developed and tailored to the target groups and specific circumstances in each Member State (...) (48) (...) In addition to assessing the overall effectiveness and performance of the Program, it is of particular importance that the interim evaluation thoroughly assess the implementation of new initiatives and of the inclusion and simplification measures put in place (...) (59) (...) It is necessary to ensure the complementarity and consistency of the Program actions, including those without a transnational or international character, with activities undertaken by the Member States and with other Union activities, in particular those relating to education and training, culture and the media, youth and solidarity, employment and social inclusion, research and innovation (...) (Article 3) (...) (a) the learning mobility of individuals and groups, and cooperation, quality, inclusion and equity, excellence, creativity and innovation at the level of organisations and policies in the field of education and training; (b) non-formal and informal learning mobility and active participation among young people, and cooperation, quality, inclusion, creativity and innovation at the level of organisations and policies in the field of youth; (c) the learning mobility of sport staff, and cooperation, quality, inclusion, creativity and innovation at the level of sport organisations and sport policies (...) (Article 15) (...) The Commission shall, by 29 November 2021, develop a framework of inclusion measures to increase participation rates among people with fewer opportunities, as well as guidance for the implementation of such measures. That guidance shall be updated as necessary over the duration of the Program. Based on the framework of inclusion measures, and with particular attention to the specific Program access challenges within the national contexts, inclusion action plans shall be developed and shall form an integral part of the national agencies' work Programs. The Commission shall monitor the implementation of those inclusion action plans on a regular basis (...) (Article 24) (...) The interim evaluation of the Program shall assess the overall effectiveness and performance of the Program, including as regards new initiatives and the delivery of inclusion and simplification measures (...) (Article 32) (...) The Program shall be implemented so as to ensure its overall consistency and complementarity with other relevant Union policies, Programs and funds, in particular those relating to education and training, culture and the media, youth and solidarity, employment and social inclusion, research and innovation (...) (Annex I) (...) Not-for-profit sport events: events held either in one country or simultaneously in several countries to raise awareness of the role of sport in diverse areas such as social inclusion, equal opportunities and health-enhancing physical activities (...) (Annex II) (...) Inclusion: 2.4.1. The number of people with fewer opportunities taking part in activities under key action 1; 2.4.2. The number of newcomer organisations and institutions taking part in the Program under key actions 1 and 2 (...)

(continued)

Table 1 (contiuned)

SUSTAINABLE DEVELOPMENT GOALS 13 (Environmental Dimension) Keyword: Climate Target number 13.3—Build knowledge and capacity to meet climate change 13.3.1 is the number of countries that have integrated mitigation, adaptation, impact reduction and early warning into primary, secondary and tertiary curricula. (59) (...) in particular those relating to education and training, culture and the media, youth and solidarity, employment and social inclusion, research and innovation, industry and enterprise, digital policy, agriculture and rural development that have a focus on young farmers, environment and climate, cohesion, regional policy, migration (...) (Article 32) (...) The Program shall be implemented so as to ensure its overall consistency and complementarity with other relevant Union policies, Programs and funds, in particular those relating to education and training, culture and the media, youth and solidarity, employment and social inclusion, research and innovation, industry and enterprise, digital policy, agriculture and rural development, environment and climate, cohesion (...) (Annex I) (...) 1.1. Learning mobility: short-term, long-term, group or individual mobility on diverse thematic areas and study subjects, including forward-looking fields such as the digital field, climate change, clean energy and artificial intelligence (...) (Annex II) (...) 2.6. Climate contribution: 2.6.1. The share of activities addressing climate objectives under key action 1; 2.6.2. The share of projects addressing climate objectives under key action 2 (...)
SUSTAINABLE DEVELOPMENT GOALS 16 (Institutional Dimension) Keyword: Peace and justice (fundamental rights) Target number 16.a—Strengthen national institutions to prevent violence and combat crime and terrorism 16.A.1 is the existence of independent national human rights institutions in compliance with the Paris principles. (28) (...) The Program should encourage the participation of young people in Europe's democratic life, including by supporting activities that contribute to citizenship education and participation projects for young people to engage and learn to participate in civic society, thereby raising awareness of European common values, including fundamental rights, as well as European history and culture (...) (64) (...) This Regulation respects the fundamental rights and observes the principles recognised in particular by the Charter of Fundamental Rights of the European Union (the 'Charter') (...)
SUSTAINABLE DEVELOPMENT GOALS 17 (Institutional Dimension) Keyword: Partnerships Target number 17.9—Enhanced SDG capacity in developing countries 17.9.1 is the dollar value of financial and technical assistance committed to developing countries. (Key action 2) Cooperation among Organisations and Institutions

8, 18, 29, and 30; to 4.7.1 (the extent to which (i) global citizenship education and (ii) education for sustainable development, including gender equality and human rights, are mainstreamed at all levels in (a) national education policies, (b) curricula, (c) teacher education, and (d) student assessment) items 12 and 33; to 4.B.1 (the volume of official development assistance flows for scholarships) items 3 and 6; and to 4.C.1 (the proportion of teachers in (a) preprimary, (b) primary, (c) lower secondary, and (d) upper secondary education who have received at least the minimum organized teacher training (e.g., pedagogical training) preservice or in-service required for teaching at the relevant level in a given country) item 32.

Concerning SDG5, the research led to the identification of indicators 5.1.1 (whether or not legal frameworks are in place to promote, enforce, and monitor equality and nondiscrimination on the basis of sex) with associated items 4, 26, and 64 and 5.C.1 (proportion of countries with systems to track and make public allocations for gender equality and women's empowerment) with related item 43.

Regarding SDG8, the indicators and items identified are 8.3.1 (proportion of informal employment in nonagricultural employment, by sex) with the related items 59, (Chapter XI, Article 32, (8), (12), (13), (33), (39), and (Key action 2) and 8.5.1 (the average hourly earnings of female and male employees, by occupation, age, and persons with disabilities) related to Article 3.

With regard to SDG9, two indicators were identified, specifically 9.5.1 (research and development expenditure as a proportion of GDP) and 9.5.2 (the number of researchers (in full-time equivalent) per million inhabitants). In this case, indicator 9.5.1 was the only one to have items under the Erasmus Regulation, namely, 12, 15, 16, 31, 32, 33, 59, Article 3, Key action 2, Article 32, Annex I, and Key action 2.

Regarding SDG 10, indicator 10.3.1 was selected (proportion of population reporting having felt personally discriminated against or harassed in the previous 12 months on the basis of a ground of discrimination prohibited by international human rights law). Thus, associated with this indicator are the following items (14), (18), (48), (59), (Article 3), (Article 15), (Article 24), (Article 32), (Annex I) and (Annex II).

SDG13 also have one indicator identified, namely, 13.3.1 (the number of countries that have integrated mitigation, adaptation, impact reduction, and early warning into primary, secondary, and tertiary curricula) linked to the items 59, Article 32, Annex I, and Annex II.

Associated with SDG16 and 17 are identified indicators for each, specifically 16.A.1 (the existence of independent national human rights institutions in compliance with the Paris Principles) and 17.9.1 (the dollar value of financial and technical assistance committed to developing countries), respectively. For indicator 16.A.1, the items of the Regulation (EU) 2021/817 are 28 and 64, and for indicator 17.9.1, the item is Key action 2.

Being aware of the contribution that the Erasmus+ program can have to the 2030 Agenda, as demonstrated by the identification of the indicators within higher education and their association with Regulation (EU) 2021/817, it is possible to realize that the SDGs do not contribute in an isolated and independent way to the desired sustainability.

Education is the basis for building citizenship and civic participation, for preserving the environment and health, and for financial sustainability, among others [4, 8, 12, 13, 28].

The awareness that systematic knowledge is more than an essential cultural resource is the starting point for the right to education. As a part of the cultural heritage, the citizen gains access to cognitive and formative patterns, increasing his chances of participating in the fate of his society and contributing to its transformation. Access to education is also a form of openness, providing the person with a key to self-construction and perceiving oneself as capable of choices. In this perspective, the right to education is an opportunity for citizen development, a path to diversified possibilities, and a tool to increase self-esteem. The specification occurs in the struggle for the universalization of schooling when new rights are recognized, such as the right of children under the age of 6 to attend a school appropriate for their age, or when the right of women, young people, and adults to attend schools is acknowledged, even today. This is also true for people with special requirements who, for whatever reason, become disabled in their motility, hearing, or other abilities and begin to demand their own educational model that accommodates their unique demands. Because social actors recognize the value of knowledge in the society in which they live, the right to education becomes a nonviolent weapon of political claim and participation. As a result, education as a right and its implementation in social practices become a tool for reducing inequities and discrimination and enabling peaceful coexistence among people all over the world. The civil presupposition of universal citizenship is the dissemination and universalization of quality school education as a citizenship right, which is part of what Kant once considered one of the conditions "of perpetual peace": the truly republican character of the States that guarantee this right of freedom and equality for all, among other things [5].

To be addressed in a way that simultaneously promotes the development of knowledge, attitude, and skills required for the preservation and improvement of environmental quality and community quality of life, environmental education must be adapted to contemporary circumstances in the city or the environment in which students live. Consequently, the practice of environmental education within the school requires much encouragement and commitment, because there is a need for immediate awareness about the problem of environmental degradation, assuming the role of social responsibility, beginning with basic actions that will become habits and practices established in society; this is the role of environmental education within the school [6]. Nonetheless, the author believes that the importance of environmental education in the school context is evident in terms of developing students' knowledge, skills, and attitudes toward environmental preservation. Thus, environmental education must be present at all educational levels, with the goal of reaching all students and reinforcing what they learned in the early grades. As a result, instructors can and should create environmental initiatives and engage with concepts and knowledge aimed at environmental preservation and sustainable resource use. Environmental education must be present at all educational levels, with the goal of reaching all students and reinforcing what they learned in the early grades [6]. Only through environmental education can a culturally complete society be achieved, and it is necessary to articulate educational actions, adequate conditions, and educator training so that they can work with environmental education themes and activities in a way that raises students' awareness and develops their criticism, generating new concepts and values about nature and contributing to environmental preservation [4].

According to Pretto [17], if we think of technologies at the service of the production of knowledge and culture, we can think of the insertion of the country in the so-called globalized world market, from another perspective, that is, of effective citizenship. Schools must be stimulating places in which the teaching of mathematics and science implies the potential to alter. Education must prepare young people to work in teams, to study on their own, to solve issues, to believe in their potential, and to have personal integrity, initiative, and the ability to invent. It must inspire innovation and provide opportunities for everyone. In this context, it must be made clear that public policy in the field of science and technology must be wide, encompassing not only innovation but, more fundamentally, scientific progress, with scientific education at all levels as a priority. It is vital to remember that scientific illiteracy will exacerbate disparities by further marginalizing already excluded majorities from the job market. To be successful, educational system reform must originate from the community, involving and valuing teachers, so that we as a society may get some perspective in knowledge society [29].

The valuing of labor is essential for the realization of human dignity, as is the execution and consolidation of decent work as a fundamental right guaranteed to all [14]. Work provides the citizen's right to feel productive, included, appreciated, and independent, which is much more than a financial concern. Adversities resulting from work relationships bring people with disabilities to a level of normalcy; that is, these are situations that any working subject can face, and this, too, provides satisfaction. Work is promoted today as a symbol of the right to citizenship and human dignity from a societal standpoint. The relationship between people with disabilities' education and qualifications has a direct impact on their salary and the employment they hold [16]. The modern world requires more schooling, qualification, up-to-date knowledge, and the learning of new technologies, but placing into the labor market is becoming more difficult every day, not only because of the new labor world demands but also because of the increase in unemployment, among the older people, the less qualified, the younger layers, and especially the young people who have not yet had their first job [9].

Higher education institutions (HEIs) promote lifelong learning, high-quality education, equity, and a variety of resources to assist empower global citizens to face today's and tomorrow's challenges. They play a critical role in ensuring a sustainable future for every citizen, which is a goal shared by all 17 Sustainable Development Goals (SDGs). [7].

Through lifelong learning, the Erasmus+ program aims to support people's educational, professional, and personal development in the fields of education and training, youth, and sport in Europe and beyond, thereby contributing to growth sustainability, quality jobs, social cohesion, driving innovation, and strengthening European identity and active citizenship. Erasmus+ intends to foster individual and group learning mobility, as well as cooperation, quality, inclusion, equity, excellence, creativity, and innovation at the organizational and policy levels in the sphere of education and training; nonformal and informal learning mobility and active participation among young people, as well as cooperation, quality, inclusion, creativity, and innovation at the organizational and policy levels in the field of youth; and sport staff learning mobility, as well as cooperation, quality, inclusion, creativity, and innovation at the organizational and policy levels in the field of sport [15].

5 Conclusions

This study concluded that SDGs 4, 5, 8, 9, 10, 13, 16, and 17 receive contributions from the Erasmus+ program. Indeed, as can be seen from the analysis of Regulation (EU) 2021/817, there is a very strong relationship between it and the sustainability underlying the SDGs associated with education, inclusion and equality, climate, decent work, innovation, peace and justice, and partnerships. The contribution that the Erasmus+ program can have to the 2030 Agenda is demonstrated by the identification of the indicators within higher education and their association with Regulation (EU) 2021/817. This conclusion allows us to realize that the SDGs do not contribute in an isolated and independent way to the desired sustainability.

In the previous section, it was possible to verify the influence or the contribution that education (SDG4) has in the achievement of other Sustainable Development Goals like the ones studied in this work. As seen, education is the basis for building citizenship and civic participation, for preserving the environment and health, and for financial sustainability, among others. The universal right to education is the guarantee of the opportunity given to citizens to make more informed and sustainable choices in various areas, improving their existence in society. Access to education is also a sort of openness, since it provides a key to self-construction and the perception of oneself as capable of making choices. In this perspective, the right to education is a chance for citizen development, a gateway to new opportunities, and a means of boosting self-esteem. Only through environmental education can a culturally complete society be achieved and generate new concepts and values about nature contributing to environmental preservation. Better education provides better and different skills, giving people the possibility to have more decent works with a better remuneration that will allow a better quality of life. Work gives citizens the right to feel productive, included, valued, and self-sufficient, which is much more than a financial concern. Given this evidence, we can consider education as the pillar of a more sustainable society and world. The Erasmus+ program, being strongly associated with education and training, allows the achievement of such noble goals linked to the 2030 Agenda as inclusion, the right to equality, access to decent work, better quality of life, and social and global peace and justice.

6 Limitations and Future Research Proposals

One limitation to this research paper has to do with the robustness of the choice of keywords that were used to select for each of the SDGs; these keywords should have been validated by EC and/or UNESCO. Another limitation is the fact that no key actions of the Erasmus+ program were analyzed, which would ensure greater robustness to the results through the treatment of European Commission reports on the program.

As future research projects, we propose the analysis and processing of participant reports to understand the real contribution of the Erasmus+ program to the 2030 Agenda, the analysis of SDGs based on the realities of each key action of the Erasmus+ program, the study of capacity building projects for higher education (CBHE) coordinated by Portuguese institutions and their alignment with Agenda 2030 and the sustainability strategy for Portugal, and a new study with the use of qualitative data analysis software to obtain more robust results and then be able to conclude with some policy recommendations.

References

1. Bañon Gomis, A. J., Guillén Parra, M., Hoffman, W. M., & McNulty, R. E. (2011). Rethinking the concept of sustainability. *Business and Society Review, 116*(2), 171–191.
2. Boeren, E. (2019). Understanding Sustainable Development Goal (SDG) 4 on "quality education" from micro, meso and macro perspectives. *International Review of Education, 65*, 277–294.
3. Boff, L. (2017). *Sustentabilidade: o que é-o que não é*. Editora Vozes Limitada.
4. Bortolon, B., & Mendes, M. S. S. (2014). A importância da educação ambiental para o alcance da sustentabilidade. Revista Eletrônica de Iniciação Científica. Itajaí. *Centro de Ciências Sociais e Jurídicas da UNIVALI, 5*(1), 118–136.
4. Charter of Fundamental Rights of the European Union (2012). OJ C 326/02, 26 October 2012, pp. 391–407. https://eur-lex.europa.eu/legal-content/PT/TXT/PDF/?uri=CELEX:12012P/TXT. Last accessed 27 May 2022.
5. Cury, C. R. J. (2006). Educação escolar e educação no lar: espaços de uma polêmica. *Educação & Sociedade, 27*, 667–688.
6. da Silva, H. O. O. (2016). A importância da educação ambiental no âmbito escolar. *Revista Interface (Porto Nacional), 12*(12), 163–172.
7. De La Torre, E. M., Perez-Encinas, A., & Gomez-Mediavilla, G. (2022). Fostering sustainability through mobility knowledge, skills, and attitudes. *Sustainability, 14*(3), 1349.
8. de Sousa, G. L., de Medeiros, A. B., Mendonça, M. J. D. S. L., & de Oliveira, I. P. (2011). A Importância da educação ambiental na escola nas séries iniciais. *Revista Eletrônica Faculdade Montes Belos, 4*(1). Pages 1, 3 and 7.
9. de Souza Nascimento, E., & Miranda, T. G. (2007). O trabalho e a profissionalização das pessoas com deficiência. *Revista Entreideias: educação, cultura e sociedade, 12*. https://doi.org/10.9771/2317-1219rf.v12i12.2761
10. Feil, A. A., & Schreiber, D. (2017). Sustentabilidade e desenvolvimento sustentável: desvendando as sobreposições e alcances de seus significados. *Cadernos Ebape. BR, 15*, 667–681.
11. Fleacă, E., Fleacă, B., & Maiduc, S. (2018). Aligning strategy with sustainable development goals (SDGs): Process scoping diagram for entrepreneurial higher education institutions (HEIs). *Sustainability, 10*(4), 1032.
12. França, L. H. D. F. P. (2011). O envelhecimento populacional e seu reflexo nas organizações: a importância da educação ao longo da vida. *Boletim Técnico do Senac, 37*(2), 49–60.
13. Loureiro, I. (2004). A importância da educação alimentar: o papel das escolas promotoras de saúde. *Revista Portuguesa de Saúde Pública, 22*(2), 43–55.

14. Miraglia, L. M. M. (2010). O direito do trabalho e a dignidade da pessoa humana pela necessidade de afirmação do trabalho digno como direito fundamental. In *XIX Encontro Nacional do CONPEDI*, pp. 9038–9047.
15. Nogueiro, T., & Saraiva, M. (2023). TQM and SDGs for Erasmus+ programmeme—Quality education, reducing inequalities, climate change, peace and justice. *Social Sciences, 12*(3), 123.
16. Passerino, L. M., & Pereira, A. C. C. (2014). Educação, inclusão e trabalho: um deba-te necessário. *Educação & Realidade, 39*, 831–846.
17. Pretto, N. (1999). Educação e inovação tecnológica: um olhar sobre as políticas pú-blicas brasileiras. *Revista Brasileira de Educação, 11*, 75–85.
18. Regulation (EU) 2021/817 of the European Parliament and of the Council of 20 May 2021 establishing Erasmus+: The Union programme for education and training, youth and sport and repealing Regulation (EU) No. 1288/2013. *OJ L, 189*, 28 May 2021, pp. 1–33. https://eur-lex.europa.eu/legal-content/EN/TXT/PDF/?uri=CELEX:32021R0817&qid=1641601177851&from=en. Last accessed 27 May 2022.
19. Sustainable Development Goal 4—Ensure inclusive and quality education for all and promote lifelong learning. https://sdg-tracker.org/quality-education. Last accessed 16 May 2022.
20. Sustainable Development Goal 5—Achieve gender equality and empower all women and girls. https://sdg-tracker.org/gender-equality. Last accessed 16 May 2022.
21. Sustainable Development Goal 8—Promote inclusive and sustainable economic growth, employment and decent work for all. https://sdg-tracker.org/economic-growth. Last accessed 16 May 2022.
22. Sustainable Development Goal 9—Build resilient infrastructure, promote sustainable industrialization and foster innovation. https://sdg-tracker.org/infrastructure-industrialization. Last accessed 16 May 2022.
23. Sustainable Development Goal 10—Reduce inequality within and among countries. https://sdg-tracker.org/inequality. Last accessed 16 May 2022.
24. Sustainable Development Goal 13—Take urgent action to combat climate change and its impacts. https://sdg-tracker.org/climate-change. Last accessed 16 May 2022.
25. Sustainable Development Goal 16—Promote just, peaceful and inclusive societies. https://sdg-tracker.org/peace-justice. Last accessed 26 Dec 2021.
26. Sustainable Development Goal 17—Revitalize the global partnership for sustainable development. https://sdg-tracker.org/global-partnerships. Last accessed 16. May 2022.
27. United Nations. *Transforming our world: The 2030 agenda for sustainable development*. https://sdgs.un.org/2030agenda. Last accessed 1 Apr 2023.
28. Vasconcelos, T. (2007). A importância da educação na construção da cidadania. *Saber e Educar, 12*, 109–117.
29. Zancan, G. T. (2000). Educação científica: uma prioridade nacional. *São Paulo em perspectiva, 14*, 3–7.

Factors and Attitudes Influencing the Implementation of Reverse Logistics (Recycling): The Case of Portuguese Consumers

Ariadne Cristina Minto and Marta da Conceição Cruz Silvério

Abstract Considering that many companies have begun to adopt reverse logistics (RL) as a strategy, consumer behavior has become a conditioning factor. The aim of this research is to identify the factors and attitudes that influence the consumer to practice RL, more specifically recycling. The investigation uses exploratory research through secondary sources and descriptive research using a questionnaire adapted from the investigation by Bezzina and Dimech (Manag Environ Qual Int J 22:463–485, 2011), among the Portuguese population over 18 years of age. In this study, a non-probabilistic convenience sampling method was chosen using the snowball sampling method. The data resulting from the study were analyzed using descriptive statistics and factor, cluster, and discriminant analysis. As results were identified four factors that influence the Portuguese consumer to perform RL: attitudes, norms, and personal recycling skills; inconvenience; satisfaction with the service provided; and knowledge on the subject. Three distinct consumer types were also identified: stimulable, disengaged, and ambassadors. The study contributes to understanding that ecological knowledge, financial incentives, purchasing advantages, and the complexity of carrying out RL (recycling) influence the consumer's attitude toward returning waste.

Keywords Reverse logistics · Recycling · Factors

A. C. Minto
Management Master, Universidade de Évora, Évora, Portugal
e-mail: ariadne.minto@poli.ufrj.br

M. da Conceição Cruz Silvério (✉)
Management Department, CEFAGUE, Universidade de Évora, Évora, Portugal
e-mail: mcs@uevora.pt

J. Reis et al. (eds.), *Driving Quality Management and Sustainability in VUCA Environments*, Springer Proceedings in Business and Economics,
https://doi.org/10.1007/978-3-031-52723-4_19

1 Introduction

As new consumption habits, competition, and legislation emerged, companies that had previously ignored the amount of postconsumer products started having to consider them [7]. They had to seek for efficiency, understanding of the relationship with the consumer, cost management, price, and better use of products that would otherwise be ruled out [22].

Reverse logistics (RL) has become an area that encompasses the needs of companies and consumers alike. That is to say, it is the area of business logistics that plans, handles, and controls the flow and the information related to the return of products after sale and postconsumption in the production cycle. It adds economic, ecological, legal, logistical, and image value [17].

For the RL to take place, some factors are of high importance in the process (distribution channels, communication, consumer participation, etc.). The present study focuses on the consumer's participation in the described process. Consumer participation in RL is fundamental as responsibility must be shared among everyone in society; only then will it be possible to transition to the circular economy, as recycling contributes to reducing impacts on the environment and avoids the need for reuse of new raw materials. In this sense, it is necessary to study the extent to which consumers recycle and what their attitudes are toward the process.

Several authors [2, 4, 8, 15] have studied consumer behavior linked to RL, thus trying to understand the factors and attitudes in the decision-making process of the consumer.

When analyzing the current situation of the market and observing the European target for 2025 (recycling 65% of all packaging), there is a noticeable need for all stakeholders involved in an RL process to understand it more thoroughly. All parts have to contribute to the improvement of this process so that it brings relevant inputs in general terms. Much of the recycling responsibility is correlated to the industry; however, the customer/consumer impact on this mechanism cannot be ignored.

In this sense, this research analyzes the RL process while focusing on the factors that lead consumers to carry out waste recycling. In addition to that, the article aims to identify the factors and attitudes that influence the consumer to practice RL, more specifically recycling.

2 Theoretical Background

2.1 Reverse Logistics

Due to the increasing globalization, market competitiveness, and high consumer demand in terms of quality of the products and services provided, the study of logistics has become a strategic element for companies. The customer has become the "main character" of consumption, and, as a result, strategic marketing and logistics decisions focus on winning over and understanding the client.

In order to understand this subject, it is necessary to understand the term reverse logistics. RL is when products are reincorporated into the production cycle. This happens as a means of reusing their constituents, demonstrating that the current logistics does not end with the final consumer, since product obsolescence has become one of the biggest environmental concerns. In this sense, it is necessary for the consumer to collaborate in the logistics process, so that the components or materials return to the production cycle [10].

At first, in the mid-1980s, the RL concept began to gain acceptance in the corporate world [19]. Shortly after, Pohlen and Farris [20] linked RL to the logistics channels, which led to the emergence of the reverse channel aiming to accumulate, transport, and process materials that would be introduced in a new production process from which a recycled product would emerge. Later, with the development of the RL concept, the definition of logistics itself was expanded, and what previously referred to a direct flow began to also include the concept of a reverse flow, being then outlined by the Council of Supply Chain Management Professionals [6].

Currently, the RL concept includes many activities, such as waste management, recycling, reuse, reprocessing, and recovery of materials, which can aid any organization converting its opportunities into profit. Therefore, the adoption of RL practices has now a greater impact on the economic and environmental performance of companies [27].

Considering that the packaging must meet the needs and values of consumers and given the growth in search for topics related to "green" consumption as well as the growing legal pressure, brands have now a greater concern for the environment and are developing packaging with recycled materials while showing to be more concerned with waste management [24].

Disposing has been rethought, and the correct disposal of solid waste developed a strong link with the consumer.

The concept of disposing was introduced in the marketing literature based on consumer behavior in the 1970s. In the beginning, disposing meant the decision of not wanting the object anymore [13]. A few years later, Young and Wallendorf [28] expanded the concept by bringing disposing into the perspective of a process, rather than a simple discrete event. Disposing was finally conceptualized as a physical and emotional event, in which consumers disconnect not only physically from the product but also from its meaning. Based on this concept, disposing is characterized by breaching any psychological and emotional links related to the symbolic aspects of ownership and consumption.

Contrary to the psychological characteristics, Roster [21] brought to the concept of disposing the fact that the consumer accepts to stop benefiting from the possession of the specific object in present or future. Thereupon, studies on disposing became increasingly important, both in terms of the environment and in terms of logistics. Once there are logistical conditions, postconsumer disposable products are returned through the reverse channel, and their components are reused as secondary raw materials, therefore returning to the production cycle through the respective market, or else they will be sent to landfills or incinerated for energy recovery [17].

Several companies started to publicize their disposing practices related to RL, encouraging the practice of returning product packaging linked to customer relationship.

According to Milichovský [18], RL incentive programs are strictly linked to consumer demands, whether due to environmental concerns or differentiation in customer loyalty. Therefore, the change in customer's consumption culture directly influences marketing and RL strategies.

2.2 Consumer Behavior

Engel et al. [9] added the disposal step as the last phase of the consumer's decision process model.

As research press ahead, as well as the addition of the concept of disposal in the purchase decision process, the definition of consumer behavior took on a new meaning. "The study of individuals, groups or organisations and the processes they use to select, protect, use or dispose products, services, experiences or ideas to satisfy the needs and the impacts that these processes have on consumers and society" [16, p. 110]. This demonstrates that studies are needed in order to understand the factors and attitudes that encourage consumers to carry out such processes.

For the sake of better understanding these factors, some studies have already been carried out, such as do Valle et al. [8] who analyzed the factors that encourage residents of Portugal to carry out recycling, converging to the following factors: "personal norms," "social norms," "adequacy of the system and information," and "location of containers"; Bezzina and Dimech [2] studied factors that seem to be significant in the recycling process for residents of Malta and found some factors such as "attitudes, norms, and personal recycling skills," "satisfaction with the service provided," "inconveniences," "awareness of consequences," "knowledge about the subject," "attitudes and norms of social recycling," "intentions of action," and "preference of structure"; Calvin et al. [4] investigated the attitudes and recycling behavior of college students and staff, and they found that "attitude," "subjective norms," "perceived behavioral control," "awareness of consequences," "moral norms," and "convenience" influence recycling; Kianpour et al. [15] carried out a survey in which they found factors such as "attitude," "subjective norm," "consumer participation and intention to dispose," and "behavioral control" that can influence the attitude of consumers in performing RL.

Based on these findings, one realizes that the environmental factor and the competitive advantages have become a relevant factor for carrying out the correct disposal of packaging, which was not observed in previous studies. This demonstrates a greater inclination of consumers to take responsibility for their acts in terms of sustainability (particularly when it comes to environmental awareness). In addition to that, the changeability of the decision-making process is in line with changes in society.

Table 1 Factors and attitudes that encourage consumers to recycle

Factors and attitudes	Authors
Environmental awareness	Bamberg and Möser [1] and Chen et al. [5]
Education and information	Thøgersen [25], Biswas et al. [3] and Chen et al. [5]
Ease of access to recycling	Thøgersen [25], Homburg et al. [12] and Chen et al. [5]
Financial incentives	Kallbekken and Sælen [14], Chen et al. [5] and Gamba and Oskamp [11]
Social responsibility	Biswas et al. [3], Chen et al. [5] and Vermeir and Verbeke [26]
Social influence	Thøgersen [25] and Chen et al. [5]

According to the literature, it is important to mention a set of factors and attitudes that encourage consumers to recycle (Table 1):

3 Research Methodology

The research method consisted of a four-stage process: exploratory research, instrument design for data collection, field work, and data analysis.

During the exploratory research stage, carried out with the aim of acquiring concepts and ideas, which would later be developed in the investigation, bibliographical research was made via books, scientific articles, and theses, to deepen knowledge about the investigated topic. This research allowed us to find a set of studies that had been already carried out (see Sect. 2.2) and develop the data collection instrument.

In the second phase of the research, a questionnaire based on the one applied by Bezzina and Dimech [2] in their study was developed. The aforementioned researchers found significant factors in the recycling process by consumers, which suited the investigation they intended to carry out. The questionnaire consisted of 26 questions, and the adaptations made took into account the Portuguese culture and their own context of RL practices. The data collection instrument was created to allow understanding consumer knowledge about the objectives of recycling and landfills, their personal awareness of the consequences of not recycling, social attitudes and norms, personal attitudes and norms, the inconvenience factors of recycling, the motivating factors, intentions to act, knowledge of strategies and capacity for action, satisfaction with the service provided, and participation in recycling. Information was collected using a five-point Likert scale.

With the data collected, we intend to identify the factors and attitudes regarding recycling and find groups of consumers with distinctive characteristics in relation to the process.

Before applying the questionnaire, a face-to-face pretest was carried out with 15 individuals in December 2021. This resulted in small improvements in terms of language use.

The population under study were individuals residing in Portugal, and in study, a non-probabilistic convenience sampling method was chosen using the snowball technique [23].

The fieldwork took place online, with the link to the questionnaire posted on social networks (Facebook, Instagram) and on the pages of university courses from January 13 to February 2, 2022.

After data collection, the data set was analyzed using descriptive statistics and factorial analysis and cluster and discriminant analysis in order to thoroughly study and investigate the central issue previously disclaimed.

4 Results Analysis and Discussions

The sample consists of 385 individuals, mostly female ($n = 245$; 63.6%), 18 and 65 years old, with the most representative group being those aged 25 and 44 years ($n = 192$; 49.9%), graduates ($n = 151$; 39.2%) or masters ($n = 119$; 30.9%) and without higher education ($n = 89$; 22.8%), with household net income higher than €1501 ($n = 174$; 45.2%) and less than €1500 ($n = 211$; 54.8%) and living in households with more than 2 people ($n = 303$; 78.7%). Respondents reside in 17 districts of which Évora (45.2%) and Lisbon (15.6%) have the greatest representation.

4.1 Factors That Influence Consumers to Carry Out Reverse Logistics

In a first analysis, 14 variables were ruled out of the study due to noncompliance with the parameters of symmetry or commonalities of variables. With regard to the studied variables, the KMO and Bartlett test assumes a value of 0.784, which indicates, according to Kaiser's parameters, a satisfactory correlation among the variables under study with a significance level of 0.00, hence rejecting the null hypothesis that there is no correlation among the variables.

The explained variance (Table 2) indicates the existence of four factors (66.56% of accumulated variance).

To perform the rotation of factors (Table 3), the Varimax method was used, and the following factors were identified as well as attitudes toward the recycling process on the part of the consumer:

- *Attitudes, norms, and personal recycling skills:* as these are matters directly related to personal attitudes toward recycling associated with norms and skills. This factor includes the following variables: I have a responsibility to separate the waste I make; I separate paper, plastic, metal, and glass for recycling; It's likely that I'll start sorting waste in the next few weeks; I always separate

Table 2 Total variance explained

Component	Initial eigenvalues			Extraction sums of squared loads			Rotation sums of squared loads		
	Total	% Variance	% Cumulative	Total	% Variance	% Cumulative	Total	% Variance	% Cumulative
1	4.016	33.466	33.466	4.016	33.466	33.466	3.188	26.567	26.567
2	1.337	11.142	44.607	1.337	11.142	44.607	1.767	14.721	41.288
3	1.278	10.647	55.255	1.278	10.647	55.255	1.630	13.581	54.869
4	1.117	9.310	64.564	1.117	9.310	64.564	1.163	9.695	64.564
5	0.990	8.248	72.812						
6	0.726	6.051	78.863						
7	0.665	5.543	84.407						
8	0.513	4.272	88.679						
9	0.437	3.639	92.318						
10	0.350	2.919	95.237						
11	0.318	2.651	97.889						
12	0.253	2.111	100.000						

Método de Extra ção: análise de Componente Principal

Table 3 Matrix rotated component[a]

	Component			
	1	2	3	4
I have a responsibility to separate the waste I make.	.795	-.191	.071	-.025
I separate paper, plastic, metal and glass for recycling.	.783	-.125	.147	-.029
It's likely that I'll start sorting waste in the next few weeksS_.	.704	-.110	.186	.177
I always separate household waste.	.695	-.201	.260	.232
I encourage other people to separate and recycle their waste.	.690	-.164	.009	.117
If a community group goes from house to house collecting plastic, paper, metal and glass, I'm willing to give them these items.	.551	.040	-.062	-.291
Sorting waste takes up too much space.	-.159	.908	-.032	-.117
Sorting waste takes up too much time.	-.252	.883	-.114	.064
My municipality has a sufficient number of places with recycling bins.	.025	.008	.900	-.009
I have plenty of opportunities to recycle my waste.	.302	-.170	.762	.079
I know the recycling targets set by the EU for Portugal.	.047	-.003	-.186	.847
I am willing to contribute to the costs involved with the waste management system.	.047	-.035	.239	.486

Extraction method: Principal Component Analysis.
Rotation method: Varimax with Kaiser Normalization.[a]

[a]Rotation converged in five iterations

household waste; I encourage other people to separate and recycle their waste; if a community group goes from house to house colleting plastic, paper, metal, and glass, I'm willing to give them these items.

- *Drawbacks:* because they associate the drawbacks of recycling. This factor includes the following variables: sorting waste takes up too much space; sorting waste takes up too much time.
- *Satisfaction with the service provided:* because services provided by municipalities and opportunities for recycling are valued. This factor includes the following variables: My municipality has a sufficient number of places with recycling bins; I have plenty of opportunities to recycle my waste.
- *Knowledge on the subject:* because it is straight linked to the personal knowledge on recycling and related agencies. This factor includes the following variables: I know the recycling targets set by the EU for Portugal; I am willing to contribute to the costs involved with the waste management system.

Table 4 shows the relationship between the results of the present study compared to studies already carried out on consumer behavior that discuss RL practices.

As in the investigation carried out by do Valle et al. [8], variables related to environmental issues still have not got any statistical expression. However, the factor associated with consumer's attitude, norms, and recycling skills has proven to be the most relevant factor for the subject in question, as in previous studies.

Factors such as "Inconveniences," "Satisfaction with the service provided," and "Knowledge on the subject" were found to be relevant in this study as well as in other previous ones. Factors such as "Awareness of consequences," "Attitudes and norms of social recycling," "Intentions of action," and "Schema preference" were not statistically significant in the present investigation.

4.2 Consumers' Profile in Regard to RL

Cluster analysis was used in order to find groups of consumers with different profiles with regard to RL. Taking into account the size of the sample, the nonhierarchical method (K-means) is the most recommended one for this type of study, which requires that the number of clusters is previously defined. To begin the procedure, the number of clusters was determined using the hierarchical method, concluding that there were three clusters in the sample.

Discriminant analysis was used to identify the variables that differ significantly between clusters. When applying Wilks' lambda, the variables passed through the assumption of equality of means showing that the difference between groups was significant. The study demonstrates the existence (Table 5) of two discriminant functions for the clusters in question.

Regarding the variables' first function, "I separate paper, plastic, metal, and glass for recycling.", "It is likely that I will start sorting waste in the coming weeks.", "I always separate household waste.", and "I have the responsibility of separating the

Table 4 Comparison table of study factors with previous studies

Present study	Bezzina and Dimech [2]	do Valle et al. [8]	Calvin et al. [4]
Personal recycling attitudes, norms, and skills	Personal recycling attitudes, norms, and skills	Personal norms	Attitude
Drawbacks	Drawbacks		Convenience
Satisfaction with service provided	Satisfaction with service provided	Containers' location	
Knowledge on the subject	Knowledge on the subject	System adequacy and information	
	Consequence awareness		Consequence awareness
	Social recycling attitudes and norms	Social norms	Moral norms
	Action intentions		
	Schema preference		
			Subjective norms
			Perceived behavioral control

Table 5 Eigenvalues | three *clusters*

Eigenvalues				
Function	Eigenvalue	% Variance	% Cumulative	Canonical correlation
1	1.938[a]	75.6	75.6	0.812
2	0.624[a]	24.4	100.0	0.620

[a]The first two canonical discriminant functions were use

Table 6 Log Determinants | three *clusters*

Log determinants		
Ward method	Position	Log determinant
1	10	−0.420
2	10	−9.006
3	10	2.784
Within pooled groups	10	−0.367

The positions and natural logarithms of the printed determinants are those of the group covariance matrices

Table 7 Box's M | three *clusters*

Box's M		
M de Box		556.247
Z	Aprox.	4.740
	dfl	110
	df2	60429.735
	Sig.	<001

Tests the null hypothesis of equal population covariance matrices

waste I make." are the ones that most contribute to discrimination. With regard to the second function, the variables are "I am more likely to recycle if recycling is linked to a rewards and incentive scheme.", "I am more likely to recycle if the solid and urban waste (SUW) fee is proportional to the waste produced by each household.", and "Separating waste takes up too much space.".

Regarding the classification of the results, 85.2% of the individuals were correctly grouped. In the first cluster, 81.3% were correctly classified, in the second cluster 92.6%, and in the third 93.8%.

From the log determinants (Table 6), it appears that the greatest dispersion occurs in the second cluster and then in the third and first cluster, respectively. Based on the Box's M Test (Table 7), which requires that various groups are equally dispersed, the null hypothesis is rejected, with no homogeneity of dispersion. Thus, it is possible to lose the quality of the discriminant functions found in the investigation. However, from the Classification of Results, one observes that, in total, 85.2% of the elements were correctly classified, resulting in its validation.

From the discriminant cluster analysis, it was concluded that the sample clusters into three distinct clusters (Figs. 1 and 2):

- The encourageable: consumers who have good ecological knowledge but are encouraged to carry out RL via financial incentives, purchase advantages, and

Fig. 1 *Clusters* Characterization

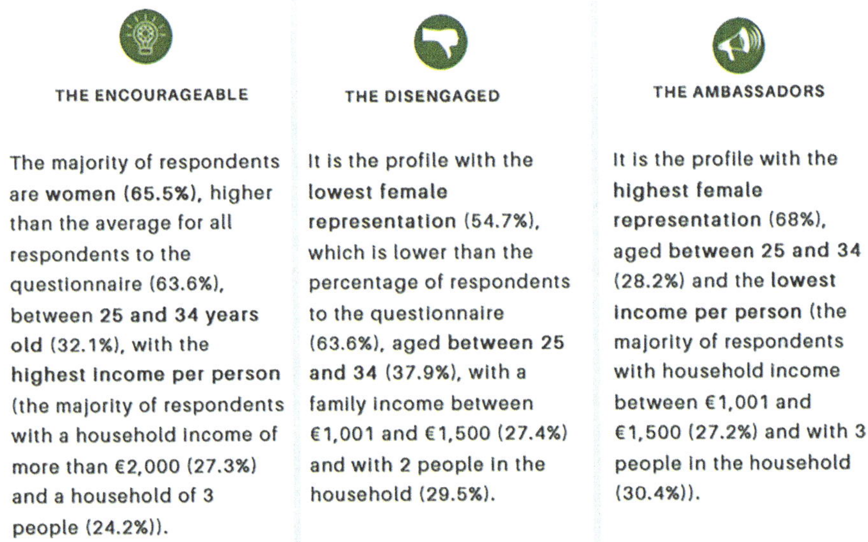

Fig. 2 Sociodemographic characterization of *clusters*

low complexity. Composed mostly by women (65.5%), reaching percentages even higher than the average of the questionnaire respondents (63.6%), between 25 and 34 years of age (32.1%), presenting the highest income per person

(most with a household income of more than €2000 (27.3%) in a three people household (24.2%)).
- The disengaged: consumers with low ecological knowledge and who demonstrate a lack of responsibility for the act of recycling, being the least engaged profile in environmental issues. Mostly composed of women (54.7%), but representing a lower female representation than the questionnaire respondents (63.6%), aged between 25 and 34 (37.9%) with family income ranging from €1001 to €1500 (27.4%) in a two people household (29.5%).
- The ambassadors: consumers with good ecological knowledge and engaged in the topic, recycling regardless of the financial incentives, purchase advantages, or complexity. It is the profile with the highest female representation (68%), aged between 25 and 34 (28.2%), with lower income per person (most respondents with a household income ranging between €1001 and €1500 (27.2%) in a three people household (30.4%)).

5 Conclusion

The study carried out along with a 385 responding sample enabled the outlining of the Portuguese consumer profile in relation to RL and resulted in relevant conclusions.

The socioeconomic profile is essentially characterized by the predominance of women (63.6%). Respondents' age ranges between 18 and 65, and they are mostly graduates or masters, with a net household income of more than €1500, live in households with more than 2 people, and reside in 17 districts, with the ones with the highest representation being Évora and Lisbon.

The analysis of the factors that influence the consumer in the process of deciding whether or not to carry out recycling as part of the RL process demonstrates that attitudes, norms, and personal recycling skills, inconveniences, satisfaction with the service provided, and knowledge on the subject are relevant. Environmental factors were not statistically observed in this analysis.

Regarding the existence of environmental awareness, there are different degrees of awareness among consumers. Some consumers who recycle are willing to participate in public information/awareness campaigns and encourage other people to separate and recycle their waste However, there are other consumers who demonstrate little environmental awareness about their own actions.

With regard to what influences consumer's attitude toward returning waste, three distinct profiles were found:

- The encourageable: consumers who have good ecological knowledge but are encouraged to carry out RL via financial incentives, purchase advantages, and low complexity.

- The disengaged: consumers with low ecological knowledge and who demonstrate a lack of responsibility for the act of recycling, being the least engaged profile in environmental issues.
- The ambassadors: consumers with good ecological knowledge and engaged in the topic, recycling regardless of the financial incentives, purchase advantages, or complexity.

It appears, therefore, that ecological knowledge, financial incentives, purchase advantages, and the complexity of carrying out RL influence the consumer's attitude toward returning their waste.

Finally, it was observed that the present study contributed to filling a shortage of studies that correlate the areas of logistics and consumer behavior. When correlated, these areas generate a positive impact on several processes, namely, on the relationship between consumers and society, being considered as a competitive advantage.

The study carried out on consumer behavior in relation to recycling has a significant contribution to management in the sense that its results, that is, knowledge about the four factors and the three consumer profiles, can help companies develop effective strategies of engagement, which includes awareness campaigns, reward programs, or partnerships with recycling organizations that are more targeted for each of the three groups identified.

The present study, according to its proposed objectives, presented some limitations during the course of investigation: difficulty in obtaining a greater heterogeneity in responses, given the restricted stay of the researcher in Portugal; sample for convenience; biased responses due to respondents' moral and social influences leading to very similar responses on certain topics such as "In our society, recycling is the right thing to do."; quantitative research, which made it impossible to establish a relationship with the surveyed sample and the thorough understanding of factors on the subject; and online data collection.

As suggestions for future investigations, the following are proposed: carrying out a qualitative and quantitative study to understand the responsibility of companies and producers on encouraging RL; conducting a qualitative study and bringing up perceptions, habits, and preferences of consumers through personal interviews; and scrutinizing the concept of irresponsible consumption in order to investigate the trends of "Reduce" and "Reuse" of the 3 Rs and not just "Recycle."

References

1. Bamberg, S., & Möser, G. (2007). Twenty years after Hines, Hungerford, and Tomera: A new meta-analysis of psycho-social determinants of pro-environmental behaviour. *Journal of Environmental Psychology, 27*(1), 14–25.
2. Bezzina, F. H., & Dimech, S. (2011). Investigating the determinants of recycling behaviour in Malta. *Management of Environmental Quality: An International Journal, 22*, 463–485.

3. Biswas, A., Roy, M., & Ganguly, S. (2014). Factors influencing recycling behavior: An integrated model. *Resources, Conservation and Recycling, 93*, 73–82.
4. Calvin, W., Ronnie, C., & Geoffrey, Q. S. (2012). Recycling attitude and behaviour in university campus: A case study in Hong Kong. *Facilities, 30*(13/14), 630–646.
5. Chen, M. F., Tung, P. J., & Chao, C. M. (2018). Factors influencing recycling intention and behavior: The mediating role of recycling knowledge and recycling attitude. *Journal of Environmental Management, 206*, 419–428.
6. Council of Supply Chain Management Professionals. (2013). *seção CSCMP supply chain management definitions and glossary*. https://cscmp.org/CSCMP/Educate/SCM_Definitions_and_Glossary_of_Terms.aspx
7. da Silva, A. A., & Leite, P. R. (2012). Empresas brasileiras adotam políticas de logística reversa relacionadas com o motivo de retorno e os direcionadores. *RGSA – Revista de Gestão Social e Ambienta, 6*, 79–92.
8. do Valle, P. O., Reis, E., Menezes, J., & Rebelo, E. (2004). Behavioral determinants of household recycling participation: The Portuguese case. *Environment and Behavior, 36*(4), 505–540.
9. Engel, J. F., Blackwell, R. D., & Miniard, P. W. (2000). *Comportamento do consumidor* (8th ed.). Copyright. Editora LTC.
10. Figueiredo, P. J. (2014). *Logística inversa no mercado de telemóveis em Portugal*. Faculdade de Economia do Porto.
11. Gamba, R., & Oskamp, S. (1994). Factors influencing community residents' participation in commingled curbside recycling programs. *Environment and Behavior, 26*, 587–612.
12. Homburg, C., Jozić, D., & Kuehnl, C. (2017, May). Customer experience management: Toward implementing an evolving marketing concept. *Journal of the Academy of Marketing Science, 45*(3), 377–401.
13. Jacoby, J., Berning, C. K., & Dietvorst, T. F. (1977). What about disposition? *Journal of Marketing, 41*(2), 22–28.
14. Kallbekken, S., & Sælen, H. (2013). Public acceptance for environmental taxes: Self-interest, environmental and distributional concerns. *Energy Economics, 36*, 536–545.
15. Kianpour, K., Jusoh, A., Mardani, A., & Streimikiene, D. (2017). Factors influencing consumers' intention to return the end of life electronic products through reverse supply chain management for reuse, repair and recycling. *Sustainability, 9*, 23.
16. Kuester, S. (2012) *Market 301: Strategic marketing & marketing in specific industry contexts* (pp. 110, 393–404). University of Mannheim.
17. Leite, P. R. (2002). Logística Reversa Nova Área da Logística Empresarial. *Revista Tecnologística*, 1–6.
18. Milichovský, F. (2016). Relationship of reverse logistics and marketing communication in Czech Republic. *Trends Economics and Management, 10*, 48.
19. Murphy, P., & Poist, R. F. (1988). Management of logistical retromovements: An empirical analysis of literature suggestions. *Journal of the Transportation Research Forum, 29*, 177–184.
20. Pohlen, T. L., & Farris, M. T. (1992). Reverse logistics in plastics recycling. *International Journal of Physical Distribution & Logistics Management, 22*(7), 35–47.
21. Roster, C. (2001). Letting go: The process and meaning of dispossession in the lives of consumers. *NA – Advances in Consumer Research, 28*, 425–430.
22. Silva, C., & Dias, O. (2020). Segmentação de mercados e diferenciação de ofertas de logística reversa. *Revista Brasileira de Marketing, 19*, 862–887.
23. Silvério, M. (2003). *Pesquisa de Marketing*. Publicações Universidade de Évora, Ciências Económicas e Empresariais.
24. Suresh, K., Mr, V., Divya, P., & Mathipurani, V. (2020). The effects of cosmetics packaging design on consumer's purchase decision. *IJETIE, 6*(3), 60–70.
25. Thøgersen, J. (2004). A cognitive dissonance interpretation of consistencies and inconsistencies in environmentally responsible behaviour. *Journal of Environmental Psychology, 24*(1), 93–103.

26. Vermeir, I., & Verbeke, W. (2006). Sustainable food consumption: Exploring the consumer "attitude – Behavioral intention" gap. *Journal of Agricultural and Environmental Ethics, 19*(2), 169–194.
27. Wang, H., Jiang, Z., Zhang, H., Wang, Y., Yang, Y., & Li, Y. (2018). An integrated MCDM approach considering demands-matching for reverse logistics. *Journal of Cleaner Production, 208*, 199–210.
28. Young, M. M., & Wallendorf, M. (1989). Ashes to ashes, dust to dust: Conceptualizing consumer disposition of possessions. *Proceedings of the AMA winter educator's conference*, pp. 33–39.

Enhancing Tourism Through Local Development Strategies (LDSs): A Study of FLAG Marche Sud Area

Federica Murmura, Guido Capanna Piscè, and Claudia Fraboni

Abstract This article focuses on the role of the community-led local development (CLLD) methodology in the promotion of the integrated development of local communities. This study explores the potential of CLLD as a tourism enhancement tool for five municipalities within the FLAG Marche Sud area. The analysis has been carried out through a literature review and in-depth interviews. To explore the relationship between CLLD and tourism, local development strategies (LDSs), and stakeholder collaboration, the research examines the methodology in terms of its definition, funding, implementation, and monitoring. The findings suggest that the CLLD approach can help local communities to maximize the opportunities offered by EU funding streams, in particular through food festivals. Also, it can help them identify their strengths and develop tourism strategies that suit their peculiarities. Anyway, all the relevant stakeholders need to cooperate to ensure a successful implementation of the methodology. The study provides a theoretical contribution to how CLLD can be an effective tool for enhancing tourism within local communities. The present article is a groundwork for the FLAG Marche Sud marketing plan.

Keywords Tourism · Community-led local development (CLLD) · Local development strategies (LDSs) · "Fisheries Local Action Groups (FLAGs)"

1 Introduction

The European Union has developed the community-led local development (CLLD) methodology to financially empower local communities and help them establish local businesses, create new services, preserve heritage, and encourage cooperation among local businesses. The CLLD approach meets the goals of integration and revitalization of communities by addressing their pressing environmental, eco-

F. Murmura (✉) · G. C. Piscè · C. Fraboni
Department of Economics, Society, Politics, Carlo Bo University of Urbino, Urbino, Italy
e-mail: federica.murmura@uniurb.it

J. Reis et al. (eds.), *Driving Quality Management and Sustainability in VUCA Environments*, Springer Proceedings in Business and Economics,
https://doi.org/10.1007/978-3-031-52723-4_20

nomic, and social needs. The FLAG Marche Sud area in Italy, which includes five municipalities (San Benedetto del Tronto, Grottammare, Pedaso, Cupra Marittima, and Porto San Giorgio) primarily reliant on tourism and fishing [1], has faced economic challenges due to the COVID-19 pandemic and financial crisis. The community-led local development approach can be used to address these challenges and enhance the potential growth of cities [2, 3]. Local development strategies are financed by EU funding streams such as the European Maritime and Fishery Fund (EMFF), the European Agricultural Fund for Regional Development (EAFRD), the European Social Fund (ESF), and the European Regional Development Fund (ERDF). Fisheries Local Action Groups (FLAGs), local enterprises, and public authorities are the main actors of the CLLD tools. Funding for CLLD projects is based on the local needs and resources of the community it aims to serve. In the FLAG Marche Sud area, tourism is a leverage for community-led local development projects. The research aims to suggest how the methodology can be used to maximize integration between local enterprises by giving priority to inputs, human resources, and other assets from the area when organizing food festivals. The research also aims to implement a marketing plan to enhance tourism in the FLAG Marche Sud area by explaining how LDSs [4] can be used to enhance tourism in coastal cities and how food festivals can lead to increased cooperation between local organizations, mainly fisheries and accommodation businesses.

Although many researchers have dealt with CLLD approaches across Europe, there has been no extensive study about how the tool can improve the tourism management of a destination. The research aims to fill this gap.

2 Literature Review

2.1 Methodology

The increasing need for community-led local development, as a tool for improving tourism management, captured the interest of researchers, with a focus on implementation, management, and implications of the methodology. The researchers' interest in CLLD multi-level tools derives predominantly from the significant number of actors involved in the process.

The main research topics include the research for CLLD in local communities, the adoption process in Italy, the guidelines for member states, the rules to follow, the implementation of community-led local development, the financing of it, the so far experiences in implementing the tool across EU, the changes in FLAGs implementation, the partnership governance, the CLLD for firms financing, and the impact of CLLD on tourism.

This work sets out to provide a systematic review and commentary along the literature about the CLLD multi-level tools. The general process of literature review adopted went through different phases, which at the end brought to outline the evolution of the literature dedicated to the use of community-led local development

(CLLD) as a tourism management approach in the European Union (EU) between 2014 and 2022. In particular, the present systematic review led to establish to what extent existing research has progressed toward clarifying our topic and identify relations, contradictions, gaps, and inconsistencies in the literature that finally allowed to describe directions for future research.

The systematic review process started, with the scoping. In fact, some scoping of the literature was needed to find out what was done before and what might make a novel and interesting scientific contribution to the literature. After some search, the review question was formulated according to our field of interest. The central research question guiding this literature review was: "what does the existing literature reveal about the efficacy of the community-led local development (CLLD) methodology in enhancing tourism, particularly in terms of its definition, funding, implementation, and monitoring within local communities such as the FLAG Marche Sud area?"

The first scoping also clarified whether the planned systematic review had already been done by other researchers. The review process continued then, with the planning phase. At this stage of the process, the research questions were broken down into individual concepts to create search terms[1] which then helped finding as many potentially relevant articles as possible to include. In the balance between sensitivity (finding as many articles as possible that may be relevant) and specificity (making sure those articles are indeed relevant), at this stage we preferred search terms to be more on the side of sensitivity so that we could trust we were not missing anything. In any case, the definition of inclusion and exclusion criteria formulated on the base of our knowledge of the literature helped in clearly defining the boundaries of the review. Having reached the searching stage, the aim was to find all available published or unpublished work which could address the research questions set through the chosen search terms. The screening stage was about reading the title and/or abstract of all work identified by the search stage and resulted in choosing all studies which were potentially eligible for inclusion. For all the work meeting the inclusion criteria, the full-text version was obtained. This first locating and sifting process continued to err on the side of sensitivity. Finally, in the eligibility stage, we sifted the full-text version of all potentially eligible articles and extracted relevant information to be included. At this point, our focus shifted from sensitivity

---

[1] **Core Keywords:** Community-led local development (CLLD); Community-based development; Local development; Grassroots development; Tourism management; Tourism strategies; Sustainable tourism; Tourism governance; European Union (EU); EU policies; EU funding; EU guidelines; Financing; Funding; Financial models; Budgeting; Adoption process; Implementation; Rollout; Uptake; Guidelines for member states; EU directives; National policy; Compliance; Partnership governance; Collaborative governance; Multi-stakeholder partnerships; Governance models; FLAGs (Fisheries Local Action Groups); Local action groups; Fisheries governance; Fisheries management; Economic growth; Economic development; GDP growth; Local economy
Additional Keywords Social inclusion; Social engagement; Social sustainability; Environmental sustainability; Green policies; Eco-friendly tourism; Stakeholder participation; Public participation; Community engagement; Policy recommendations; Best practices; Policy implications

to specificity, making sure that potentially eligible studies were indeed relevant and appropriate for inclusion.

Studies were identified and gathered from two of the most popular online databases, Scopus and Web of Science, as well as from relevant governmental reports.

2.2 Literature Review

Having gone through all the stages illustrated in the methodology, we got all the potentially eligible work and extracted all relevant information to be included in the review.

We found a total of 96 publications of which we read the abstracts and/or titles to decide which to exclude and which to keep since useful and relevant to the research questions. After the screening process, a total of 49 articles remained. Of all the publications which were considered eligible for inclusion, we read the full text (eligibility process), excluded 29 articles due to the lack of empirical data, and extracted the relevant information from a total of 20 publications.[2]

So, at the end, we got a total of 20 peer-reviewed articles, published between 2014 and 2022, meeting the criteria, which will be the core of the following descriptive literature review.

The present descriptive review highlights how the main literature stream is considering the community-led local development as an effective tool for promoting economic growth mainly through the development of local communities and cities in the EU.

2

Criteria type	Search criteria	Description
Inclusion	Database sources	Scopus and Web of Science, as well as relevant governmental reports
Inclusion	Temporal scope	Published between the years 2014 and 2022
Inclusion	Relevance	Articles must be directly pertinent to the research questions
Inclusion	Empirical data	Articles must contain empirical data
Inclusion	Geographic focus	European Union, with specific attention to Italy
Inclusion	Stage of search	Scoping, planning, searching, screening, and eligibility
Exclusion	Database sources	Other than Scopus, Web of Science or governmental reports
Exclusion	Temporal scope	Articles published outside the range of 2014–2022
Exclusion	Empirical data	Articles lacking empirical data

In the study "The Marche: Italy's New Frontier for International Retirement Migration" [10], the Marche region in Italy is analyzed in terms of its international retirement migration, focusing on the age groups, activities, and impact on the territory of elderly tourists. The paper suggests policies for improving the tourist experience and addressing challenges. However, the study is limited to elderly tourists only.

The paper "Community-Led Local Development Cohesion Policy 2014–2020" [5] defines the CLLD methodology, which is an area-based approach to solving specific social, economic, and environmental needs of a community. It promotes community, efficiency in governance, integrated bottom-up approaches, sustainability, and innovation. The EU established the Lead Fund to manage strategy development, funding, implementation, oversight, and reporting of the integrated LDSs.

The work "Guidance for Member States and Programme Authorities on Community-led Local Development in European Structural and In-vestment Funds" [6] provides guidance on how to implement the complicated CLLD methodology efficiently. It examines the Common Provisions Regulation (CPR), which outlines rules for the use of funds to achieve goals consistent with EU sustainable development goals. Funding should only be granted to LDSs that tackle pressing economic, environmental, and social needs of local communities, and projects need to promote efficiency and innovation in organizing markets and creating new products and services.

The study "From LEADER to CLLD" [7] highlights the increase in established FLAGs due to the community-led local development methodology and their role in tackling issues such as economic development, demographic retention, social inclusion, and environmental protection.

The study "The Bottom-Up Development Model as a Governance Instrument for the Rural Areas" [8] examines the cases of two Italian FLAGs and suggests that highly autonomous local authorities perform better in using EU funds to meet pressing needs in their communities.

The paper "Stakeholder Participation in Planning Rural Development Strategies" [4] emphasizes the importance of stakeholder participation and feedback in the CLLD methodology and suggests that involving all stakeholders can improve democracy at the local community level.

The work "The Unaccomplished Quest for Urban Policies in Italy" [9] examines the need for a clear strategic policy design and vision for urban development and suggests that adopting a common and clear strategic policy can enhance cohesion and boost economic performance in areas such as the FLAG Marche Sud.

In "Tailored Polities in the Shadow of the State's Hierarchy" [3], the author discusses the innovative features of CLLD financing, including its multi-fund approach, which allows local communities to engage in a wider range of activities. The author also emphasizes the importance of a Lead Fund to manage financial budgeting, implementation, control, and reporting of LDS projects under community-led local development, which could lead to more efficient achievement of sustainable development goals.

"Four Ideal Types for Leaderability" [10] identifies four leadership models that FLAGs should implement, including high strategic capability and limited involvement in decision-making. The paper emphasizes the importance of FLAGs as local actors with expertise on complex tourism and coastal-city issues, who guide local stakeholders in cooperation and sourcing funds to stimulate market opportunities through innovation or network creation.

In "Are FLAGs Contributing to Solving Social Exclusion Problems?" [11], the authors suggest that EU financing streams like EMF, ESF, and ERDF are essential in providing financial support to FLAGs. They argue that a thorough diagnosis of the level of social exclusion is necessary before implementing LSDs to meet the needs of less privileged people in the territory.

In the work "Implementation of the Leader Programme in the Observed EU Countries" [12], the authors examined the implementation of the CLLD methodology in four EU member states. They found that focusing on local self-sustainability through teaching, sourcing inputs, and learning from the local community helps in networking and increasing cooperation among the local businesses.

"The role of Food Festivals in Branding Culinary Destinations" [13] emphasizes the potential of enhancing the food and gastronomy sector by holding unique culinary festivals tailored to the tastes and preferences of frequent visitors. The paper suggests that culinary festivals should meet all visitors' needs and interests to create added value for the community and that FLAGs should work together with stakeholders to ensure high-quality cuisine and food culture.

In "Sustaining Rural Areas, Rural Tourism Enterprises, and EU Development Policies" [14], the authors suggest that local governments should stabilize the economic environment, provide information, invest in specific expertise, and involve local representatives and FLAGs in identifying business downsides to overcome challenges in implementing LDSs. Additionally, the paper suggests that organizations participating in FLAG Marche Sud should implement programs to teach locals some of the most widely spoken foreign languages.

In "Implementing ERDF Through CLLD" [15], the authors investigate community-led local development implementation experiences in some EU countries and suggest that FLAGs can activate a broad range of actions in their territories. However, more funds need to be directed toward LDSs to improve the recipients' experience. The paper suggests that experiences of other EU member states can be used as a benchmark for implementing strategies and connections aimed at enhancing tourism in the FLAG Marche Sud area.

The aim of this review was to provide an objective discussion on the ongoing adjustment of CLLD approaches to the diverse European standards. The review highlighted the need for more quantitative, long-term studies to fully evaluate impacts over time. In general, this systematic review of academic research contributes to build a critical knowledge base for stakeholders interested in strengthening integrated local tourism management through community-driven models.

3 Data and Method

The in-depth analysis of the 52 semi-structured interviews conducted during the month of September 2021 in the FLAG Marche Sud region offered a deep understanding of CLLD approaches in their connection with tourism and socioeconomic factors. Face-to-face interviews have been carried out involving different stakeholders such as local representatives, organizations, municipal entities, tourists, and locals, ensuring a wide range of viewpoints.

Data Collection Details

- Interviews conducted: 52
- Interviewees: local representatives, organizations, municipal authorities, tourists, and other locals
- Breakdown of interviews: 18 FLAG Marche Sud actors, 2 municipal authorities, 20 tourists, 5 local representatives, 2 local citizens
- Interview locations: hotels, villages, municipal offices, and other locations within the FLAG Marche Sud area
- Eligibility criteria for interviewees: specific criteria based on employment and involvement in the management or formulation of an LDS in the area
- Sample size: people of different ages, socio-professional backgrounds, and length of residence in the FLAG territory

The interviews, which were carried out in various settings like hotels, villages, and municipal offices, aimed to assess the efficacy of the CLLD approach in driving social and economic development. The interview results showed the ability of food festivals to attract visitors to the region and set out the travel behavior of tourists in the area, thus offering a thorough overview of tourism in the FLAG Marche Sud region. To be considered for an interview, local governments' officers and private companies' employees had to meet certain criteria, such as a minimum of 10 years employment and involvement in LDS management or formulation in the FLAG area.

Also, tourists within the FLAG region were interviewed, in order to track the frequency and duration of their visits and their interest in food festivals. The interview sample included individuals of different ages, socio-professional backgrounds, and lengths of residence in the FLAG area. The research aimed to explore the CLLD approach performance across all of the economic sectors, with a particular emphasis on organizations fostering social inclusion through tourism in the FLAG Marche Sud area.

To protect data and privacy, every interviewee was requested to sign an informed consent form. None of the 52 individuals invited to be interviewed declined. Transparency was guaranteed by keeping the interview transcripts for future reference. The interviews were tailored to the respondent's needs, including the omission or modification of undesired questions to preserve a friendly atmosphere. All interviews were recorded to ensure seamless communication with the respondents.

The responses analysis was focused on the connection between tourism and the CLLD approach. The analysis also covered issues like the ability of food festivals to attract visitors, the local stakeholders' perception of the CLLD methodology, and tourists' travel habits. To protect the respondents' privacy and let them feel at ease during the interview, pseudonymisation was guaranteed.

The results underscore a robust understanding and acceptance of the CLLD approach among stakeholders, especially among municipal authorities and organizations involved in LDS management. Said stakeholders perceive CLLD as a significant driving force for fostering social inclusion and for enhancing economic growth.

Eventually, the interviews' analysis offers a deeper understanding of the region's management dynamics:

- Stakeholder Cooperation: the interviews emphasized the importance of stakeholder collaboration in the FLAG Marche Sud area. The participating organizations are interlinked, actively partnering with diverse stakeholders, including actors from the fishing sector, local government, and cultural associations. This collaborative approach is not just an asset but a necessity, given the diverse needs of the municipality. Future research could examine the structure and management of these collaborations and their contribution to the overall development of the area.
- Tourist Preferences: the interviews showed that a good part of the FLAG Marche Sud area tourists are culinary enthusiasts and are attracted by the region's history and valuable reputation. This implies that the area's gastronomic heritage is a valuable resource that could be used to enhance tourism. Future research could deeper investigate which aspects of the region's history and reputation tourists are most interested in.
- SWOT Analysis: the SWOT analysis identified the strengths, weaknesses, opportunities, and threats for the FLAG Marche Sud area. A more nuanced analysis, though, could have highlighted the factors underlying these dynamics. For example, which specific aspects of the region's social capital and professionalism are regarded as strengths? What are the causes of the identified weaknesses, such as the lack of innovation and the limited foreign language skills? How can the identified opportunities be exploited, and the threats mitigated?
- Social Inclusion Through Tourism: the interviews suggested that tourism plays a critical role in the FLAG Marche Sud area. However, the data did not return information on how tourism is promoting social inclusion in the area. Future research could explore how local communities benefit from tourism and what strategies are being employed to ensure that the advantages of tourism are equally distributed.

4 Results

The present research illustrates how EU CLLD methodology can be used to enhance tourism in the FLAG Marche Sud area. Also, it highlights the competitive

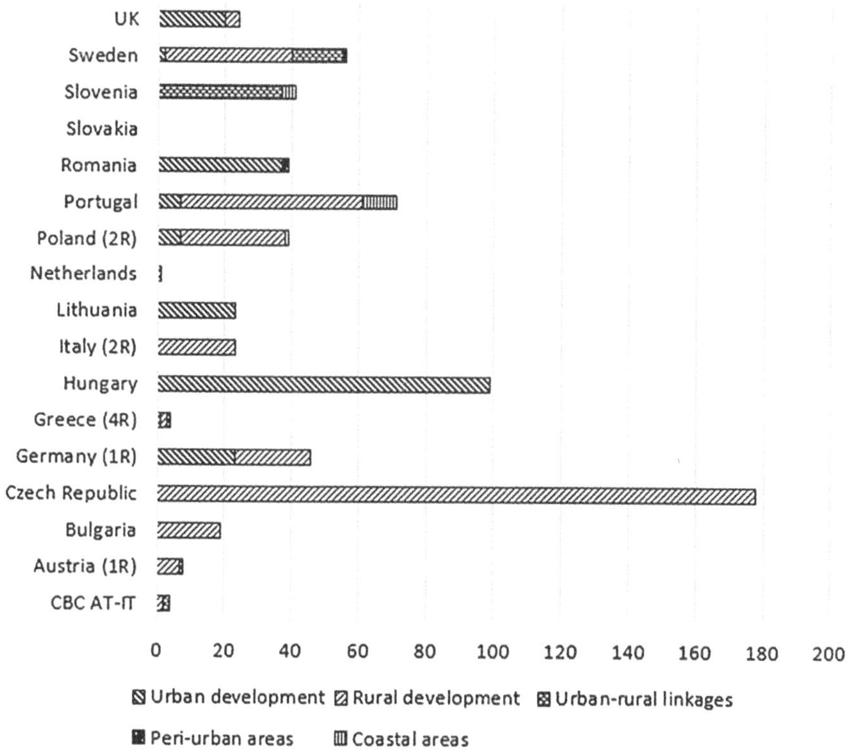

Fig. 1 Territorial focus of FLAGs in EU member states [7]

advantages of promoting LDSs such as food festivals. With regard to the research question "how can CLLD enhance tourism?", the data analysis and the above literature review led to the following conclusions: most organizations participating in the FLAG Marche Sud have tourism or fishing as core businesses. Figure 1 shows that countries adopting the CLLD methodology have five different territorial focuses: rural development, urban development, urban-rural linkages, coastal areas, and peri-urban areas. For Italy, the focus is on the rural areas [16]. The graph and the present study demonstrate that EU member states have adopted a bottom-up approach to achieve integrated development within the local communities. Regional integration has a leading role in fostering sustainable economic growth within the areas of implementation.

However, it is important to note that Italy has many coastal cities that have established LDSs. The multi-funding approach adopted by the CLLD methodology has the potential for achieving integrated development in coastal cities. In the FLAG Marche Sud area, fishing and fish processing is one of the main activities along with tourism. This means that organizations participating in the FLAG Marche Sud can obtain funds for fishing-related tourism, so that local fishermen can produce

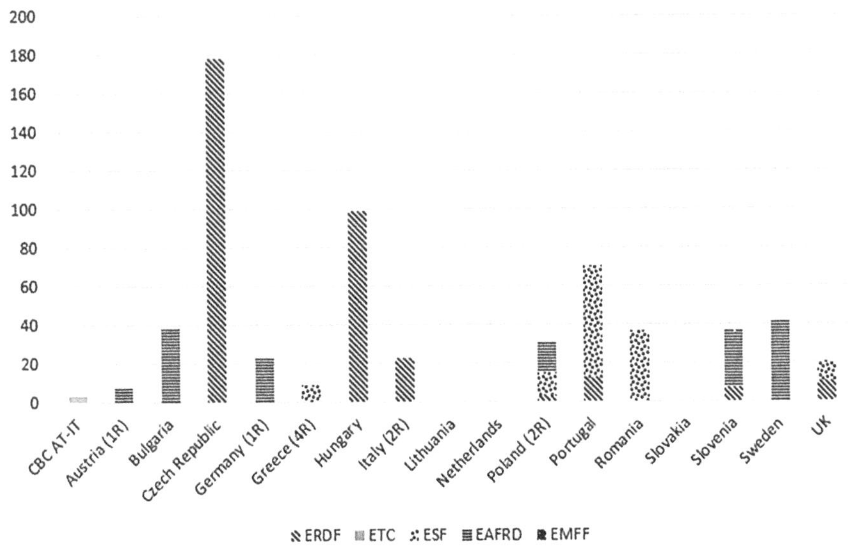

Fig. 2 Lead Funds in multi-funded FLAGs of EU member states [7]

inputs for local accommodation businesses. The EMFF is the perfect funding stream for fisheries and tourism activities in the coastal areas [3]. Figure 2 shows that the ERDF is considered the Lead Fund in Italy. The organizations participating in the FLAG Marche Sud can clearly show how food festivals, financed by EU funding streams, can lead to an increase in tourist flows. Accordingly, the empowerment of local fishing actors can be reached fostering cooperation between these and the accommodation businesses. A way to realize this kind of cooperation may be through the supply of raw materials from the local fishing businesses to the accommodation firms involved in the food festivals.

The present study also emphasizes the need for LDSs that meet the pressing social, environmental, and economic needs of the local communities in which they operate. For the FLAG Marche Sud area, the most pressing need is to enhance social inclusion by providing job opportunities for the neediest individuals. Figure 3 shows how, according to the literature, social inclusion is the main thematic approach for the CLLDs in Italy [7]. Thematic approaches declare the aim of FLAGs in the implementation of LDSs. Social inclusion involves the implementation of new opportunities for other people, as well, to engage with the local community. Apart from social exclusion, theoretical approaches, adopted by other EU member states, list economic development, demographic support, environmental protection, and access to services.

The analysis of data collected through interviews shows that the organizations involved in the FLAG Marche Sud area are effectively using the CLLD methodology. In fact, they cooperate with various stakeholders: actors from the fishing sector, local government, and local cultural associations. The cooperation with

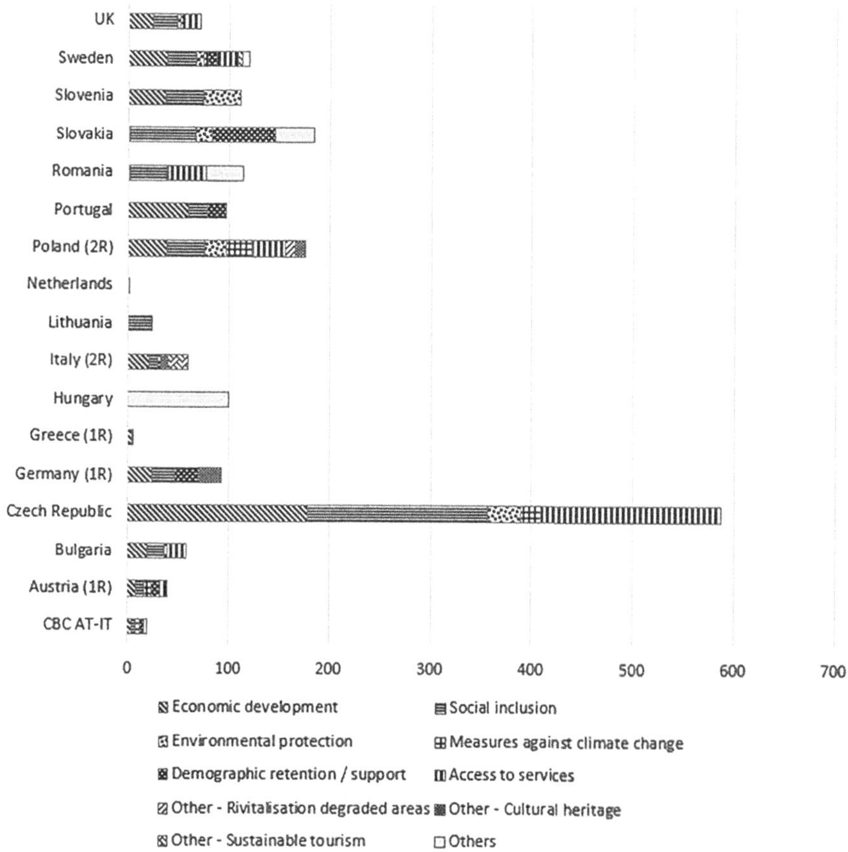

Fig. 3 Thematic approaches adopted by various EU member states [7]

different stakeholders is seen as an effective way to address the pressing needs of the municipality. The interviews also revealed that half of the tourist respondents are food enthusiasts who are attracted by the history and reputation of the five FLAG municipalities.

Moreover, the collected data was used to create a SWOT analysis that determines the competitive position of the FLAG Marche Sud area. The interviews provided information on the current and future potential of the area, as well as the internal and external threats affecting the municipality. The SWOT analysis identified the following, as strengths of the five municipalities: high social capital and professionalism, rich history and cultural excellence, and good accessibility in terms of rail and roads. However, lack of innovation, seasonality of raw materials, and low foreign language skills were identified as weaknesses. Opportunities include connecting with other coastal cities along the Adriatic Sea whose economies are similarly based on tourism, as well as safeguarding the environment to foster unique tourist and cultural attractions. The threats faced by the five municipalities include the declining availability of fish and the increase of fish imports from abroad.

5 Discussion

In this study, the researchers analyzed existing literature to explore the potential of CLLD to improve tourism management in the FLAG Marche Sud area. The findings align with most of the reviewed literature, indicating that CLLDs have the potential to promote integration and foster economic growth within the local communities. However, the successful implementation of CLLD requires careful analysis of the pressing needs in the area but also about active participation and support from local governments and about the involvement of all stakeholders in the policy formulation. These factors are crucial in ensuring that CLLD initiatives are effective and sustainable. Overall, this study highlights the importance of cooperation and stakeholder engagement in CLLD and emphasizes the need for strong leadership and effective communication to achieve successful outcomes.

Tourism organizations participating in the FLAG Marche Sud can engage in market analysis to identify ways for improving the tourists' experience while also tackling the pressing needs of the local community. This can be done by creating connections between local businesses and municipalities to ensure sustainability. But, the elaboration of an effective strategy, based on the market analysis, is critical. In the present analysis, a literature review and in-depth individual interviews have been carried out to assess how the FLAG Marche Sud communities use the CLLD methodology to enhance tourism management. The research shows how the organizations play a critical role in fostering cooperation among all the stakeholders in the community. Thus, local communities and municipalities should consider FLAGs as the main identity of the coastal cities' development agency.

The research also illustrates how the marketing strategy of the FLAG Marche Sud area has been focused on tourism management enhancement through the CLLD methodology. Organizations participating in the FLAG can foster integration between local fishing actors and accommodation businesses by organizing food festivals which showcase the food culture and the heritage of the local communities. Besides that, food festivals are valuable ways to promote the values and local identity of the territory among tourists. Thus, the marketing strategy should focus on the showcasing of local cuisines during food festival programs. This strategy also helps triggering curiosity of tourists who are food enthusiasts and promote authentic traditional cuisines in a particular region. Generally, food festivals give a chance to improve the livelihood of local fisheries actors while also improving the general quality of local cuisines across the whole region. Eventually, food festivals can play a part in creating a positive brand for the FLAG Marche Sud area as a must-go tourist destination.

Anyway, the main effect of empowering local fisheries actors and organizing food festivals is the creation of employment opportunities for locals. In fact, organizing frequent food festivals could create employment opportunities for locals who work for the festivals. The use of local fish products in food festivals also promotes the development of local fishing actors by creating demand for their products. This shows the im-portance of establishing LDSs which are tailored to solve the pressing needs of the community they serve.

6 Conclusion

The EU employs the CLLD methodology to drive change in local communities by financially empowering them to establish LDSs. Funding and oversight are provided by EU financial streams such as ESF, ERDF, EAFRD, and EMFF, which also establish guidelines for managing the operations and funding of LDSs. Important stakeholders such as FLAGs formulate strategies, manage funding, implement projects, and monitor operations while ensuring that all partners are involved in the decision-making process. Other stakeholders in the CLLD methodology include organizations participating in the FLAG, local representatives, the local community, and the municipality. In the case of FLAG Marche Sud, stakeholders include the local municipality, local representatives, private organizations, and other public bodies.

The community's strength lies in its reputation as a popular tourist destination. Therefore, organizations participating in the FLAG Marche Sud can successfully organize food festivals that use locally produced food to promote integration among local businesses. Integration leads to the transfer of skills and an overall improvement in the quality of food offered by accommodation businesses, creating a unique culinary culture that tourists can appreciate. It is important not to forget that 50% of tourists visiting the FLAG Marche Sud area are food enthusiasts who travel to explore local cuisines. Food festivals organized by FLAG Marche Sud actors empower the local communities by creating demand for local fish products and new employment opportunities.

While this study provides useful insights into the potential benefits of food festivals, it has some limitations that suggest avenues for future research. For example, the model exploring the link between food festivals and tourism enhancement has not yet thoroughly been investigated. Further studies should aim to quantify the tourism demand for culinary culture and food festivals. This could involve surveying potential tourists about their interest in food festivals in coastal areas like the FLAG Marche Sud region. Also, future research should estimate the number of new visitors that food festivals attract to a region. Conducting market research in connection with tourism boards and monitoring tourism statistics before and after food festivals could shed light on this. A deeper understanding of the impact of food festivals on tourism could give to regions, like the FLAG Marche Sud one, the tools to strategically develop these events in ways that maximize the economic opportunities for local fishermen and communities. Finally, while this study provides evidence of the benefits of food festivals on tourism, future research should dig deeper into the effects of those events on tourism in order to provide actionable recommendations for managers and policymakers.

References

1. King, R., Cela, E., Morettini, G., & Fokkema, T. (2019). The Marche: Italy's new frontier for international retirement migration. *Population, Space and Place, 25*(5), e2241.
2. Courades, J. M., & Brosei, P. (2018). CLLD/LEADER: Applying the partnership principle to local development. *European Structural & Investment Funds Journal, 6*(3), 210–222.
3. Servillo, L. (2019). Tailored polities in the shadow of the state's hierarchy. The CLLD implementation and a future research agenda. *European Planning Studies, 27*(4), 678–698.
4. Sisto, R., Lopolito, A., & Van Vliet, M. (2018). Stakeholder participation in planning rural development strategies: Using backcasting to support local action groups in complying with CLLD requirements. *Land Use Policy, 70*, 442–450.
5. European Commission. (2014). *COHESION POLICY 2014-2020. Community-led local development*. European Commission.
6. European Commission. (2018). Guidance for member states and programme authorities on community-led local development in European structural and investment funds. *European Structural and Investment Funds, 4*, 1–62.
7. Servillo, L., & De Bruijn, M. (2018). From LEADER to CLLD: The adoption of the new fund opportunities and of their local development options. *European Structural and Investment Funds Journal, 6*(3), 223–233.
8. Gargano, G. (2021). The bottom-up development model as a governance instrument for the rural areas. The cases of four local action groups (LAGs) in the United Kingdom and in Italy. *Sustainability, 13*(16), 9123.
9. Fedeli, V. (2021). The unaccomplished quest for urban policies in Italy. Waiting for Godot in the country of one hundred cities. In *A modern guide to National Urban Policies in Europe*. Edward Elgar Publishing.
10. La Sala, P., De Pascale, G., Dicecca, R., & Fiore, M. (2017). Four ideal types for leaderability: Cases from local action groups. *Rural Areas and Development, 14*, 93–106.
11. Kola-Bezka, M. (2020). Are FLAGs contributing to solving social exclusion problems? The case of Kujawsko-Pomorskie Voivodship. *Bulletin of Geography. Socio-economic Series, 48*(48), 35–45.
12. Roncevic, A., Lez, T., & Gregoric, M. (2020). Implementation of the leader programme in the observed EU countries. In *Economic and social development (book of proceedings), 50th international scientific conference on economic and social development*, p. 99.
13. Yang, F. X., Wong, I. A., Tan, X. S., & Wu, D. C. W. (2020). The role of food festivals in branding culinary destinations. *Tourism Management Perspectives, 34*, 100671.
14. Apostolopoulos, N., Liargovas, P., Stavroyiannis, S., Makris, I., Apostolopoulos, S., Petropoulos, D., & Anastasopoulou, E. (2020). Sustaining rural areas, rural tourism enterprises and EU development policies: A multi-layer conceptualisation of the obstacles in Greece. *Sustainability, 12*(18), 7687.
15. Kah, S. (2019). Implementing ERDF through CLLD: Experiences so far. *European Structural & Investment Funds Journal, 7*(1), 47–57.
16. Pollermann, K., Aubert, F., Berriet-Solliec, M., Laidin, C., Lépicier, D., Pham, H. V., et al. (2020). LEADER as a European policy for rural development in a multi-level governance framework: A comparison of the implementation in France, Germany and Italy. *European Countryside, 12*(2), 156–178.

Knowledge Base for Innovations in Higher Education Institutions (HEIs) Toward Smart and Sustainable Future: A Study on EUDRES

Agita Livina, Gavinolla Mahender Reddy, and Līga Horsta

Abstract An ecosystem where innovation is the goal becomes an important proponent to promote a smart and sustainable future (SSF). In doing so, it is imperative to provide an enabling environment where innovative ideas leverage existing issues by combining various resources—intellectual, physical, human, and financial—for innovation to be applied in real life. In this regard, higher education institutions (HEIs) are considered as a driver for regional innovation, as they provide enabling environment for idea generation, foster creation of new knowledge and expertise, and, in order to teach students, are in contact with the industry that has a vital role in the development SSF. Studies in the past examined the need for promoting innovation in the HEIs to create SSF. However, research about small cities and regions is scarce. With this background, Engaged, European, entrepreneurial university as driver for smart and sustainable regions (E^3UDRES^2 or EUDRES) is a European universities alliance located in the small cities and regions, working together on research, teaching, and innovation for creating the SSF. In this regard, the study aims to understand how the HEI in regions provides the enabling environment for innovation to create SSF, using the case of Vidzeme University of Applied Sciences, Latvia.

Further, the study throws light on knowledge base and resources on best practices related to innovation for the SSF. The study is conducted by applying multiple research methods such as content analysis, the Delphi method, and a case study, and data mainly includes policy documents, written records and reports, and expert opinion. The outcome of this descriptive and exploratory study provides significant information on the implications of innovation at HEIs for SSF.

A. Livina (✉) · L. Horsta
Vidzeme University of Applied Sciences, Valmiera, LV, Latvia

G. M. Reddy
Vidzeme University of Applied Sciences, Valmiera, LV, Latvia

National Institute of Tourism & Hospitality Management, Hyderabad, India

J. Reis et al. (eds.), *Driving Quality Management and Sustainability in VUCA Environments*, Springer Proceedings in Business and Economics,
https://doi.org/10.1007/978-3-031-52723-4_21

Keywords Higher education · Innovation · Entrepreneurship · Ecosystem · Smart and sustainable future

1 Introduction

Sustainable development (SD) is widely considered as an important approach to all that is good and desirable in society [24], with the equitable management of society, economy, and environment by providing appropriate institutional framework. Innovation, know-how, the progress of information and communication technologies (ICT) and their applications, and entrepreneurial activities are considered as the most important proponents to lead the path of SD [47]. This has further led to visioning the cities and regions with the idea of smart and sustainable growth, leading to the broader aim of SD [21]. Meanwhile, sustainable growth is considered an "economic growth that is fair, equitable and just, and that does not result in loss of environmental capital" [1, 12], and smart growth is regarded as the "development that makes cities or the regions more livable, economically efficient and more effective by providing just access to infrastructure and services through technological innovations" [1, 7]. In other words, going smart is all about the intelligent management of man-made environments [1].

In this regard, it is widely acknowledged that innovation and entrepreneurship (IE) play an important role to provide accurate solutions to the problems [13] for the best quality of life [31] to create smart and sustainable future (SSF). For instance, IE has been widely used by several countries in developing smart cities [17]. Universities are deemed to act as an imperative to provide knowledge ecosystem for the IE activities to create SSF [30, 40]. There are several studies that elucidated the need for promoting IE in the universities to create a smart and sustainable future. For instance, the triple helix model of university-industry-government interactions emphasizes the need for the IE and creates knowledge for the SSF [2, 13]. Most importantly, regional universities are considered as important ones in the regional context, as they produce knowledge and expertise that has an important role for the SSF regions [46]. At the same time, Andersons and Bushati [2] point out that focusing on the needs of the region in specific and narrow sectors, especially in the less developed regions with sectors with less added value, is not conducive to development. Livina and Rozentale [33] conducted research on talent attraction in small and medium urban areas based on case study of Vidzeme University of Applied Sciences, Latvia. Livina and Rozentale concluded that for young people who start their family life and career, the most crucial factor for selecting a place for residence is economic considerations, including job and carrier opportunities (Livina & Rozentale, [33]). It can also be inferred from this that active innovation, which promotes and expands the creation of businesses in the region, is important.

It is believed that the knowledge base or information resources on existing practices, new and trending approaches and models, and entrepreneurial ecosystem by the various stakeholders of IE create an immense impact on overall idea of entrepreneurial university and their contribution to the SSF.

To the author's knowledge, however, there hasn't been any research done or models developed that offer significant knowledge standardized data on the resources already available that are beneficial for the IE ecosystem from the viewpoint of an entrepreneurial institution. At the same time, innovation and entrepreneurship are a changing and dynamic environment. Policies, planning and strategy documents, networks of innovation and incubation centers, and scientific works are examples of these resources and best practices who are stimulating SSF. In this regard, the study's goal is to present a knowledge base model on tools and best practices for IE in the context of a university, with a focus on SSF. Since EUDRES is regarded as one of the significant university consortiums that focus on SSF, it was chosen as a case study. The purpose of creation of a "European, entrepreneurial university as driver for smart and sustainable regions" (E^3UDRES[2] and here after referred as EUDRES) with a consortium of universities is to create an IE ecosystem that focuses on the European small cities and regions and the small cities and their surroundings [15].

2 Literature Review-Conceptual Framework

2.1 Innovation and Entrepreneurship for Smart and Sustainable Future

IE is considered as continuous and complementary processes, in which innovation serves as the foundation for entrepreneurship, which in turn encourages innovation to flourish and recognize its economic and societal benefits [50]. Given the significance of entrepreneurship, it is feasible to help people develop their latent abilities to put ideas into practice and utilize entrepreneurship to find opportunities with the potential to create value in the areas of the economy, society, and the environment [4, 11]. Therein, innovations that are aimed at integrating, harnessing, and advancing physical and social infrastructures for environmental protection, economic regeneration, and improved public services and well-being are the focus of SSF [5]. Entrepreneurs provide new ideas and profit from utilizing them to build new technology required for SSF [35]. The growth of regions depends on entrepreneurial endeavors. In less developed areas, entrepreneurial thought and behavior can assist in improving the lives of those who are unemployed or underprivileged while also promoting social cohesiveness [32]. Integrating IE for the SSF becomes common practice while contributing to the global agenda of sustainable development goals [5, 11, 36]. However, research on SSF is mainly dominated in the context of cities and urban areas, ignoring the rural areas and small cities. Accordingly, several definitions elucidated the concept in the context of cities. SSF will become more prominent in the coming years; this fact supports the current trend as it becomes more crucial to reduce the amount of energy, water, waste, heat output, air pollution, and all together carbon dioxide that is produced [5]. SSF becomes an important global agenda for the different stakeholders in the process of planning, development, and management of cities, regions, and rural areas [4, 22, 36]. Universities are expected to take a chance to actively shape our future and be taught to students to embrace personal responsibility for both their own well-being and the well-being of future generations by inculcating SSF aspects [22]. With this thought, the proceeding part of the review reveals about entrepreneurial universities and SSF.

2.2 Entrepreneurial Universities Toward Smart and Sustainable Future

At the core of any entrepreneurial culture, "Entrepreneurial Universities have the ability to innovate, recognize and create opportunities, work in teams, take risks and respond to challenges" [23]. Entrepreneurial universities are believed to add value for a SSF by creating and disseminating information, skills, and capabilities to minimize risk and be change-adaptive, by challenging the established value of the outcomes of the universities [14, 38]. Universities in several parts of the world have included the concepts of IE into their curricula and methods of instruction. By combining with other academic disciplines, university-wide entrepreneurship training programs are greatly expanded beyond business programs or business schools, by creating cross-sector study programs and launching projects not related to specific study course or even university, in association with companies and cooperation partners. Funding for these activities comes from the national or global level. In the recent past, the idea has been applied to the different stakeholders of the university [34]. It is evident in several studies [10, 37] that the entrepreneurial universities contribute significantly to the SSF. There have also been other innovative methods for integrating universities with regional and community development that aim to have broader effects on communities than just high-tech start-ups, such as addressing opportunities for diverse populations and directing research at current issues with the economy, society, and the environment [19].

Already on year 2008, there have been researchers that described that to foster EI, university should act as a platform that offers institutional stability, credibility, access to a variety of knowledge, a setting for interdisciplinary collaboration, and the capacity to secure financial and human capital resources in a knowledge-hub environment where all of these initiatives involve multiple organizations [48].

2.3 Knowledge Base on Smart and Sustainable Future

A knowledge base is a collection of linked facts and information that is maintained and shared for use in product design or developing something new [49]. Recent years have seen a significant expansion of science and innovation, making it possible to acquire and apply a wealth of continuously updated scientific achievements that contain innovative knowledge [8], useful for the entrepreneurship and SSF.

Understanding how entrepreneurial universities are addressing the knowledge platforms or intellectual capacities to successfully generate societal impacts is complex [18]. A survey made the point that SMEs are unsure of which intelligent solutions perfectly complement their systems and processes and how they ought to leverage new technology. They therefore require knowledge bases that are compatible with current databases and procedures in terms of issues, problems, policies, and existing technological solutions at universities and so on [25]. This study substantiates the need for the knowledge base for providing smart solutions, and most importantly, universities can play an important role in this context. In this regard, few studies focused on knowledge base of SSF, mainly the use of big data for the citizens and start-ups for the IE to provide several public services. For instance, cities in the USA and Denmark developed website to access the open data [9, 42]. Other models include the interventions of municipalities, such as T-Hub in Hyderabad, India, that provides information on problems to be addressed in the city to the citizens for the creation of SSF of cities (Kummitha & Crutzen, [28]). In a study conducted by Chi et al. [8], the creation of a scientific publication management model based on knowledge graph and data analysis technologies is useful for the students in understanding the sustainability issues and providing solutions. However, this model is built based on scientific production and not considering other aspects of the ecosystem of SSF such as policies, best practices, and case studies and other knowledge platforms and so on. In the Vidzeme regional context, study was examined on knowledge in microenterprises [26]. Another study examined combinatorial knowledge for the improved innovation performance in organizations; however, study was limited to ICT sector [45]. The abovementioned studies are mainly limited to specific area or domain of study. However, in order to create SSF through IE, it is important to have comprehensive and structured knowledge base for several reasons.

3 Methodology

A review of the literature shows that there is a connection between entrepreneurship, innovation, and universities and also their role toward the SSF. Several studies also provided knowledge based on various aspects of SSF focusing on specific sector, industry, or discipline or sometimes specific stakeholder involved in SSF. In general, there are no previous studies that provided knowledge base consisting of various aspects of SSF and also stakeholders. This study fundamentally aimed at providing a comprehensive knowledge on existing practices and resources required for the SSF. Most importantly, it should provide a base for the EUDRES stakeholders to understand the existing knowledge, resources, and practices that are required for the creation of SSF. Essentially there are two aspects connected to the study, first developing a structured format or model with various components that are connected to the major themes of SSF in general and EUDRES in particular. Next was to segregate the available practices, resources, and knowledge according to the theme or relevance. In doing so, study adopted mixed methods and it consists of several steps. In the first phase of the study, extensive literature review was conducted in order to understand the concept [43], on IE, SSF, and related aspects, followed by the content analysis of websites, policy, planning, and vision documents to know the context. Content analysis was applied to thoroughly analyze the data and provide themes [20, 44]. In the process of developing tools, Delphi method was followed; in other words, format for the knowledge base, requesting opinions from experts based on their knowledge and experience as subject-matter experts [3], from the EUDRES consortium was considered. In the latter stage, as a case study, it was applied in the context of Vidzeme University of Applied Sciences main influence zone in Latvia—Vidzemes Planning Region—to provide realistic and topical output of the situation in the region in regards of SSF. Case study was applied to contribute to the existing knowledge, in this case Vidzeme University of Applied Sciences, Latvia [37]. Derived structured model was validated with the senior academicians and administrators of the university. The outcome of this descriptive and exploratory study provides significant information on the implications of innovation at HEIs for SSF.

4 Contextual Framework-EUDRES

The study's geographical parameters are defined by the contextual framework. EUDRES with nine institutions from Austria, Portugal, Hungary, Belgium, Romania, the Netherlands, Germany, Finland, and Latvia has been selected as a case study for the current study. EUDRES is a network of higher education institutions in Europe that have made the decision to closely cooperate and share ideas in the areas of research, instruction, and innovation, among other things. The goal is to establish a model multi-university campus throughout Europe, cocreate outstanding ideas and concepts for future universities, and integrate challenge-based learning, mission-oriented research, human-centered innovation, as well as open and engaged knowledge exchange, as interrelated core areas. Initiatives from the bottom up that have a regional impact and take into account regional conditions will determine the SSF future of the European Union through IE [16]. The Vidzeme University of Applied Sciences (VUAS) is considered as a regional leader in higher education, science, knowledge transfer, and skill development. It provides a digital era ecosystem with a variety of options for pursuing professional higher education as well as interdisciplinary research that addresses societal issues. VUAS is a consortium of EUDRES striving to contribute to the SSF of the Vidzeme region and beyond.

5 Findings for Model on Structured Knowledge Base

Proposed structured knowledge base consists of mainly three levels of best practices or resources such as regional, national, and university level. Regions, country, and university resources were looked at in the deviation of physical, intellectual, human, network, and financial resources to be accessed and combined for the development of the IE ecosystem. The latter part of the study discussed the best practices at the university level realized and started projects and other sources of inspiration. After look at scientific communities, innovation networks and hubs, conferences and other sources of knowledge, experience, and inspiration and possible co-operations, authors analyzed the three most important aspects as follows:

1. The most relevant, unique assets of the regional/national innovation ecosystem; most influential and core stakeholders in the regional innovation ecosystem.
2. Possibility and accessible resources on coaching and training courses, methods, and formats.
3. Entrepreneurship and innovation skills, followed by articles/scientific publications related IE and entrepreneurial university.

The abovementioned practices are mainly suggested by the staff members of the consortium of the entrepreneurial universities, representing the countries that are promoting the consortium of the entrepreneurial university. The idea is to enhance the knowledge of the existing practices that are useful for the member universities to focus on their respective regions, in order to provide solutions to create a smart and sustainable future. Information on the entrepreneurship activities and support networks and start-ups includes the objective of the organization, need and importance, focused area or targeted beneficiaries of innovation or innovative services, and most importantly relevance for the entrepreneurial university. Similarly, policies, strategies, and vision and planning documents are selected mainly from the consortium member representative regions with a national level context, as they will be helpful to know the government vision or support for the entrepreneurial ecosystem. The existing landscape of scientific production related to innovation, entrepreneurship, smart and sustainable future, and entrepreneurial university for the SSF includes scientific knowledge base. Although the best practices are covered from the consortium of universities, the information is dominated by Latvia and VUAS, followed by other members. However, due to the space constraints, detailed formats were not discussed. Table 1 shows the international, national, and university level practices or resources.

6 Discussion and Conclusion

Cities are creating the ecosystems that will allow different parties to participate in technological innovations and entrepreneurship, while polices provide an enabling atmosphere that will be advantageous to entrepreneurs to create SSF [29]. In this regard, a detailed knowledge on existing policies and practices and enabling factors of the universities to create IE would be an advantage for the stakeholders. IE and

Table 1 Format for structured knowledge base for the SSF

International level (Europe) practices/resources	National level practices/resources	University level practices/resources
Incubation centers, networks, start-ups Best practices at university level—projects and other sources of inspiration Scientific communities, innovation networks or hubs, conferences, and other sources of knowledge, experience, and inspiration Accessible resources on coaching and training courses, methods, formats Empowering entrepreneurship and innovation skills	Incubation centers, networks, start-ups Best practices at university level—Projects and other sources of inspiration Scientific communities, innovation networks or hubs, conferences and other sources of knowledge, experience and inspiration Accessible resources on coaching and training courses, methods, formats; empowering entrepreneurship and innovation skills Most relevant and core stakeholders of the regional innovation ecosystem	Best practices at university level—Projects and other sources of inspiration Scientific communities, innovation networks or hubs, conferences and other sources of knowledge, experience and inspiration Physical, intellectual, and human networks Most relevant, unique assets of the regional innovation ecosystem Accessible resources on coaching and training courses, methods, formats; empowering entrepreneurship and innovation skills Scientific knowledge on innovation and entrepreneurship (publications) by VUAS

Source: Authors elaboration from the focus group discussion and content analysis

technological advancement without understanding the regional or local relevance would create more complex problems and become irrelevant for the SSF [27]. Thus, understanding the citizen concerns and needs is an important aspect for SSF [39]. Therein, the research outcomes of the applied universities and universities in regions are an appropriate way to understand the regional and local problems, as they focus more on local issues and challenges. In this regard, proposed knowledge base would act as a source for the information not only on local problems but also relevant development policies. Particularly the proposed model would have greater implications on the EUDRES stakeholders that are involved in creating SSF. Sensing the needs of the local and grassroots levels before inventing or promoting technology requires extensive learning from scratch. An appropriate method to increase effectiveness is to match technological platforms available at universities to solve the social problems and by bringing them together [6]. At this stair, EUDRES is considered as an important concept that allows to make it possible and for which the knowledge bases provide an enabling platform to understand the existing technological advantages and research outcomes on citizen concerns. EUDRES have opportunity to provide knowledge base platform in situations where there are similar regional issues with stakeholder involved on cross-sector scientific knowledge and for the other resources through the proposed platform. Particularly, universities must forge alliances with decision-makers, business, and the general public to provide the required foundation for the development of inclusive SSF in which the general public plays an active role [29]. A thorough understanding based on SSF would reinforce this much more. Knowledge bases, information, and resources on current methods of work have a significant impact not only on the total teaching-learning environment for innovation and entrepreneurship but also on our general conception of the entrepreneurial university.

There is room to include more features of best practices or resources in the knowledge base's formats, which are now restricted to main SSF and IE components. Despite this, study tried to provide format for knowledge base on existing practices. Further, there is a scope for EUDRES alliance to carry forward the future studies. Future academics may examine additional facets of organized knowledge bases in universities and nations outside of the EUDRES group. Despite certain drawbacks, the study makes a substantial contribution to the production of a structured knowledge base, which opens up the possibility for ontological analysis and helps turn complex issues into solutions that are easily understood.

Acknowledgments Authors would like to thank the Entrepreneurship and Innovation Network for Smart and Sustainable European Regions (EINS) of EUDRES at Vidzeme University of Applied Sciences, Valmiera, Latvia, for funding this research.

A.1 Annexure (Table 2)

Table A.1 Format for structured knowledge base for the SSF

International level practices—Europe	Country wide practices—Latvia	University level practices—Vidzeme University
Incubation centers, networks, start-ups	*Incubation centers, incubation centers, networks, start-ups*	*Best practices at university level-projects and other sources of inspiration*
European Institute of Innovation and Technology (EIT), EU	LIAA Valmiera Business and Innovation Incubator (VBII), Latvia	B-LAB, Vidzeme University of Applied Sciences, Latvia
Institute of Entrepreneurship Development (iED), EU	The Latvian Rural Consultation and Education Center (LLKC), Latvia	I-LAB & Champions' Dinner, Vidzeme University of Applied Sciences, Latvia
EIT Climate-KIC (Knowledge and Innovation Community), EU	The association ZINIS, Latvia	S-LAB, Vidzeme University of Applied Sciences, Latvia
European Association of Research and Technology Organisations (EARTO), EU	Most relevant and core stakeholders of the regional innovation ecosystem	Bachelor in Business Administration (Start-up Management), Ventspils University of Applied Sciences, Latvia
European Social Entrepreneurship and Innovative Studies Institute, EU	Latvian digital accelerator	Rezekne Academy of Technologies (RTA), Latvia
European Consortium of Innovative Universities (ECIU), EU	Latvian IT cluster, Latvia	*Scientific communities, innovation networks or hubs, conferences and other sources of knowledge, experience and inspiration*
European Association of Innovation Consultants (EAIC), EU	Creative Hives (NGO developed under Vidzeme Tourism Association), Latvia	Vidzeme Innovation Week, Latvia
European Association for Innovation and Growth, EU	LIKTA—Latvian Information and Communications Technology Association, Latvia	International Week, Vidzeme University of Applied Sciences, Latvia
European Association for Creativity and Innovation (EACI), EU	LETERA—Latvian Electrical Engineering and Electronics Industry Association, Latvia	International Scientific Conference SOCIETY. TECHNOLOGY. SOLUTIONS, Vidzeme University of Applied Sciences, Latvia
European Business and Innovation Centre Network (EBN), EU	Latvia Convention Bureau, Latvia	Share IT—conference, Latvia
European Start-up Association (ESA), EU	Latvian Chamber of Commerce and Industry (LTRK), Latvia	The World Association of Free Latvians (PBLA), Latvia
European Innovation Academy, EU	Best practices at university level-projects and other sources of inspiration	Student Group Consultancy Module, Vidzeme University of Applied Sciences, Latvia
The Young Entrepreneurs Organization of the European Union (JEUNE), EU	B-LAB, Vidzeme University of Applied Sciences, Latvia	*Physical, intellectual, and human networks*
YES for Europe—The European Confederation of Young Entrepreneurs, EU	I-LAB & Champions' Dinner, Vidzeme University of Applied Sciences, Latvia	Virtual and Augmented Reality Laboratory, Vidzeme University of Applied Sciences, Latvia
European Family Businesses (EFB), EU	S-LAB, Vidzeme University of Applied Sciences, Latvia	Mobile Technology Laboratory, Vidzeme University of Applied Sciences, Latvia
European Small Business Alliance (ESBA), EU	Bachelor in Business Administration (Start-up Management), Ventspils University of Applied Sciences, Latvia	Multimedia Laboratory, Vidzeme University of Applied Sciences, Latvia
	Rezekne Academy of Technologies (RTA), Latvia	

International level practices—Europe	Country wide practices—Latvia	University level practices—Vidzeme University
ANJE—National Association of Young Entrepreneurs, EU The European Researchers' Night, Europe *Best practices at university level-projects and other sources of inspiration* Demola projects ForEMLink, Denmark The Third Way The COPERNICUS Alliance (CA) GRASP-A network of knowledge creation through artful thinking Female Founders	*Scientific communities, innovation networks or hubs, conferences, and other sources of knowledge, experience, and inspiration* Vidzeme Innovation Week, Latvia International Week, Vidzeme University of Applied Sciences, Latvia International Scientific Conference SOCIETY. TECHNOLOGY. SOLUTIONS, Vidzeme University of Applied Sciences, Latvia Share IT—conference, Latvia The World Association of Free Latvians (PBLA), Latvia Student Group Consultancy Module, Vidzeme University of Applied Sciences, Latvia *Physical, intellectual, and human networks* Virtual and Augmented Reality Laboratory, Vidzeme University of Applied Sciences, Latvia Mobile Technology Laboratory, Vidzeme University of Applied Sciences, Latvia Multimedia Laboratory, Vidzeme University of Applied Sciences, Latvia Spatial Research Laboratory, Vidzeme University of Applied Sciences, Latvia Computer Network Laboratory, Vidzeme University of Applied Sciences, Latvia E-learning Management and Technology, Vidzeme University of Applied Sciences, Latvia	Spatial Research Laboratory, Vidzeme University of Applied Sciences, Latvia Computer Network Laboratory, Vidzeme University of Applied Sciences, Latvia E-learning Management and Technology, Vidzeme University of Applied Sciences, Latvia Mechatronics Laboratory, Vidzeme University of Applied Sciences, Latvia Electrical Engineering Laboratory, Vidzeme University of Applied Sciences, Latvia HESPI, Vidzeme University of Applied Sciences (ViA HESPI), Latvia Socio-Technical Systems Engineering Institute (SSII), Vidzeme University of Applied Sciences, Latvia Sciences Knowledge and Technology Centre (ZTC), Vidzeme University of Applied Sciences, Latvia EURAXESS contact point at Vidzeme University of Applied Sciences, Latvia *Most relevant, unique assets of the regional innovation ecosystem* Smart transformation of function (services) of small places (villages), Latvia Circular economy hubs in project Change CornEr (internal project of EUDRESS project) Circular economy hub in project "circulairwerkt" regio Hasselt

(continued)

Table A.1 (continued)

International level practices—Europe	Country wide practices—Latvia	University level practices—Vidzeme University
Accessible resources on coaching and training courses, methods, formats, empowering entrepreneurship and innovation skills EntreComp: The Entrepreneurship Competence Framework Transforming education with community-developed teaching materials: evidence from direct observations of STEM college classrooms *Appreciative inquiry* LIFT European Network of Learning Factories (LIFT) *Scientific communities, innovation networks or hubs, conferences, and other sources of knowledge, experience, and inspiration* *Accessible resources on coaching and training courses, methods, formats, empowering entrepreneurship and innovation skills*	Mechatronics Laboratory, Vidzeme University of Applied Sciences, Latvia Electrical Engineering Laboratory, Vidzeme University of Applied Sciences, Latvia HESPI, Vidzeme University of Applied Sciences (ViA HESPI), Latvia Socio-Technical Systems Engineering Institute (SSII), Vidzeme University of Applied Sciences, Latvia Sciences Knowledge and Technology Centre (ZTC), Vidzeme University of Applied Sciences, Latvia EURAXESS contact point at Vidzeme University of Applied Sciences, Latvia *Most relevant, unique assets of the regional innovation ecosystem* Smart transformation of function (services) of small places (villages), Latvia Circular Economy hubs in project Change CornEr (internal project of Eudress project) Circular economy hub in project "circulairwerkt" regio Hasselt *Accessible resources on coaching and training courses, methods, formats, empowering entrepreneurship and innovation skills* *Handbook "VIA I-LAB and ALC-PBL Approach Integration in the Study Courses"* *Scientific knowledge on innovation and entrepreneurship by Latvian universities*	*Accessible resources on coaching and training courses, methods, formats, empowering entrepreneurship and innovation skills* Handbook *VIA I-LAB and ALC-PBL Approach Integration in the Study Courses* *Scientific knowledge on innovation and entrepreneurship by Vidzeme University (publications)*

References

1. Ahvenniemi, H., Huovila, A., Pinto-Seppa, I., & Airaksinen, M. (2017). What are the differences between sustainable and smart cities? *Cities, 60*, 234–245.
2. Andersons, A., & Bushati, J. (2019). Smart specialisation concept application in universities: E-business online studies model development. In: *Environment. Technology. Resources. Proceedings of the 12th international scientific and practical conference*, 20–22 June 2019, Vol. II, pp. 190–194.
3. Barrett, D., & Heale, R. (2020). What are Delphi studies? *Evidence-Based Nursing, 23*(3), 68–69.
4. Belz, F. M., & Binder, J. K. (2014). *A process view on sustainability entrepreneurship: Towards a model.* Entrepreneurship Summer University at ISCTE-IUL, 2013, p. 146.
5. Bibri, S. E. (2021). Data-driven smart sustainable cities of the future: An evidence synthesis approach to a comprehensive state-of-the-art literature review. *Sustainable Futures, 3*, 100047.
6. Carvalho, L. (2015). Smart cities from scratch? A socio-technical perspective. *Cambridge Journal of Regions, Economy and Society, 8*, 43–60.
7. Cavada, M., Hunt, D. V. L., & Rogers, C. D. F. (2016). Do smart cities realise their potential for lower carbon dioxide emissions? *Proceedings of the Institution of Civil Engineers: Engineering Sustainability, 169*(6), 243–252.
8. Chi, Y., Qin, Y., Song, R., & Xu, H. (2018). Knowledge graph in smart education: A case study of entrepreneurship scientific publication management. *Sustainability, 10*(4), 995.
9. Cohen, B., Almirall, E., & Chesbrought, H. (2016). The city as a lab: Open innovation meets the collaborative economy. *California Management Review, 59*(1), 5.
10. Cunningham, J. A., Guerrero, M., & Urbano, D. (2017). Chapter 1: Entrepreneurial universities—Overview, reflections, and future research agendas. In *The world scientific reference on entrepreneurship* (pp. 3–19). https://doi.org/10.1142/9789813220591_0001
11. Dana, L. P., Salamzadeh, A., Hadizadeh, M., Heydari, G., & Shamsoddin, S. (2022). Urban entrepreneurship and sustainable businesses in smart cities: Exploring the role of digital technologies. *Sustainable Technology and Entrepreneurship, 1*(2), 100016.
12. De Jong, M., Joss, S., Schraven, D., Zhan, C., & Weijnen, M. (2015). Sustainable-smart resilient-low carbon-eco-knowledge cities; making sense of a multitude of concepts promoting sustainable urbanization. *Journal of Cleaner Production, 109*, 25–38.
13. Etzkowitz, H., & Zhou, C. (2017). *The triple helix: University–industry–government innovation and entrepreneurship*. Routledge.
14. Etzkowitz, H., Webster, A., Gebhardt, C., & Terra, B. R. C. (2000). The future of the university and the university of the future: Evolution of ivory tower to entrepreneurial paradigm. *Research Policy, 29*(2), 313–330.
15. EUDRES. (2023a). *Everything you need to know about the E^3UDRES^2*. European University Alliance. Retrieved from https://eudres.eu/who-we-are
16. EUDRES. (2023b). *Long-term vision mission statement* (p. 2). Retrieved from https://eudres.eu/documents-and-materials
17. Ferraris, A., Santoro, G., & Pellicelli, A. C. (2020). "Openness" of public governments in smart cities: Removing the barriers for innovation and entrepreneurship. *International Entrepreneurship and Management Journal, 16*, 1259–1280.
18. Fischer, B., Guerrero, M., Guimón, J., & Schaeffer, P. R. (2021). Knowledge transfer for frugal innovation: Where do entrepreneurial universities stand? *Journal of Knowledge Management, 25*(2), 360–379.
19. Forrant, R. (2001). Pulling together in Lowell: The university and the regional development process. *European Planning Studies, 9*(1), 613–628.
20. Gavinolla, M. R., Livina, A., & Swain, S. K. (2022). State of the research on teacher education and sustainability: A bibliometrics analysis. *Journal of Teacher Education for Sustainability, 24*(2), 147–165. https://doi.org/10.2478/jtes-2022-0022

21. Gazzola, P., Del Campo, A. G., & Onyango, V. (2019). Going green vs going smart for sustainable development: Quo vadis? *Journal of Cleaner Production, 214*, 881–892.
22. Gupta, B. B., Gaurav, A., & Panigrahi, P. K. (2023). Analysis of the development of sustainable entrepreneurship practices through knowledge and smart innovative based education system. *International Entrepreneurship and Management Journal, 19*(2), 923–940.
23. Hisrich, R. D., Peters, M. P., & Sheperd, D. A. (2002). *Entrepreneurship*. McGraw-Hill.
24. Holden, E., Linnerud, K., & Banister, D. (2014). Sustainable development: our common future revisited. *Global Environmental Change, 26*, 130–139.
25. Khosravani, M. R., Nasiri, S., & Reinicke, T. (2022). Intelligent knowledge-based system to improve injection molding process. *Journal of Industrial Information Integration, 25*, 100275.
26. Krumina, I., Krumins, G., & Rozentale, S. (2015). Vidzeme region of Latvia: Knowledge management in micro enterprises. *Procedia-Social and Behavioral Sciences, 213*, 781–786.
27. Kummitha, R. K. R. (2018). Entrepreneurial urbanism and technological panacea: Why smart city planning needs to go beyond corporate visioning. *Technological Forecasting and Social Change, 137*, 330–339.
28. Kummitha, R. K. R., Crutzen, N. (2019). Smart cities and the citizen-driven internet of things: A qualitative inquiry into an emerging smart city. *Technological Forecasting and Social Change*, 44–53, 140.
29. Kummitha, R. K. R. (2019). Smart cities and entrepreneurship: An agenda for future research. *Technological Forecasting and Social Change, 149*, 119763.
30. Kutty, A. A., Abdella, G. M., Kucukvar, M., Onat, N. C., & Bulu, M. (2020). A system thinking approach for harmonizing smart and sustainable city initiatives with United Nations sustainable development goals. *Sustainable Development, 28*(5), 1347–1365.
31. Lee, S. M., & Trimi, S. (2018). Innovation for creating a smart future. *Journal of Innovation & Knowledge, 3*(1), 1–8.
32. Lindner, J. (2018). Entrepreneurship education for a sustainable future. *Discourse and Communication for Sustainable Education, 9*(1), 115–127.
33. Livina, A., & Rozentale, S. (2020). Challenge of talent attraction in small and medium urban areas: Case of Valmiera City, Latvia. In M. Rehm, J. Saldien, & S. Manca (Eds.), *Project and design literacy as cornerstones of smart education. Smart innovation, systems and technologies* (Vol. 158). Singapore: Springer. https://doi.org/10.1007/978-981-13-9652-6_15
34. Morris, M. H., Kuratko, D. F., & Cornwall, J. R. (2013). *Entrepreneurship programs and the modern university*. Edward Elgar Publishing.
35. Munoz, P., & Cohen, B. (2016). The making of the urban entrepreneur. *California Management Review, 59*(1), 71–91.
36. Nižetić, S., Šolić, P., González-De, D. L. D. I., & Patrono, L. (2020). Internet of things (IoT): Opportunities, issues and challenges towards a smart and sustainable future. *Journal of Cleaner Production, 274*, 122877.
37. OECD. (2022). *Advancing the entrepreneurial university: Lessons learned from 13 HEInnovate country reviews* (OECD SME and Entrepreneurship Papers, No. 32). OECD Publishing. https://doi.org/10.1787/d0ef651f-en
38. Peterka, S. O., & Salihovic, V. (2012). What is entrepreneurial university and why we need it. *Economy of Eastern Croatia: Yesterday, Today, Tomorrow, 1*(1), 98–107.
39. Rashid, Y., Rashid, A., Warraich, M. A., Sabir, S. S., & Waseem, A. (2019). Case study method: A step-by-step guide for business researchers. *International Journal of Qualitative Methods, 18*, 1609406919862424.
40. Rinaldi, C., Cavicchi, A., Spigarelli, F., Lacchè, L., & Rubens, A. (2018). Universities and smart specialisation strategy: From third mission to sustainable development co-creation. *International Journal of Sustainability in Higher Education, 19*(1), 67–84. https://doi.org/10.1108/IJSHE-04-2016-0070
41. Shahrokni, H., Lazarevic, D., & Brandt, N. (2015). Smart urban metabolism: Towards a real-time understanding of the energy and material flows of a city and its citizens. *Journal of Urban Technology, 22*(1), 65–86.

42. Snow, C. C., Hakonsson, D. D., & Obel, B. (2016). A smart city is a collaborative community: Lessons from smart Aarhus. *California Management Review, 59*(1), 92–108.
43. Snyder, H. (2019). Literature review as a research methodology: An overview and guidelines. *Journal of Business Research, 104*, 333–339.
44. Stough, T., Ceulemans, K., Lambrechts, W., & Cappuyns, V. (2018). Assessing sustainability in higher education curricula: A critical reflection on validity issues. *Journal of Cleaner Production, 172*, 4456–4466. https://doi.org/10.1016/j.jclepro.2017.02.017
45. Tödtling, F., & Grillitsch, M. (2015). Does combinatorial knowledge lead to a better innovation performance of firms? *European Planning Studies, 23*(9), 1741–1758.
46. Wise, G., & Wilkinson, S. (2016). *How regional universities drive regional innovation*. Available at https://www.industry.gov.au/sites/g/files/net3906/f/May%202018/document/other/how_regional_universities_drive_regional_innovation.docx
47. Wu, J., Guo, S., Huang, H., Liu, W., & Xiang, Y. (2018). Information and communications technologies for sustainable development goals: State-of-the-art, needs and perspectives. *IEEE Communications Surveys & Tutorials, 20*(3), 2389–2406.
48. Youtie, J., & Shapira, P. (2008). Building an innovation hub: A case study of the transformation of university roles in regional technological and economic development. *Research Policy, 37*(8), 1188–1204.
49. Zhang, L., Tan, R., Peng, Q., Shao, P., Dong, Y., & Wang, K. (2022). Construction and application of enterprise knowledge base for product innovation design. *Applied Sciences, 12*(13), 6358.
50. Zhao, F. (2005). Exploring the synergy between entrepreneurship and innovation. *International Journal of Entrepreneurial Behavior and Research, 11*(1), 25–41.

Index

J. Reis et al. (eds.), *Driving Quality Management and Sustainability in VUCA Environments*, Springer Proceedings in Business and Economics,
https://doi.org/10.1007/978-3-031-52723-4

The manufacturer's authorised representative in the EU is Springer Nature Customer Service Centre GmbH, Europaplatz 3, 69115 Heidelberg, Germany. If you have any concerns regarding our products, please contact ProductSafety@springernature.com

Printed and bound by CPI Group (UK) Ltd, Croydon, CR0 4YY
28/09/2026
02224073-0009